17th International Conference on Calorimetry in Particle Physics (CALOR 2016)

Journal of Physics: Conference Series Volume 928

Daefu, South Korea
15 – 20 May 2016

ISBN: 978-1-5108-5312-6
ISSN: 1742-6588

Printed from e-media with permission by:

Curran Associates, Inc.
57 Morehouse Lane
Red Hook, NY 12571

Some format issues inherent in the e-media version may also appear in this print version.

Printed by Curran Associates, Inc. (2018)

For permission requests, please contact the Institute of Physics at the address below.

Institute of Physics
Dirac House, Temple Back
Bristol BS1 6BE UK

Phone: 44 1 17 929 7481
Fax: 44 1 17 920 0979

techtracking@iop.org

Additional copies of this publication are available from:

Curran Associates, Inc.
57 Morehouse Lane
Red Hook, NY 12571 USA
Phone: 845-758-0400
Fax: 845-758-2633
Email: curran@proceedings.com
Web: www.proceedings.com

Table of contents

Volume 928

17th International Conference on Calorimetry in Particle Physics (CALOR2016)
15–20 May 2016, Daegu, South Korea

Accepted papers received: 8 November 2017
Published online: 27 November 2017

Preface

Papers

1.Opening Session and Invited talk

2.Large Hadron Collider

6.High Luminosity LHC

7.Challenges for calorimeters

9.Poster session

IOP Conf. Series: Journal of Physics: Conf. Series **928** (2017) 011001 doi:10.1088/1742-6596/928/1/011001

Preface

Calorimetric particle detection with scintillation counters started in nuclear physics after World War II, and in the following seventy years calorimeters have become essential detectors in particle physics experiments providing measurements of the energy and position of particles and jets, energy flow information, particle identification, and fast trigger information used in real-time event selection. The first CALOR conference was held at Fermilab in 1990 and this series of conferences has become the main meeting to exchange research findings in all areas of calorimetry, including recent results from large running calorimeter systems, beam test results, ideas for the design of new particle physics experiments, underground experiments, experiments in space, and so on.

The XVIIth International Conference on Calorimetry in Particle Physics, CALOR 2016, was held in Daegu, South Korea from 15-20 May, 2016. This conference supported presentations and profound discussions on calorimeters at the Large Hadron Collider, experience with current calorimetric systems, calorimeters for future accelerator experiments such as the High Luminosity LHC, including discussion of the challenges for future calorimeters, and new concepts in calorimetry.

On May 16, 2016, CALOR2016 opened with four interesting invited talks on new physics searches at CMS and ATLAS, on dark matter searches, on neutrinoless double beta decay experiments, and on the history of calorimetry over the last seventy years. There was good harmony that, in the most recent physics results, calorimeters played important roles in accelerator and non-accelerator experiments and in the summary of the history of calorimetry. During the afternoon of the first day, the calorimeters at the Large Hadron Collider were discussed, mainly CMS and ATLAS experimenters presented talks on calibration, monitoring, trigger, energy measurement, and the performance of their calorimeter systems, displaying the role of calorimeters in important new physics searches.

The second day brought interesting studies performed in various experiments not at the LHC, for example, in the morning session, experience with the current calorimeter systems for NA62, SND, CMD-3 at the VEPP-2000 collider, ISS-CREAM, AMoRE, and CaloCube were discussed. It was a good opportunity to witness calorimeters from deep underground up to calorimeters in experiments in space. In the afternoon, the conference hall was full of enthusiasm for calorimeter studies for future accelerator physics experiments. These studies involved much effort in worldwide particle physics communities on calorimeters at sPHENIX, a future electron-ion collider, CEPC, Mu2e, Belle II, etc.

The main topic of the third day was the upgrades of the ATLAS and CMS calorimeters soon to be exposed to a High Luminosity LHC. The afternoon session was reserved for an excursion to Gyeongju, the capital of the Shilla Kingdom 1300 years ago. We visited Gyeongju with its well preserved ancient Korean heritage and its natural environment, and took time to feel the atmosphere of this historic site in Korea.

The morning of the fourth day was devoted to discussions of calorimeters for the High Luminosity LHC experiments, mainly the upgrade of the endcap calorimeter of CMS and the tile and liquid argon calorimeters of ATLAS. The rest of the fourth day detailed the endeavors to solve known calorimeter issues such as radiation-hardness, precise energy measurement, precision timing to mitigate pile-up effects, calibration for a calorimeter in a space station experiment, and so on.

The last day was devoted to new concepts in calorimetry, in particular new concepts for future lepton collider experiments which require excellent electromagnetic and hadronic performance. All time slots were filled with talks on the particle flow algorithm calorimeters.

The five days of CALOR 2016 were a precious time when we could discuss various studies of calorimeters and calorimetry in particle physics experiments. It was a great honor to host and organize CALOR 2016 where world calorimeter experts got together and shared their knowledge and experience. Most of all, I would like to express my gratitude to the members of International Advisory Committee who allowed Kyungpook National University to host CALOR 2016 on behalf of the local organizing committee. I should not forget to thank the session conveners who made the sessions go smoothly and drew meaningful discussions. I am also grateful to the institutions and industrial partners who provided generous support, namely Kyungpook National University, Center for High Energy Physics, Radiation Science Research Institute, NOTICE, TPS, and AutoTech Korea. I specially thank John Hauptman for reviewing the proceedings. Without the happy collaboration of the members of the local organizing committee and DEXCO, CALOR 2016 would not have been successful. I would like to express my gratitude for their efforts to host this successful conference. I sincerely appreciate all speakers and participants who enriched CALOR 2016 with their precious talks, discussions and creativity. Calorimetry and calorimeters will become more and more important in future particle physics experiments to obtain better physics data and meaningful results. I am sure that the CALOR conference will prosper and be a main place to develop future world high energy physics experiments.

Sehwook Lee
on behalf of the Local Organizing Committee

IOP Conf. Series: Journal of Physics: Conf. Series **928** (2017) 011002 doi:10.1088/1742-6596/928/1/011002

International Advisory Committee

Nural Akchurin, (Texas Tech Univ.)

James Brau, (Univ. of Oregon)

Marcella Diemoz, (INFN Roma)

Antonio Ereditato, (Univ. of Bern)

Tao Hu, (IHEP Beijing)

Michele Livan, (Univ. of Pavia)

Steve Magill, (Argonne Natl. Lab.)

Amelia Maio, (LIP and FC-University of Lisbon)

Sven Menke, (MPI Munich)

Rainer Novotny, (Univ. of Giessen)

Iouri Tikhonov, (Budker INP)

Richard Wigmans, (Texas Tech Univ.)

Ren-yuan Zhu, (CALTECH)

Local organizers

General Chair: *Hong Joo Kim, (Kyungpook National Univ.)*

Scientific secretary: *Sehwook Lee, (Kyungpook National Univ.)*

Guinyun Kim, (Kyungpook National Univ.)

Dong Hee Kim, (Kyungpook National Univ.)

Sun Kee Kim, (Seoul National Univ.)

Youngduk Kim, (Institute of Basic Science)

Ilheung Park, (Sungkyunkwan Univ.)

Hwanbae Park, (Kyungpook National Univ.)

Byung Gu Cheon , (Hanyang Univ.)

IOP Conf. Series: Journal of Physics: Conf. Series **928** (2017) 011002 doi:10.1088/1742-6596/928/1/011002

Session Conveners

Opening Session:

Seh Wook Lee (Kyungpook National University)

Large Hadron Collider:

Francesca Nessi-Tedaldi (ETH Zurich)

Nural Akchurin (Texas Tech University)

Experience with current calorimetric systems:

Byung Gu Cheon (Hanyang University)

Calorimetric systems at non-accelerator:

Tao Hu (IHEP, CAS)

Calorimeters for future accelerator experiments:

Richard wigmans (Texas Tech University)

Renyuan Zhu (California Institute of Technology)

High Luminosity LHC:

Ana Maria Henriques Correia (CERN)

Adam Para (Fermi National Accelerator Lab)

Boris Shwartz (Budker Institute of Nuclear Physics)

Challenges for calorimeters:

John Michael Hauptman (Iowa State University)

Jim Brau (University of Oregon)

Craig Woody (Brookhaven National Laboratory)

New concepts for calorimetry:

DongHee Kim (Kyungpook National University)

Rainer Willi Novotny (Justus-Liebig-University)

IOP Conf. Series: Journal of Physics: Conf. Series **928** (2017) 011002 doi:10.1088/1742-6596/928/1/011002

List of Participants

AKCHURIN, Nural Texas Tech University (US)

AKHMETSHIN, Ravel Budker INP (RU)

ALIBERTI, Riccardo Johannes-Gutenberg-Universitaet Mainz (DE)

ARYAL, Pabitra Kyungpook National University (KR)

BAE, Hanwook Kyungpook National University (KR)

BENDAVID, Josh California Institute of Technology (US)

BENITEZ, Jose Feliciano University of Iowa (US)

BORNHEIM, Adolf California Institute of Technology (US)

BOYARYNTSEV, Andriy Inst. for Scint. Materials of Nat. Acad. of Science of Ukraine (UA)

BRAU, James University of Oregon (US)

BUNDOCK, Aaron Imperial College Sci., Tech. & Med. (GB)

BUTANOV, Khakimjan Kyungpook National University (KR)

C. R., Kesavulu Kyungpook National University (KR)

CHLEBANA, Frank Fermi National Accelerator Lab. (US)

CHOMONT, Arthur Univ. Blaise Pascal Clermont-Fe. II (FR)

CIRIOLO, Vincenzo Universita & INFN, Milano-Bicocca (IT)

DAVIDEK, Tomas Charles University (CZ)

FISCHER, Cora Universitat Autònoma de Barcelona (ES)

GECSE, Zoltan Fermi National Accelerator Lab. (US)

HA, Daehoon Kyungpook National University (KR)

HAN, Daehee Kyungpook National University (KR)

HARTBRICH, Oskar Deutsches Elektronen-Synchrotron (DE)

HAUPTMAN, John Michael Iowa State University (US)

HENRIQUES, Ana CERN (CH)

HU, Tao IHEP, CAS (CN)

HUH, Changgi Kyungpook National University (KR)

JANG, Jonghun Kyungpook National University (KR)

JANG, Wooyoung Kyungpook National University (KR)

JENG, Geng-Yuan University of Maryland (US)

JEON, Hyebin Kyungpook National University (KR)

JEON, JinA Sungkyunkwan University (KR)

JO, Hyunsuk Kyungpook National University (KR)

KANG, Kookhyun Kyungpook National University (KR)

KANG, Sinchul Kyungpook National University (KR)

KARKI, Sujita Kyungpook National University (KR)

KAZANIN, Vassili Budker Institute of Nuclear Physics (RU)

KESHELASHVILI, Irakli Forschungszentrum Jülich GmbH (DE)

IOP Conf. Series: Journal of Physics: Conf. Series **928** (2017) 011002 doi:10.1088/1742-6596/928/1/011002

KHAN, Arshad Kyungpook National University (KR)
KIM, Bobae Kyungpook National University (KR)
KIM, DongHee Kyungpook National University (KR)
KIM, Geon-Bo Institute for Basic Science (KR)
KIM, Guinyun Kyungpook National University (KR)
KIM, HongJoo Kyungpook National University (KR)
KIM, Hyelim Kyungpook National University (KR)
KIM, Sung-Hyun Hanyang University (KR)
KIM, Taehun Kyungpook National University (KR)
KOTCHETKOV, Dmitri Ohio University (US)
LAKTINEH, Imad Universite Claude Bernard-Lyon I (FR)
LEE, Hye Young Sungkyunkwan University (KR)
LEE, Hyun Su CUP, Institute for Basic Science (KR)
LEE, Jooyoung Kyungpook National University (KR)
LEE, Seh Wook Kyungpook National University (KR)
LEE, Seungcheol Kyungpook National University (KR)
LEE, Jongho Kyungpook National University (KR)
LUCCHINI, Marco Toliman CERN (CH)
MARCH, Luis Universite de Geneve (CH)
MARTINI, Matteo Istituto Nazionale Fisica Nucleare Frascati (IT)
MELNIKOVA, Natalya Budker Institute of Nuclear Physics (RU)
MERIDIANI, Paolo Universita e INFN, Roma I (IT)
MÜLLER, Fabian Forschungszentrum Jülich GmbH (DE)
MUNOZ CAMACHO, Carlos IPN-Orsay, CNRS/IN2P3 (FR)
NESSI-TEDALDI, Francesca ETH Zurich, Switzerland (CH)
NEUBUSER, Coralie Deutsches Elektronen-Synchrotron (DE)
NOVOTNY, Rainer Willi Justus-Liebig-University (DE)
OCHANDO, Christophe Centre National de la Recherche Scientifique (FR)
OCHESANU, Silvia Crytur spol. s.r.o (CZ)
PACINI, Lorenzo Universita e INFN, Firenze (IT)
PAK, Sang Il Kyungpook National University (KR)
PANDEY, Indra Raj Kyungpook National University (KR)
PARA, Adam Fermi National Accelerator Lab. (US)
PARK, Hwanbae Kyungpook National University (KR)
PARK, Jeongmin Kyungpook National University (KR)
PHAN, Vuong Kyungpook National University (KR)
PIGAZZINI, Simone Universita & INFN, Milano-Bicocca (IT)
POESCHL, Roman Laboratoire de l'Accelerateur Lineaire (FR)
POLIFKA, Richard University of Toronto (CA)

IOP Conf. Series: Journal of Physics: Conf. Series **928** (2017) 011002 doi:10.1088/1742-6596/928/1/011002

SAUVAN, Jean-Baptiste CERN (CH)

SCHOENENBERGER, Myriam Eidgenoessische Tech. Hochschule Zuerich (CH)

SHWARTZ, Boris Budker Institute of Nuclear Physics, Novosibirsk (RU)

SMITH, Jacob University of Maryland College Park (US)

SON, Jukyung Kyungpook National University (KR)

SONG, Jihwan Kyungpook National University (KR)

STÄRZ, Steffen CERN (CH)

STEEN, Arnaud National Taiwan University (TW)

STREBLER, Thomas LLR - Ecole Polytechnique (FR)

SUEHARA, Taikan Kyushu University (JP)

TAPAN, Ilhan Uludag University (TR)

TEIXEIRA DE LIMA, Rafael Northeastern University (US)

VALDES SANTURIO, Eduardo Stockholm University (SE)

VAN DE KLUNDERT, Merijn University of Antwerp (BE)

WANG, Zhigang Chinese Academy of Sciences (CN)

WIGMANS, Richard Texas Tech University (US)

WOODY , Craig Brookhaven National Laboratory (US)

YANG, Un-Ki Seoul National University (KR)

YE, Ryonghae Kyungpook National University (KR)

YOO, Hwi Dong Seoul National University (KR)

YOON, Young Soo Institute for Basic Science (KR)

YOON, Young Soo University of Maryland College Park (US)

ZHU, Ren-Yuan California Institute of Technology (US)

IOP Conf. Series: Journal of Physics: Conf. Series **928** (2017) 011003 doi:10.1088/1742-6596/928/1/011003

Peer review statement

All papers published in this volume of *Journal of Physics: Conference Series* have been peer reviewed through processes administered by the proceedings Editors. Reviews were conducted by expert referees to the professional and scientific standards expected of a proceedings journal published by IOP Publishing.

IOP Conf. Series: Journal of Physics: Conf. Series **928** (2017) 012001 doi:10.1088/1742-6596/928/1/012001

Seventy Years of Calorimetry

Richard Wigmans

Department of Physics
Texas Tech University
Lubbock, TX 79409-1051, U.S.A.

E-mail: wigmans@ttu.edu

Abstract. In this opening talk of the CALOR 2016 conference, I briefly review some milestones in the history of calorimetry as a detection technique in particle physics. I also discuss common misconceptions, which are unfortunately widespread, and give you my personal outlook on the future.

1. Introduction

We are gathered here in Daegu for the 17th edition of the CALOR conference series, which started in 1990 at Fermilab. Calorimeters have become the heart and soul of practically all modern experiments in high-energy accelerator experiments. They are also playing a very important role in experiments that study natural phenomena, such as high-energy cosmic rays and neutrinos of solar, atmospheric and cosmic origin. In this opening talk, I will briefly review 70 years of calorimetry as a particle detection technique, by discussing some of the milestones in this history. The rest of my talk is devoted to a discussion of a few misconceptions that are unfortunately all too common. I'll finish with my personal outlook on the future of this exciting field.

2. Milestones

2.1. Nuclear γ-ray detectors

The use of calorimeters as particle detectors was pioneered in nuclear physics, where crystals such as NaI(Tl) were already used as γ-ray detectors in the early 1950s. A first milestone occurred around 1965, when semiconductor detectors made their appearance. Figure 1 illustrates how the much larger number of signal quanta available in such detectors improved the energy resolution and led to a revolution in experimental nuclear physics. Nowadays, one can do even much better with certain cryogenic devices, which are based on the detection of Cooper pairs, whose binding energy is typically three orders of magnitude smaller than the energy needed to create an electron-hole pair in silicon or germanium. The resolution of all these calorimeters is dominated and limited by Poissonian signal quantum fluctuations, and thus scales like $E^{-1/2}$. This is typically not the case in calorimeters used for particle physics experiments, where the energies are measured in units of GeV, and increasingly in TeV. Sampling fluctuations, which also lead to a $E^{-1/2}$ energy dependence, typically dominate the energy resolution of electromagnetic sampling calorimeters, but especially in hadron calorimeters the energy resolution is limited by other factors.

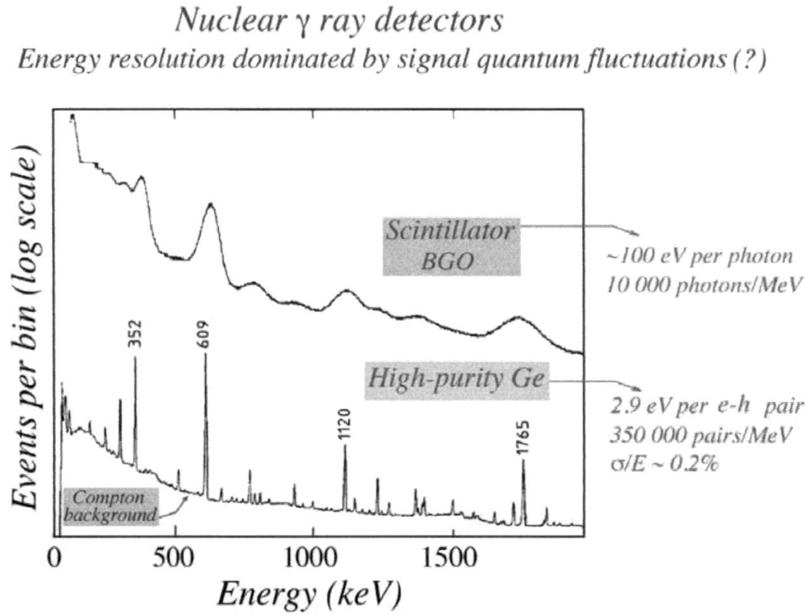

Figure 1. Nuclear γ-ray spectrum of decaying uranium nuclei, measured with a BGO scintillation counter (*upper curve*) and with a high-purity Ge crystal (*lower curve*).

2.2. *Shower counters*

The first application of calorimeters in particle physics experiments occurred around 1970. They were called *shower counters* and were used in magnetic spectrometers to detect the particles that produced no signals in the tracking detectors, mainly γs from π^0 decay. Figure 2 shows an example of such a spectrometer, used in Fermilab experiment E-70, which discovered the Υ family of $b\bar{b}$ resonances. The calorimeters used in this experiment were made of lead glass, a popular material in this type of application, where the calorimeters had a very limited and specific task.

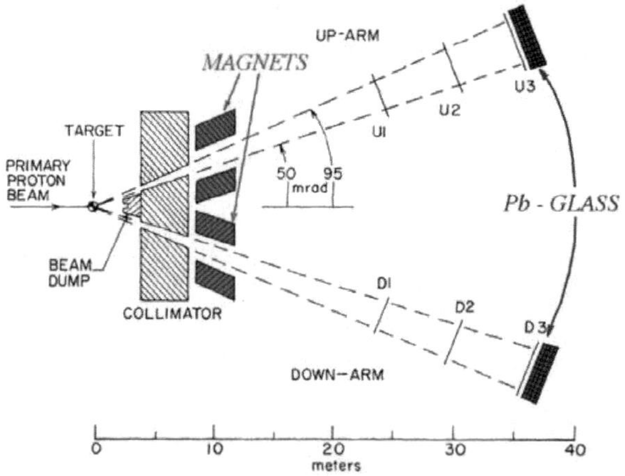

Figure 2. The magnetic spectrometer of Fermilab experiment E-70 [1].

IOP Conf. Series: Journal of Physics: Conf. Series **928** (2017) 012001 doi:10.1088/1742-6596/928/1/012001

2.3. Instrumented targets

During the mid-1970s, both Fermilab and CERN started experiments with neutrino beams. The detectors used for these experiments were essentially calorimeters, but they had very different and more elaborate tasks than the shower counters mentioned above. They acted at the same time as the target for the neutrino interactions and as the detector for the reaction products. Because of the very small cross section of the ν-induced interactions, these detectors had to be very massive (kilotons). They were typically sampling calorimeters, where the target and detector tasks were carried out by separate materials, *e.g.,* iron as absorber and scintillating plastic slabs as active material. It turned out that a properly chosen readout structure could also provide very effective triggers, which separated background events (*e.g.,* caused by cosmic rays) from the rare neutrino induced beam events.

2.4. 4π calorimeters

Because of the successful performance of these instrumented neutrino targets, calorimeters also became crucial components of detectors at colliding beam experiments. Figure 3 shows the 4π calorimeter of the UA2 experiment at CERN's proton-antiproton collider, where the intermediate vector bosons (W, Z) were discovered in 1983. This calorimeter was longitudinally segmented into a $17X_0$ deep em section

Figure 3. The calorimeter used in the UA2 experiment at the CERN $Sp\bar{p}S$ collider [2].

(lead/plastic-scintillator), followed by two hadronic sections (iron/plastic-scintillator), each $2\lambda_{\rm int}$ deep. The central calorimeter shown in this figure was segmented into 240 independent cells, each covering $10°$ in the polar angle and $15°$ in azimuth, and built in a tower structure pointing toward the center of the interaction region. It served as a model for many calorimeters that would follow in years to come.

2.5. Compensation

On a fundamental level, one of the main mysterious aspects of calorimetry was the poor performance for hadronic particle detection. While sampling calorimeters could achieve energy resolutions at the 1% level without too much difficulty for detecting electrons and photons, hadronic resolutions were worse by a considerable factor in the same instruments. In addition, calorimeters were typically nonlinear for hadron detection, gave a different response to protons and pions of the same energy and had asymmetric, non-Gaussian response functions. In an attempt to solve this issue by brute force, eliminating all possible instrumental effects, Benvenuti *et al.* built a homogeneous calorimeter consisting of 60 tonnes of liquid scintillator (Figure 4a) and tested it with beams of hadrons. However, the energy resolution of this

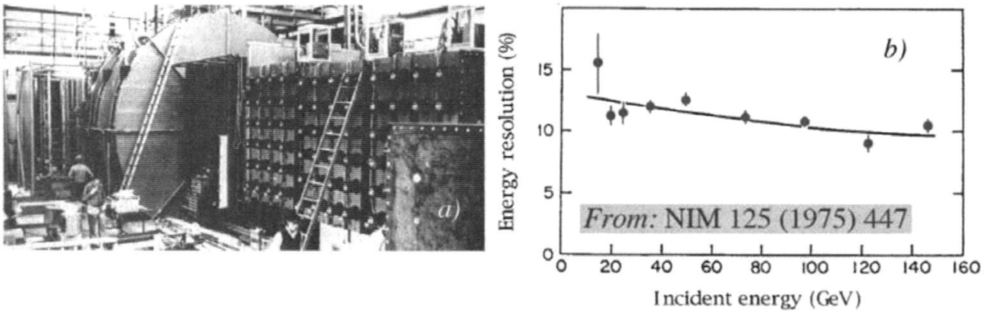

Figure 4. A calorimeter consisting of 60 tonnes of liquid scintillator (a), and the hadronic energy resolution measured with this detector (b).

detector did not get better than $\sim 10\%$, at any energy (Figure 4b) [3]. Clearly, fluctuations in the number of scintillation photons produced in the shower development did not contribute significantly to the energy resolution of this instrument.

The reasons for the mentioned problems with hadron shower detection, and the solution to these problems, were discovered/developed in the second half of the 1980s. Key to the solution was to design the calorimeter in such a way that the response (average signal per unit deposited energy) was the same for the em and non-em components of hadron showers: $e/h = 1$. One detector in which this condition was realized is shown in Figure 5, which also shows the response function for pions at three different energies. In terms of hadronic energy resolution, this detector still holds the world record. The detector was also very linear for hadron detection, and the response functions were well described by Gaussian functions.

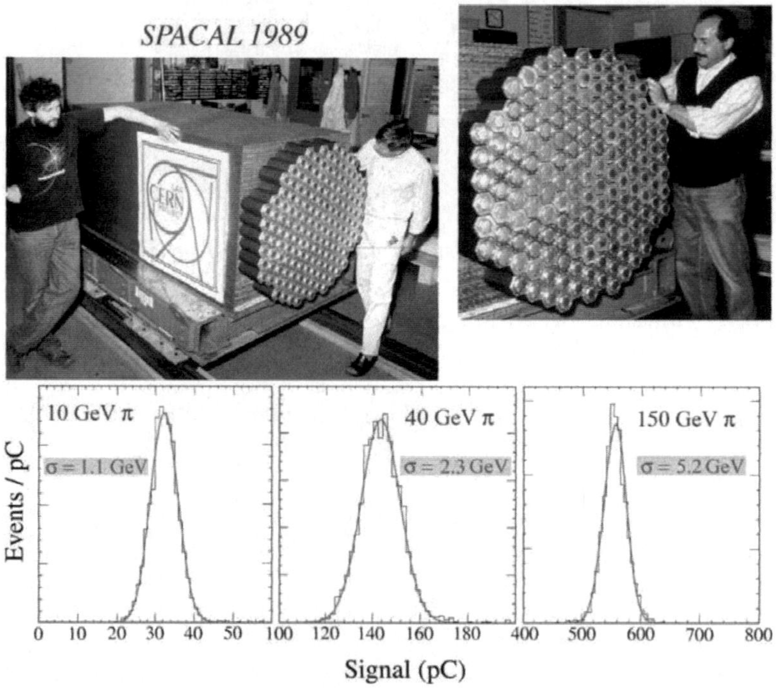

Figure 5. The Spaghetti calorimeter built at CERN (1989), and hadronic response functions measured with this detector [4].

IOP Conf. Series: Journal of Physics: Conf. Series **928** (2017) 012001 doi:10.1088/1742-6596/928/1/012001

2.6. Dual-readout calorimetry

Compensating hadron calorimeters depend crucially on the contributions of the neutrons that are abundantly produced in hadronic shower development to the signals. This implies a small sampling fraction, and a relatively large signal integration time and volume. In order to crcumvent these limiting conditions, while still benefitting from the important advantages offered by compensation, dual-readout calorimetry was developed. Dual-readout calorimeters provide two signals, *e.g.,* light produced by scintillation and Čerenkov processes. These signals provide complementary information about the shower development, and make it possible to measure the em fraction of hadron showers *event by event.* This in turn makes it possible to eliminate the effects of event-to-event fluctuations in that em component on the calorimeter performance. The principle is illustrated in Figure 6, and was experimentally demonstrated in [5]. Until now, no full scale calorimeters of this type have been constructed, but

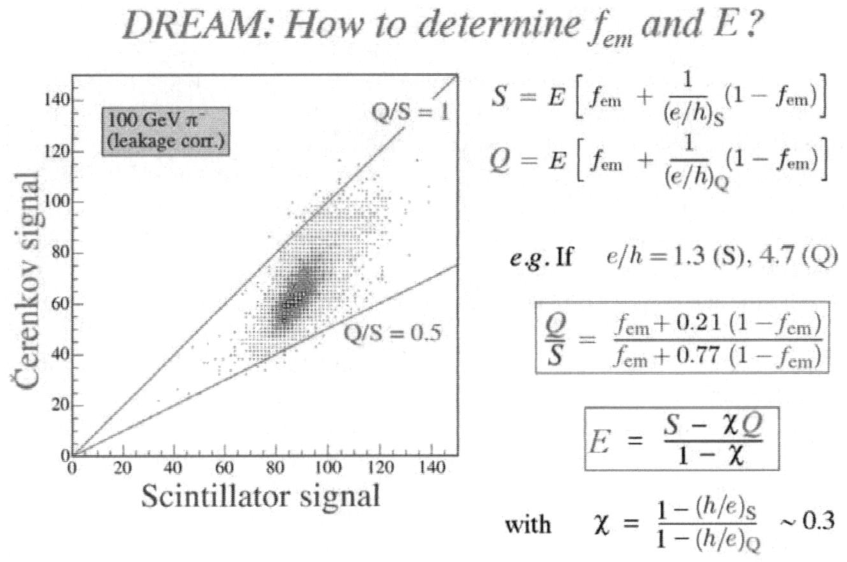

Figure 6. Measurement of the em energy fraction ($f_{\rm em}$) and the energy of showering hadrons (E) on the basis of the measured ratio of the Čerennkov and scintillation signals (Q/S) from the DREAM copper/fiber calorimeter [5].

results obtained with 1-ton detectors, combined with Monte Carlo simulations, have indicated that the performance may be at least as good as that of the best compensating calorimeters, with better energy resolution for em showers.

2.7. Imaging calorimeters

Inspired by the development of the Time Projection Chamber, which allows tracking of charged particles in a three-dimensional volume, imaging calorimeters have been developed. These detectors, in which the particles are absorbed in a large volume of liquid argon, provide information about developing showers that rivals that of the bubble chambers of several decades ago. The difference is that this information is now provided fully electronically and does not have to be obtained by laborious scanning of photographs of the events. Figure 7 shows an example of an event in the imaging calorimeter of the ICARUS experiment, which pioneered this technique [6]. It is important to realize that, since calorimeters of this type only register charged shower particles, one should not expect hadronic energy resolutions at the level of that achieved with compensating or dual-readout calorimeters, which are specifically designed with high-quality calorimetric hadron detection in mind.

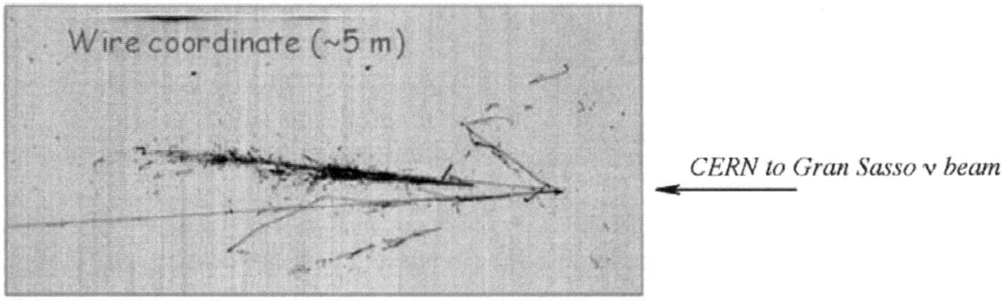

Figure 7. Example of a neutrino interaction in the ICARUS T600 calorimeter, installed in the Gran Sasso lab (Italy).

3. Misconceptions

3.1. Shower particles contributing to the calorimeter signals

The most common, important and consequential misconception about calorimetry is that a shower is a collection of minimum ionizing particles (mips). Already in the early days, it was realized that the signal from a high-energy electron absorbed in a sampling calorimeter was substantially different from that of a muon that traversed this calorimeter and deposited the same energy in it as the showering electron. This is due to the fact that the composition of the em shower changes as a function of depth, or age. In the late stages, most of the energy is deposited by soft γs which undergo Compton scattering or photoelectric effect, and the sampling fraction for this shower component (*i.e.,* the fraction of the energy that contributes to the calorimeter signals) may be very different from that of the mips that dominate the early stages of the shower development. This causes major complications for the intercalibration of the different sections of a longitudinally segmented calorimeter, as is well known from the experiences of, for example, ATLAS [7] and AMS [8], and will also have important consequences for applications of Particle Flow Analysis (PFA), which relies on calorimeters with many longitudinal segments.

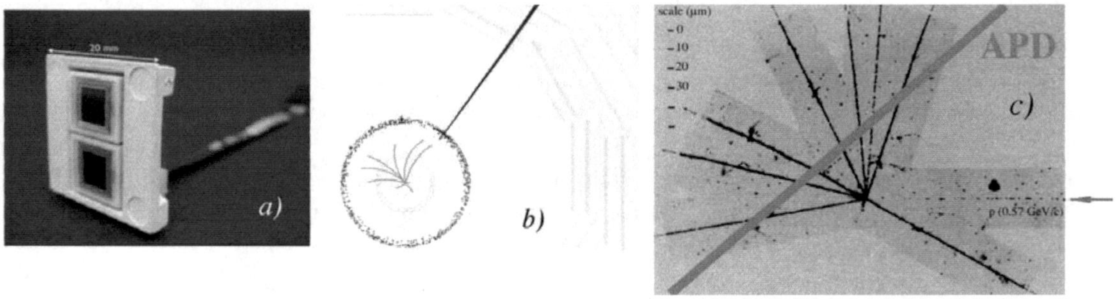

Figure 8. Arrangement of the APDs used to read the signals from the CMS ECAL (a). A "spike" event recorded by the CMS calorimeter system (b). A typical nuclear interaction in a developing hadron shower, induced by a 0.57 GeV/c proton (c).

Another aspect of this problem is the fact that *a single* shower particle may cause catastrophic effects for the calorimeter performance. This is particularly true for hadron showers, and may be illustrated by a recent example taken from the CMS experiment. The CMS calorimeter system consists of a crystal based em calorimeter, followed by a brass/plastic-scintillator hadronic compartment. Each $PbWO_4$ crystal is read out by two APDs (Figure 8a). When hadrons are sent into this calorimeter system, it produces sometimes events in which an anomalously large signal is recorded in one individual crystal. Such events are referred to as "spikes" (Figure 8b). Figure 8c shows an example of a nuclear interaction that is a typical feature in hadron shower development. A proton with a momentum of 0.57 GeV/c interacts with a nucleus of the absorber structure, and produces seven even lower-energy charged particles, presumably protons and/or nuclear aggregates such as α particles in this process. Presumably, there are also at least as many neutrons produced in this reaction, but these do not ionize the material and are thus invisible in this figure. The charged fragments are all heavily ionizing, with typical dE/dx values of 100 - 1000 times that of a mip. If such an event happens close to an APD, these charged fragments may create a very large signal. The APDs are intended to detect scintillation photons produced in the $PbWO_4$ crystals, and the energy scale of the calorimeter signals is set by the production rate of such photons. However, the APDs produce much larger signals when traversed by a charged particle. The densily ionizing fragments of an event such as the one shown in Figure 8c may produce signals that are interpreted as an energy deposit of several hundred GeV inside the scintillating crystals, and this is precisely what causes these spikes. As an aside, I mention that this phenomenon should be very easily recognizable if the two APDs connected to each crystal would be read out separately, since the described phenomenon would only occur in one of them. However, in order to save money, CMS had ganged them together and treated the two APDs as one readout cell.

3.2. Scaling of the energy resolution

Another widespread misconception concerns the energy resolution. Frequently, the relative energy resolution (σ/E) of a particular calorimeter is expressed as $x\%/\sqrt{E}$. However, this is rarely a correct description of reality, since in practice other factors, which are not governed by Poisson statistics, contribute to the energy resolution, and such factors often dominate the performance, especially at the low and high ends of the energy spectrum for which the detector is intended. As an example, Figure

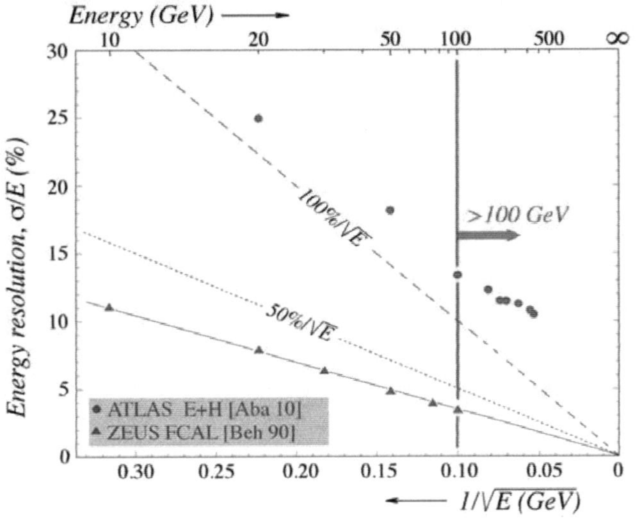

Figure 9. Hadronic energy resolutions of the ZEUS [9] and ATLAS [10] calorimeters.

9 shows the hadronic energy resolutions of the ZEUS and the ATLAS calorimeters. The experimental data points are plotted on a scale that is linear in $E^{-1/2}$ and runs from right to left. Scaling with $1/\sqrt{E}$ implies that the data points should be located on a straight line through the bottom right corner in this plot. This is indeed the case for the compensating ZEUS calorimeter, for which the resolution is quoted as $35\%/\sqrt{E}$. However, the resolution of ATLAS does not at all scale with $1/\sqrt{E}$. As a matter of fact, the data points are at all energies located well above the $100\%/\sqrt{E}$ line in this plot, and for energies larger than 100 GeV, the resolution is more than a factor of four worse than for ZEUS. Yet, in talks about ATLAS, the hadronic energy resolution is often quoted as $0.6 - 0.7/\sqrt{E}$.

Another common mistake with regards to energy resolution has to do with its very definition. The energy resolution is the precision with which the energy of an unknown object can be determined from the signals it produces in the calorimeter. Typically, this resolution is determined as the relative width of the signal distribution measured for a beam of mono-energetic particles from an accelerator. However, this is only correct if the average value of that measured signal distribution corresponds indeed to the correct energy of these particles. Response non-linearities tend to invalidate that assumption.

3.3. Signal (non)linearity

Calorimeters may be non-linear for a variety of reasons. Intercalibration of longitudinal sections, signal saturation and the energy dependence of the em shower fraction (in hadron showers) are the most common causes. Many calorimeters are non-linear, even though their owners often pretend otherwise.

A common misconception is that a calorimeter is linear if the average signals plotted versus the deposited energy can be described with a straight line. **This is incorrect**. The straight line has to extrapolate through the origin of the plot. Signal linearity means that the average calorimeter signal is *proportional* to the deposited energy, *i.e.*, the *response is constant*. Figure 10 illustrates this issue.

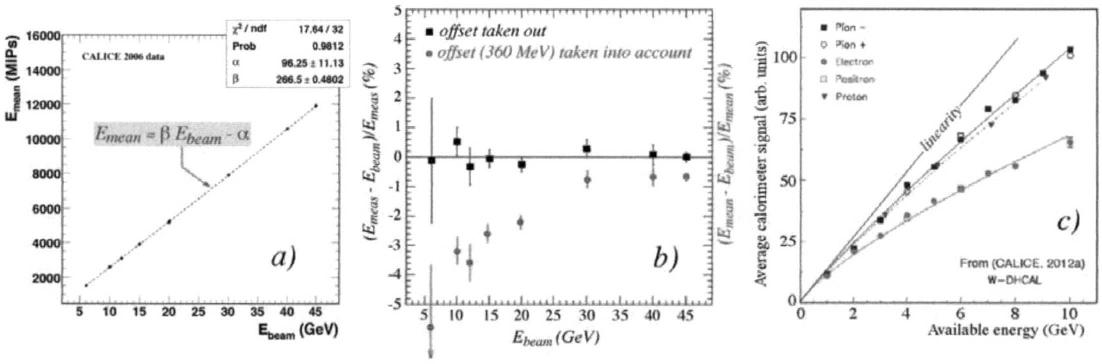

Figure 10. Average signal as a function of electron energy for the W/Si ECAL built by CALICE (*a*) [11]. Residual signals from this detector, before and after taking out a 360 MeV offset (*b*). Non-linearity as a result of saturation effects in a digital CALICE calorimeter [12].

Diagrams *a* and *b* concern data obtained with a W/Si em calorimeter built by CALICE [11]. The authors fit the measured signals with the following expression:

$$E_{\text{mean}} = \beta\, E_{\text{beam}} - 360 \text{ MeV} \tag{1}$$

Then, they define

$$E_{\text{meas}} = E_{\text{mean}} + 360 \text{ MeV} \tag{2}$$

and plot

$$(E_{\text{meas}} - E_{\text{beam}})/E_{\text{meas}}$$

as a function of the beam energy. The result is represented by the (black) squares in Figure 10b. They conclude that "*the calorimeter response is linear to within approximately 1%*." This is highly misleading. When the calorimeter signals they actually *measured* are used to check the linearity, *i.e.*, when

$$(E_{\text{mean}} - E_{\text{beam}})/E_{\text{mean}}$$

is plotted as a function of the beam energy, the results, represented by the (red) full circles in Figure 10b, look quite different. I conclude from these results that the authors measured a response non-linearity of 5% over one decade in energy.

Whereas this type of nonlinearity is the result of the intercalibration of the numerous longitudinal segments of this PFA calorimeter, Figure 10c shows non-linearity with a different origin. It concerns data obtained with a digital hadron calorimeter built by CALICE [12]. This calorimeter contains 500,000 readout cells (1×1 cm^2 RPCs), which produce "digital" signals ("yes" or "no") in response to charged particles. However, this type of cell produces the same signal, regardless whether it is caused by 1, 3 or 29 shower particles. This leads to signal non-linearity, more so in em showers which are narrower than hadronic ones and where the shower particle density is thus larger. In this calorimeter, an important source of fluctuations is suppressed, and therefore also the energy resolution measured with it is meaningless.

3.4. Other misconceptions
Other widespread misconceptions concern

- The practical importance of compensation and its consequences, which are often regarded as limited to a *constant term in the hadronic energy resolution* (which is incorrect by itself, since the effects of non-compensation are *not* energy independent),
- The presumed equivalence of showers induced by γs and electrons/positrons of the same energy
- The presumed equivalence of showers induced by protons, pions and kaons of the same energy, and
- The assumption that all calorimeter problems can be solved offline

For details on these issues, the reader is referred to the new edition of my book [13].

4. Outlook

Calorimetry has come a long way in the past seventy years. Much of what has been learned about the inner workings of these somewhat mysterious detectors has been the result of dedicated generic research & development projects, although the important contributions of work carried out in the ZEUS prototype phase definitely deserve an honorary mention in this context.

Unfortunately, times have changed. There are no longer significant resources available for this type of R&D. New experiments are designed based on somebody's concept of what the detectors should look like, and prototype work primarily concentrates on technical aspects of that concept. This approach is followed, for example, in projects carried out in the CALICE framework, which is geared towards application of PFA at future electron-positron colliders [12]. Based on my observations, calorimetry research in this new era is characterized by ignorance, misconceptions (such as the ones discussed in the previous section) and a general lack of interest in fundamental issues, combined with a strong belief that all eventual problems can be solved with technology: the use of tungsten, silicon, RPCs must of course lead to better results, because this is a more modern approach than lead + plastic.

One experiment that experiences the consequences of this modern approach is CMS. Every component of the CMS calorimeter system suffers from design flaws that were preventable:

- The $PbWO_4$ crystals composing the ECAL are each read out with two APDs, which were ganged together and treated as one readout element. If these APDs had been read out individually, spike events (Figure 8), which will become an even more serious problem than they are already now when the luminosity and collision energy are further increased, could have been avoided.
- The signals from the hadron calorimeter are read out by "Hybrid Photon Detectors", a technology that had never before been used on a comparable scale, and which turns out to have so many problems (*e.g.,* with noise) that this calorimeter is at the present time barely useable, and the decision has been made to completely overhaul this readout system. These problems could have been avoided in the first place with several available low-tech options.
- The Very-Forward calorimeter (HF) has such a small sampling fraction that each photoelectron produced by Čerenkov light in the quartz fibers corresponds to 5 GeV energy deposit. As a result, this calorimeter is extremely sensitive to background effects, *e.g.,* coming from beam halo muons traversing the PMTs. These produce signals consisting of dozens of photoelectrons and are thus interpreted as sources of very large energy deposits.

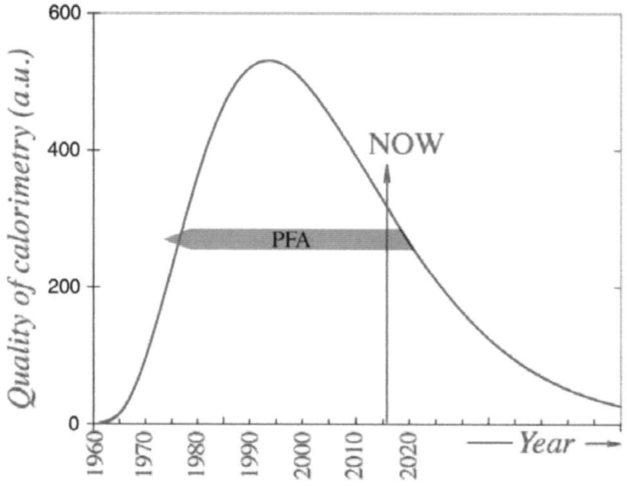

Figure 11. The quality of calorimetry for particle physics experiments as a function of time. See text for details.

In a new development, CMS has decided to replace the endcap section of its calorimeter system, where the supposedly very radiation hard $PbWO_4$ crystals have already been seriously damaged by radiation after $\sim 1\%$ of the total expected lifetime dose, by a W/Fe/Si PFA calorimeter. Also this decision was not based on any serious prototype studies, and there are good reasons to believe that this is a disaster waiting to happen, comparable in scope (and cost) to the "warm-liquid upgrade" that meant the end of the UA1 experiment 30 years ago. Interestingly, some of the characters involved in this decision are the same.

Because of these developments, I think the future of calorimetry is bleak. Figure 11 depicts the quality of the calorimeters used in particle physics experiments as a function of time. This curve remarkably resembles a longitudinal shower profile. We have clearly passed the peak, and I am afraid that large-scale applications of PFA, such as envisaged in CMS, will set us back half a century, to the days of the magnetic spectrometers such as the one shown in Figure 2. PFA systems will actually be 4π versions of the magnetic spectrometers from those days (Figure 2), albeit on a scale that is an order of magnitude more compact (4 m instead of 40 m), with all the problem associated with that, since showers develop irrespective of where the absorbing calorimeters are located.

References

[1] Appel, J A *et al.*(1975). Nucl. Instr. and Meth. **127**, 495.
[2] Bagnaia, P *et al.*(1984). *Z. Phys.* **C24**, 1.
[3] Benvenuti, A *et al.*(1975). Nucl. Instr. and Meth. **125**, 447.
[4] Acosta, D *et al.*(1991). Nucl. Instr. and Meth. **A308**, 481.
[5] Akchurin N *et al.*(2005) Nucl. Instr. and Meth. **A537**, 537.
[6] Almerio, S. *et al.*(2004). Nucl. Instr. and Meth. **A527**, 329.
[7] Aharrouche M *et al.* 2006, Nucl. Instr. and Meth. **A568**, 601.
[8] Cervelli F *et al.* 2002, Nucl. Instr. and Meth. **A490**, 132.
[9] Behrens U *et al.* 1990, Nucl. Instr. and Meth. **A289**, 115.
[10] Abat, E *et al.*(2010). Nucl. Instr. and Meth. **A621**, 134.
[11] Adloff, C *et al.*(2009). Nucl. Instr. and Meth. **A608**, 372.
[12] Sefkow, F *et al.*(2016). *Rev. Mod. Phys.* **88**, 1.
[13] Wigmans R 2017, *Calorimetry - Energy Measurement in Particle Physics*, 2nd ed., Oxford University Press

IOP Conf. Series: Journal of Physics: Conf. Series **928** (2017) 012002 doi:10.1088/1742-6596/928/1/012002

ATLAS Tile calorimeter calibration and monitoring systems

Arthur CHOMONT, on behalf of the ATLAS collaboration
LPC Université Blaise Pascal, 4 avenue Blaise Pascal, 63178 AUBIERE CEDEX
E-mail: arthur.chomont@cern.ch

Abstract. The ATLAS Tile Calorimeter (TileCal) is the central section of the hadronic calorimeter of the ATLAS experiment and provides important information for reconstruction of hadrons, jets, hadronic decays of tau leptons and missing transverse energy. This sampling calorimeter uses steel plates as absorber and scintillating tiles as active medium. The light produced by the passage of charged particles is transmitted by wavelength shifting fibres to photomultiplier tubes (PMTs), located on the outside of the calorimeter. The readout is segmented into about 5000 cells (longitudinally and transversally), each of them being read out by two PMTs in parallel. To calibrate and monitor the stability and performance of each part of the readout chain during the data taking, a set of calibration systems is used. The TileCal calibration system comprises cesium radioactive sources, Laser and charge injection elements, and allows for monitoring and equalization of the calorimeter response at each stage of the signal production, from scintillation light to digitization. Based on LHC Run 1 experience, several calibration systems were improved for Run 2. The lessons learned, the modifications, and the current LHC Run 2 performance are discussed.

1. Introduction

The TileCal Calorimeter of ATLAS [1] is a sampling calorimeter made of scintillating tiles as active medium and steel plates as absorbers. The light produced by charged particles through scintillating tiles are collected by wavelength shifting fibres and transmitted to the PMTs. The signal from the PMTs is shaped and amplified with two gains. The sampling and the digitization is realised by ADCs. During collisions, if an event is selected by the trigger system, the digitized signals are collected and processed by a ReadOut Driver. In parallel to this, integrators measure the integrated current from the PMTs.
The reconstructed energy of each channel, E(GeV), is derived from the raw response, A(ADC), as follows:

$$E(GeV) = A(ADC)C_{ADC \to pC}C_{pC \to GeV}C_{Cesium}C_{Laser}$$

This relation is established for each TileCal channel. The factors can evolve in time because of instabilities in PMTs response induced by high light fluxes or optics aging. The calibration systems are used to monitor the stability of these elements, and provide per-channel calibration. The conversion constant from pC to GeV, $C_{pC \to GeV}$, was determined during dedicated test beam campaigns. The remaining calibration constants are provided by individual systems:

- A movable cesium gamma-source to calibrate the optical components and the PMT gains.
- A Laser system to monitor the PMTs and the electronics components.
- A Charge Injection System (CIS) to monitor the electronics and extract the conversion factors from ADC counts to pC.

A monitoring of beam conditions and Tile optics with the so-called integrator system (minimum bias) can also be used in combination with Laser and cesium results to understand channel gain deviations, because the calibration tools follow different and partially overlapping readout paths allowing for easier identification of potential failure and for crosschecks.

2. The cesium calibration system

The cesium calibration system [2] uses three movable ^{137}Cs-sources, emitting 0.662 MeV photons to illuminate scintillators. The signal generated by the Cs source is collected through a special readout that integrates the analog PMT signals. This system is used to calibrate the optical components of the calorimeter as well as to monitor the PMTs. The precision of the system is at the order of 0.3% with cesium calibration scans spaced by one to three months.
From Run 1 to Run 2 of the LHC, a new water storage system, lower pressure in the hydraulics, and more precise water level metering during the scans have been achieved to improve stability and safety of the operation.
The deviation of C_{Cs} calibration constant from 1 for A cells (in %) over the period of six years (2009-2015) is shown in Fig.1.

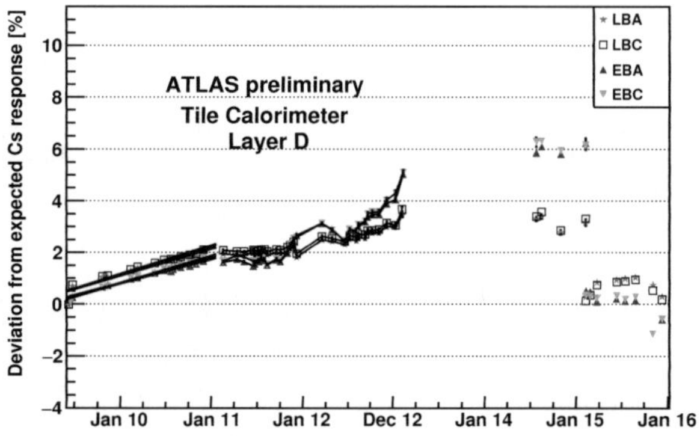

Figure 1. Deviation of TileCal response to cesium sources per partition, in the inner radial layer (A layer), from 2009 to 2016. Figures are taken from Ref.[5].

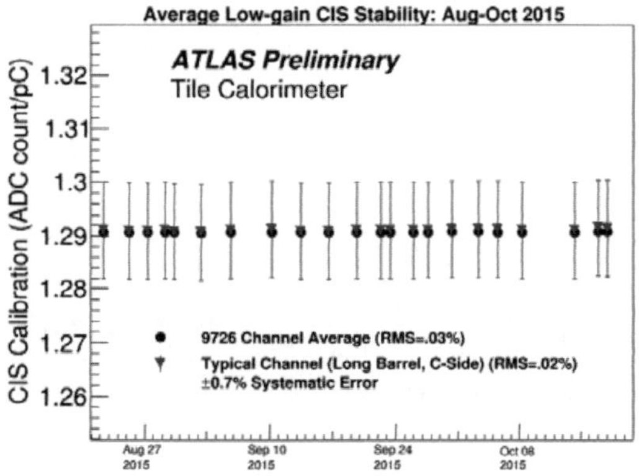

Figure 2. Time stability of the average high-gain readout calibration constants from August 2015 to October 2015 for 19421 ADC channels. The calibrations are measured in-situ with the Charge Injection System (CIS). The RMS values printed in the legend represent a measure of the fluctuations present in calibrations. Figures are taken from Ref.[5].

3. The Laser calibration system

The Laser calibration system [3] is used to measure the gain of each channel sending a known amount of light to the photocathode of the PMTs. The Laser system is used to monitor both High- and Low-gain regimes of the calorimeter readout by modifying the light amplitudes. Deviations of channel responses are considered with respect to their nominal responses (at the

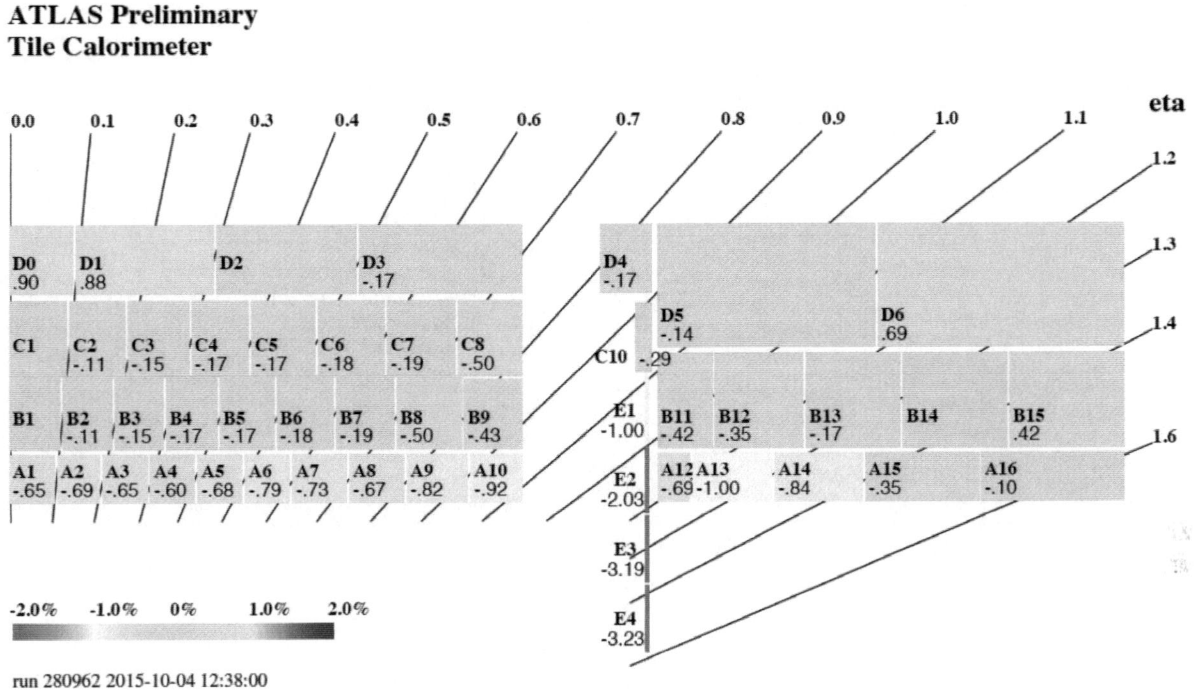

Figure 3. Difference in PMT response in % (averaged over 64 modules) measured in two Laser calibration runs in October 2015. Figures are taken from Ref.[5].

time of the latest cesium calibration). The Laser calibration runs are usually taken twice a week to monitor the individual PMT gain variations between the cesium monthly scans.
The typical precision on the gain variation during Run 1 of the LHC was better than 0.5% per channel. During the Long Shutdown from 2012 to 2015 of the LHC, a new Laser II system has been developed to correct shortcomings in electronics and light monitoring of the first system with the consequence to have an improved resolution. The new system has been used since the beginning of the Run 2 in 2015 to monitor the channel gain deviations as shown on Figure 3.

4. The Charge Injection System

The CIS [4] is used to calibrate the readout electronics by injecting a known charge in the ADC and measuring the electronic response. The conversion factor from pC to ADC counts, $C_{ADC \to PC}$, is extracted from this measurement for both High-Gain and Low-Gain regimes of the readout. This system is also used to monitor the front-end electronics and to correct for non-linearities.
The overall stability of the calibration factor is at the level of 0.02% (Fig.2) and usually less than 1% of the channels exhibit large fluctuations. Their calibration constants are updated if the deviation from previous measured value is above the CIS systematic error (0.7%),

5. The Minimum Bias system

The Minimum Bias system can also be used for calibration. The first purpose of this system is the monitoring of ATLAS instantaneous luminosity.
Both the Minimum Bias and the cesium systems measure the signal coming from scintillators and the variation in PMT response over time is expected to be the same. The comparison

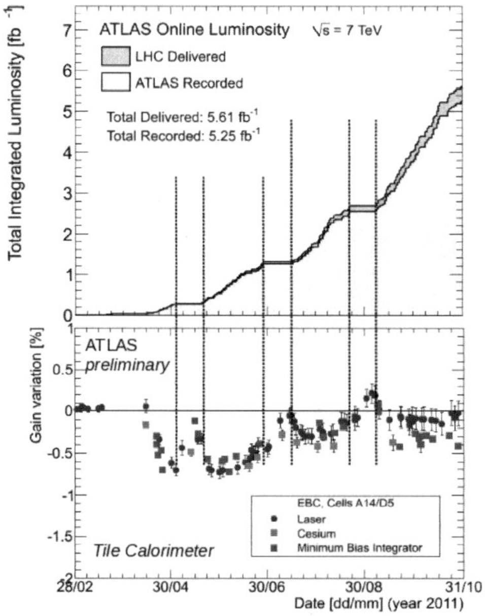

Figure 4. The upper part described the increase in the total integrated luminosity delivered by the LHC and recorded by ATLAS during 2012. The plateaus described period without collision. The bottom part described the variation of the response in cell A13 ($1.2 < |\eta| < 1.3$) during 2012 measured by 3 calibration systems: Minimum Bias (red squares), cesium (green squares), Laser (blue circles). The agreement between the three measurements could indicate that the origin of the deviation is the PMT. Figures are taken from Ref.[5].

between those systems and the Laser system can give insights on origins of channel drifts as illustrated in Figure 4.

6. Conclusion

The calibration systems of the TileCalorimeter were operated susccessfully throughout Run 1 of the LHC. The stability in time achieved using these systems during Run 1 was within 1%. During the LS1 of the LHC from 2012 to 2015, shortcomings of Laser and cesium systems have been corrected. Since the beginning of Run 2 of the LHC the calibration systems have been used attaining an improved resolution with respect to Run 1.

References

[1] The ATLAS Collaboration, The ATLAS experiment at CERN Large Hadron Collider, JINST 3 (2008) S08003.
[2] E.A. Starchenko et al., Cesium monitoring system for ATLAS Tile Hadron Calorimeter, Nucl. Instrum. Meth.A494 (2002) 381.
[3] S. Viret, et al., The TileCal Laser monitoring, Nucl. Instrum. Meth. A617 (2010) 120122.
[4] K.J. Anderson, et al., Design of the Front-end Analog Electronics for the ATLAS Tile Calorimeter, Nucl.Instrum. Meth. A551 (2005) 469.
[5] TileCal Public Results, https://twiki.cern.ch/twiki/bin/view/AtlasPublic/ApprovedPlotsTileCalibration .

ATLAS Tile Calorimeter time calibration, monitoring and performance

T Davidek[1] **on behalf of the ATLAS Collaboration**

[1] Charles University in Prague, Faculty of Mathematics and Physics, IPNP, Czech Republic

E-mail: `Tomas.Davidek@mff.cuni.cz`

Abstract. The Tile Calorimeter (TileCal) is the hadronic calorimeter covering the central region of the ATLAS experiment at the LHC. This sampling device is made of plastic scintillating tiles alternated with iron plates and its response is calibrated to electromagnetic scale by means of several dedicated calibration systems.

The accurate time calibration is important for the energy reconstruction, non-collision background removal as well as for specific physics analyses. The initial time calibration with so-called splash events and subsequent fine-tuning with collision data are presented. The monitoring of the time calibration with laser system and physics collision data is discussed as well as the corrections for sudden changes performed still before the recorded data are processed for physics analyses. Finally, the time resolution as measured with jets and isolated muons is presented.

1. Introduction

The Tile Calorimeter is part of the ATLAS experiment [1], which explores the proton-proton (and also lead-lead) collisions at unprecedented high energies achieved at the LHC. TileCal surrounds the liquid-argon electromagnetic and hadronic end-cap calorimeters and it spans over the ATLAS central region ($|\eta| < 1.7$). Mechanically, the calorimeter is divided into a central part (Long Barrel) and two Extended Barrels. Each part consists of 64 wedges (modules), see Fig. 1.

The readout cells are organized into 3 radial layers of the total depth of $7.4\,\lambda_{\rm int}$. The cell segmentation in pseudorapidity is 0.1 ($\Delta\eta = 0.2$ in the outermost radial layer), while in azimuthal direction the cell size is defined by the module geometry ($\Delta\phi = 2\pi/64 \approx 0.1$). Each cell is readout by two PMTs and the signals are further processed by the readout electronics.

1.1. Calibration systems

TileCal exploits three dedicated calibration systems. Each of them monitors different stage of the signal processing:

- Cesium system measures the signal induced by ^{137}Cs radioactive source that passes through all tiles in dedicated calibration runs. It calibrates optics and PMTs, the signal is read out through dedicated slow electronics used for Cs system and minimum bias monitoring.
- Laser pulses are used to measure the PMT response combined with the fast readout electronics that is also used for physics data.

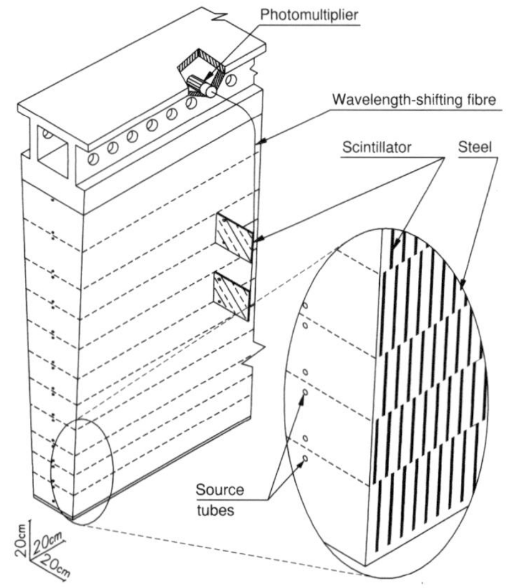

Figure 1. The sketch of one TileCal module shows its geometry and the principle of the signal readout [2]. The scintillating tiles are oriented parallel to the incoming particles. They are readout on both edges by wavelength-shifting (WLS) fibers, which route the light to the photomultipliers (PMTs) housed in the outer part together with the front-end electronics. The readout cells are defined by grouping the fibers onto a PMT.

- Charge injection system calibrates the fast readout electronics by injecting pulses of specified charge into the fast readout electronics.

The combination of the results from individual calibration systems allows to calibrate the detector response at the electromagnetic scale as well as to spot where the problem is in case of unexpected behaviour [2]. The calibration systems and their performance are also detailed in another contribution presented at the same conference [3].

1.2. Signal reconstruction

The analog pulse from the PMT is shaped and split into two branches (high- and low-gain, gain ratio 64:1) to ensure both good signal-to-noise ratio for small signals (e.g. from muons) and large dynamic range up to $\sim$ 800 GeV in each channel.

Each pulse is then sampled every 25 ns and the amplitude A (proportional to the deposited energy), time t_0 and quality factor are reconstructed. Currently, the Optimal Filter (OF) is used in physics data reconstruction [4]:

$$A = \sum a_i S_i \qquad (1)$$

$$t_0 = \frac{1}{A}\sum b_i S_i \qquad (2)$$

where S_i represent individual values of the sampled pulse. The OF constants a_i, b_i, derived from the known pulse shape, depend on the pulse arrival phase, which is determined in the time calibration.

2. Time calibration

As already discussed in Section 1.2, the time calibration is important for the energy measurement. Nevertheless, the precise time synchronization is also relevant for the removal of signals which do not originate from the proton-proton (pp) collisions as well as for the time-of-flight measurement of hypothetical heavy slow particles that would enter in the calorimeter [5].

The aim of the time calibration is to set the phase in each channel so that a particle travelling from the ATLAS interaction point at the speed of light produces the signal with measured time equal to zero. The time calibration is performed in the two following steps.

2.1. Splash events
Special single beam data-taking sessions (runs) used by various detectors for calibration purposes are usually organized at the beginning of the LHC data-taking period. During these runs, only one bunch of about $5 \cdot 10^8$ protons is circulating in LHC and it hits the closed collimator approximately 40 m upstream of the ATLAS detector.

Many high-energy particles are produced in these so-called splash events. Since TileCal is located in the central part of the ATLAS experiment, high energy muons produced parallel to the beam axis are the main component of the particle flux entering the calorimeter. Large signals are observed during splash events in all calorimeter cells, allowing a precise time measurement of the pulses. Time constants are derived in each channel after accounting for the time of flight of the particles crossing the calorimeter.

2.2. Calibration with jets
The pp collisions are also used to calibrate the timing of the calorimeter cells. Events with jets pointing to the TileCal η-range and satisfying standard quality criteria are used for this purpose. In order to avoid possible bias from non-collision background, only cells/channels being a part of reconstructed jets are considered in the given event. Since the average cell time slightly depends on the energy deposited in a cell (see Section 4), a specific energy range has to be used. In each channel, the condition $2 < E_{\text{channel}} < 4$ GeV is further imposed as a trade-off between the statistics and the Gaussian shape of the time response.[1]

The time constants are determined as a Gaussian mean of the time distribution in each channel, obtained with the above mentioned procedure.

3. Time monitoring
Laser calibration events are shot during physics runs in the empty bunch-crossings[2] with a frequency of about 1 Hz. Since all PMTs are illuminated at the same time, these data provide good statistics to monitor the time calibration. Laser events are not synchronized with the LHC clock, therefore a phase correction (average of the reconstructed time in all TileCal channels) is applied in every event. A dedicated tool analyzing these events was set up in order to spot sudden changes in the time synchronization. An example is shown in Fig. 2. The laser-based results are available soon after the physics run is over, therefore the corresponding time constant correction can be applied before the collision data are reconstructed for the physics analyses.

The impact of the time constant correction was investigated with jet collision data, the results are displayed in Fig. 3. While the Gaussian core of the time distribution remains unchanged (it corresponds to non-affected channels), the tails are reduced significantly.

A complementary time monitoring tool exploiting the jet collision data was developed for Run-2. It is based on the same principles as the time calibration with jets (see Section 2.2). These results are used as a cross-check to that from laser.

4. Time performance
As already noted in Section 2.2, the mean cell time measured with jets depends on the energy deposited in the cells. Figure 4 shows the mean cell time as a function of the cell energy after correcting for the run-to-run differences due to changes in the beam phase with respect to the LHC clock.

[1] The lower energy bins are more populated, but the corresponding time spectra exhibit high-time tails due to slow hadronic component of the shower.

[2] When LHC is filled with proton bunches, some consecutive positions are left empty. This translates into short interval (5.6 μs) without collisions in every LHC orbit. These intervals are used by individual sub-detectors for taking calibration events.

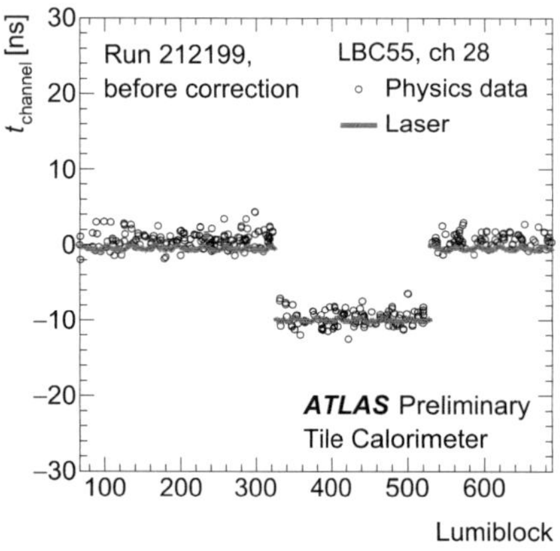

Figure 2. Example of sudden change in the time calibration as detected by the laser-based tool in one channel [6]. Superimposed are the results obtained with the jet collision data.

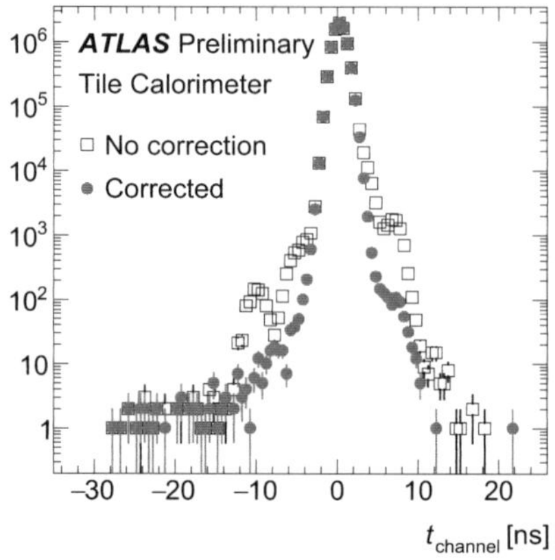

Figure 3. The impact of the time constant corrections for sudden changes in the time calibration, as observed for one data-taking period in Run-1 [6]. The RMS of the measured time distribution improves from 0.90 ns to 0.82 ns.

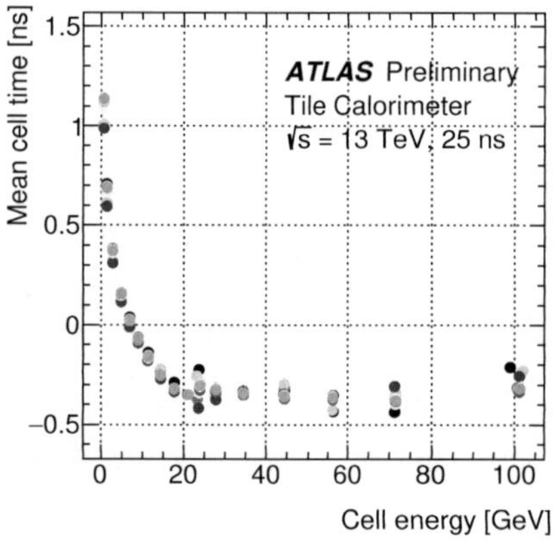

Figure 4. The mean cell time as measured with jet collision data as a function of energy deposited in the cell, after correcting for the run-to-run differences [6]. Symbols of different colours correspond to different runs.

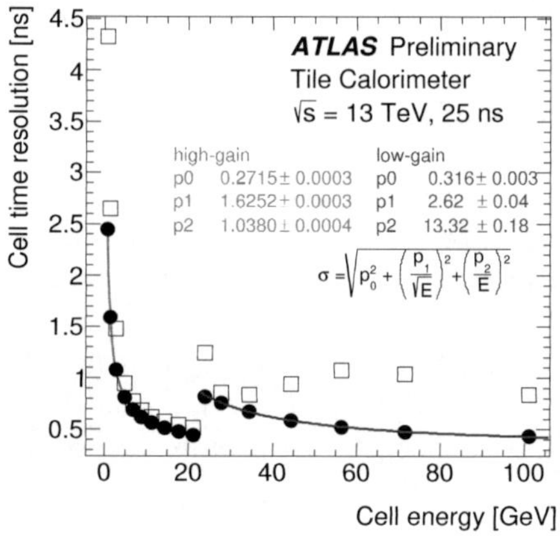

Figure 5. The cell time resolution as measured with jets [6]. Full circles denote Gaussian widths and open squares indicate the RMS of the underlying time spectra. The discontinuity around 25 GeV corresponds to the transition between the high- and low-gain.

Similar studies were performed already with Run-1 data [6]. While the mean time in jet events shows the same behaviour, the mean time obtained with isolated muons is independent from the cell energy, as no hadronic component is involved.

The time resolution was also studied and the results obtained with jets in Run-2 are shown in Fig. 5. The RMS values are systematically higher than Gaussian widths, indicating non-Gaussian tails in the measured time distribution that originates from the concurrent interactions in neighbouring bunch-crossing.

The time resolution was also compared between jet and muon events. The results were found to be compatible with one another [6].

5. Conclusions

Two methods for the time calibration in the ATLAS Tile Calorimeter were presented: one with single beam data and another with proton-proton collisions. The time synchronization is monitored during the physics data taking by means of the laser calibration events as well as with jet data, allowing for time constant correction before the data are processed for the physics analyses. Due to the frequency of sudden changes in the time calibration during Run-1, these corrections resulted in significant tail reduction and improvement of the RMS of the measured time distribution.

The mean cell time as a function of the deposited energy in jet data was presented as well as the time resolution, where the constant term approaches 300 ps.

Acknowledgments

The support from the Ministry of Education, Youth and Sports of the Czech Republic is acknowledged.

References

[1] Aad G et al 2008 The ATLAS Experiment at the CERN Large Hadron Collider *J. of Instrum.* **3** S08003
[2] Aad G et al 2010 Readiness of the ATLAS Tile Calorimeter for LHC collisions *Eur. Phys. J.* C **70** 1193
[3] Chomont A 2016 ATLAS Tile Calorimeter calibration and monitoring systems, talk presented at CALOR2016 conference
[4] Valero A 2012 Implementation and Performance of the Signal Reconstruction in the ATLAS Hadronic Tile Calorimeter *Physics Procedia* **37** 1765-71
[5] Aad G et al 2015 Searches for heavy long-lived particles with the ATLAS detector in proton-proton collisions at $\sqrt{s} = 8$ TeV *J. High Energy Phys.* JHEP1501(2015)068
[6] Tile Calorimeter public plots, `https://twiki.cern.ch/twiki/bin/view/AtlasPublic/TileCaloPublicResults`

IOP Conf. Series: Journal of Physics: Conf. Series **928** (2017) 012004 doi:10.1088/1742-6596/928/1/012004

The CMS Calorimeter Trigger for LHC Run II

Dr. Aaron Bundock [1]

Research Associate, Imperial College Sci., Tech. Med. (GB)

E-mail: aaron.bundock@cern.ch

Abstract. The Compact Muon Solenoid (CMS) experiment has implemented a sophisticated two-level online selection system that achieves a rejection factor of nearly 10^5. During Run II, the LHC has increased its centre-of-mass energy up to 13 TeV and will progressively reach an instantaneous luminosity of $2\times10^{34}\text{cm}^{-2}\text{s}^{-1}$. In order to guarantee a successful and ambitious physics programme under this intense environment, the CMS Trigger and Data acquisition (DAQ) system has been upgraded. A novel concept for the L1 calorimeter trigger is introduced: the Time Multiplexed Trigger (TMT). In this design, which is similar to the CMS DAQ or High Level Trigger (HLT) architecture, nine main processors each receive all of the calorimeter data from an entire event via 18 pre-processors. The advantage of the TMT architecture is that a global view and full granularity of the calorimeters can be exploited by complex algorithms. The goal is to maintain the current thresholds for calorimeter objects and improve the performance for their selection. The introduction of new triggers based on the combination of calorimeter objects is also foreseen.

1. Introduction

During Run II at the Large Hadron Collider at CERN, proton beams are collided with a centre-of-mass energy of 13 TeV at a bunch crossing rate of 40 MHz. The CMS experiment [1] is a general purpose detector built to perform precision measurements of Standard Model processes and searches for new physics at the high energy frontier. The high rate of collisions and large amounts of data produced by each of the sub-detectors require a sophisticated hardware trigger system to rapidly select which events may contain interesting physics processes and should be saved for offline analysis. The CMS Level 1 trigger must reduce the event rate from 40 MHz to 100 kHz within a timeframe of 3.2μs, passing events to the software-based High Level Trigger, which further reduces the rate to 1 kHz for recording to disk.

Run II provides a substantial increase in instantaneous luminosity, resulting in a challenging environment in which the trigger must quickly identify physics objects using trigger primitives constructed from the detector readout and generate a trigger decision. A significant increase in the number of pileup vertices makes it difficult to maintain the low energy thresholds required to measure electroweak scale physics whilst also providing necessary trigger bandwidth for new physics searches. The Run I trigger would produce a factor of six increase in trigger rates in the beam conditions of Run II. To maintain both low trigger thresholds and a Level 1 rate of 100 kHz, the trigger was completely replaced with a new system with a novel architecture and greater logic resources, allowing more complex identification algorithms introducing online pileup subtraction at Level 1 and improving object selection performance.

[1] On behalf of the CMS Collaboration

2. The CMS Calorimeter Trigger

The upgraded CMS calorimeter trigger has adopted a novel time-multiplexed architecture consisting of two layers, with higher granularity calorimeter data for the full detector being processed in a single main processor (MP), providing extra latency for more sophisticated algorithms (Figure 1). Layer 1 consists of 18 CTP7 pre-processor cards that receive trigger primitives optically from regions of ECAL and HCAL, sum the transverse energies and further format the data ready for processing at Layer 2, which consists of 10 MP7 cards that perform the object identification and compute global energy sums. Both cards are built around high performance Xilinx Virtex 7 FPGAs, which provide the large IO and logic resources that are required [3]. The main processors receive data on 72 links, each link containing 2 out of 72 ϕ segments with alternate links carrying data from the top and bottom η halves of the calorimeters. Input data are pipelined in η such that processing can commence once a minimum amount of data is received. A Molex Flexplane patch panel provides the interface between Layer 1 and Layer 2, fanning out data from the pre-processor cards for all detector regions to each main processor. The time required for a given MP to receive full detector information is around 7 bunch crossings, therefore at least 8 MPs are required. However, 9 MPs are used such that any given bunch crossing in an orbit will be sent to the same MP.

The data protocol used between Layer 1 and Layer 2 is a 16-bit word comprising the sum of ECAL and HCAL transverse energies, their ratio, and various feature flags, thus allowing the separate calorimeter energies to be utilized. Each link carries two of these words, with the full 32-bit word computed at a clock frequency of 240 MHz. The MP identifies up to twelve jets, e/γ, and tau candidates, and calculates the total transverse energy, total jet energy, along with their x and y components, both with and without the forward hadronic calorimeter data. The transverse momenta and η-ϕ co-ordinates of the particle candidates and the energies of the sums are sent from each MP over 6 links to the demux board (also an MP7), which calculates the vector sums and forwards data to the global trigger.

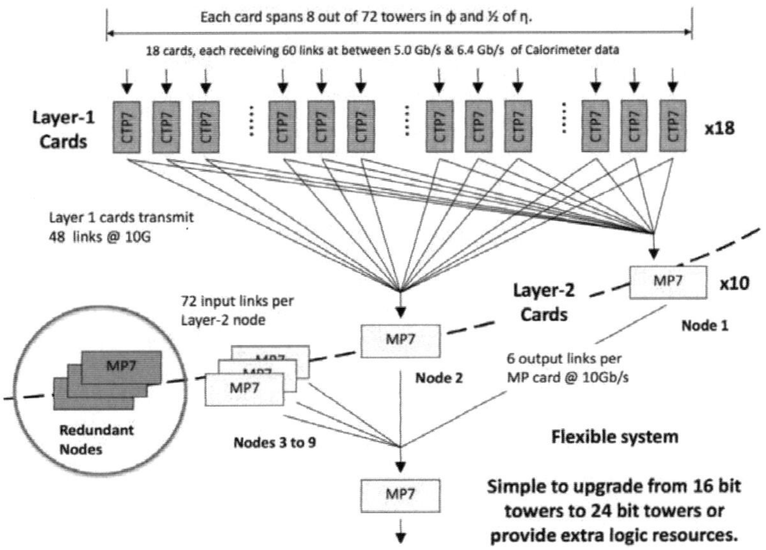

Figure 1. CMS calorimeter trigger architecture.

3. Algorithms

3.1. Jet and energy sums algorithms

Jet candidates are identified starting with a seed tower, which must have an energy above a programmable threshold of around a few GeV. A jet is then constructed by taking a 9×9 sliding window centred on the local maxima. The window size is chosen to correspond with the cone radius of 0.4 used in the offline jet reconstruction algorithm. An inequality mask is applied comparing seed tower energy with the rest of the tower energies within the window to avoid double counting of jets. The jet candidate is then defined as the sum of all tower energies within the 9×9 window. Comparisons between firmware algorithm output and software *anti-kt* jet finder algorithm used in offline reconstruction show excellent agreement [4].

Dynamic pile-up subtraction is applied to each jet, operating on an event-by-event basis and estimating the local energy density to provide an appropriate response to fluctuating pileup conditions. The "chunky donut" algorithm sums the tower energies in four 3×9 strips on each edge of the 9×9 window, and subtracts the sum of the lowest three energy strips from the total jet energy. Studies have shown a strong correlation between the chunky donut estimate of the pileup energy density and number of interactions [4].

Total transverse energy is calculated by summing the energies of all towers in the event, with the x and y components also calculated and used as input to a cordic function to produce the missing transverse energy. The scalar sum of all jet energies, H_T, is also calculated, along with its corresponding vector sum, H_T^{miss}.

3.2. e/γ algorithm

e/γ candidates are identified by first forming clusters seeded by a local energy maxima above a programmable threshold. As much as possible of the e/γ candidate energy is included whilst minimising the effect of pile-up by limiting the maximum size of clusters to 8 trigger towers. Due to the magnetic field, electron and photon showers spread mostly in the ϕ direction, and thus better containment of the shower is achieved by extending the region in ϕ. The improved calorimeter resolution now available allows the candidate position to be calculated as an energy-weighted average centred on the seed tower, which provides a factor of four improvement with respect to the legacy algorithm. Shape vetoes are utilized with dynamic clustering to improve background rejection by discriminating against jet clusters and reducing the fake rate, whilst maintaining high efficiency. Isolation criteria can be applied to significantly reduce e/γ rates for a small loss in efficiency.

3.3. τ algorithm

τ candidates that decay hadronically are identified using the same dynamic clustering implementation used for e/γ. Several clusters may be produced from the τ decay products that are spatially separated along ϕ due to the magnetic field, and these can be merged. Isolation can also be applied in a similar way to e/γ candidates. The footprint of the τ candidate is subtracted from the isolation energy, which is compared to an η dependent threshold. Background rejection is improved by utilising a shape veto LUT to remove background-like clusters.

The algorithms described above are also implemented in a C++ bit-level emulator. This provides a cross-check of the algorithm performance that can be used online for data quality monitoring. Figure 2 shows a comparison between firmware and emulator outputs produced using a sample of $t\bar{t}$ Monte Carlo events. Excellent agreement is observed, demonstrating that the firmware algorithms are well understood.

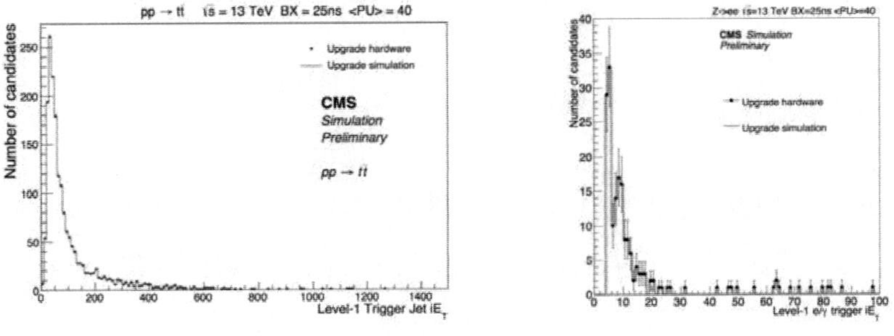

(a) (b)

Figure 2. Excellent agreement for transverse energies between firmware and emulator for (a) jets and (b) e/γ candidates.

4. Performance

The upgraded Level-1 jet algorithm with pile-up subtraction provides significantly improved performance. Figure 3 shows rate vs. efficiency curves comparing the expected performance of the Run I and Run II triggers. Simulated $t\bar{t}$ events were used to estimate efficiency, and minimum bias MC was used to estimate the rate. Efficiency is computed with respect to events fulfilling a corresponding generator-level requirement, and the curves produced by varying the Level 1 transverse momenta threshold. The upgraded jet algorithm maintains high performance in efficiency whilst providing a significant reduction in rate of multi-jet triggers. Similar levels of improvement in performance are achieved for energy sums and τ leptons [5] . The improved performance of the e/γ algorithms is discussed in [6].

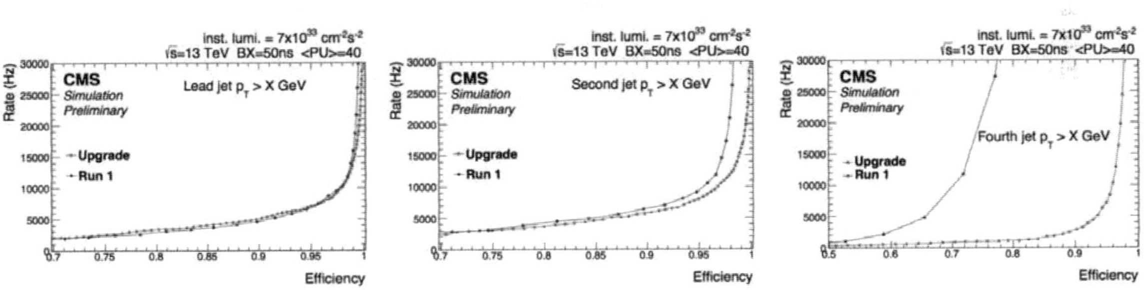

Figure 3. Rate vs. efficiency curves comparing the expected performance of the Run 1 and Run II (upgrade) Level-1 jet trigger using simulated $t\bar{t}$ events.

References

[1] CMS Collaboration, *The CMS Experiment at the CERN LHC*, 2008 JINST 3 S08004
[2] CMS Collaboration, *CMS Technical Design Report for the Level-1 Trigger Upgrade* CERN- LHCC-2013-011
[3] K. Compton et al., *The MP7 and CTP-6: multi-hundred Gbps processing boards for calorimeter trigger upgrades at CMS* 2012 JINST 7 C12024
[4] A. Zabi et al., *Triggering on electrons, jets and tau leptons with the CMS upgraded calorimeter trigger for the LHC RUN II*, CMS CR -2015/297
[5] B. Kreis et al., *L1 calorimeter trigger upgrade: jet and energy sum performance and commissioning status*, CMS DP -2015/051
[6] T. Strebler et al. *The role of the CMS electron and photon trigger in the study of the Higgs boson and other resonances*, CMS CR -2016/129

IOP Conf. Series: Journal of Physics: Conf. Series **928** (2017) 012005 doi:10.1088/1742-6596/928/1/012005

Overview of Energy Reconstruction, and Electron and Photon Performances with the CMS ECAL in Run II

Rafael Teixeira de Lima, on behalf of the CMS Collaboration
Northeastern University, Boston - MA, USA

E-mail: rafael.teixeira.de.lima@cern.ch

Abstract.
The electromagnetic calorimeter (ECAL) of the Compact Muon Solenoid (CMS) Experiment is crucial for achieving high resolution measurements of electrons and photons. Maintaining and possibly improving the excellent performance achieved in Run I is vital for measurements of the Standard Model Higgs boson and searches for new higher mass resonances in final states with electrons and photons. Since spring 2015, the ECAL has operated with proton-proton collisions at 13 TeV center-of-mass energy and at a reduced bunch spacing of 25 ns. The instantaneous luminosity delivered by the LHC during Run II is expected to exceed the levels previously attained. The average number of concurrent proton-proton collisions per bunch-crossing (pileup) is expected to reach up to 40 interactions. In this summary we present new crystal energy reconstruction algorithms and clustering techniques that have been developed to maintain the excellent performance of the CMS ECAL throughout Run II. We will show first performance results from 2015 data, achieved through energy calibrations using electrons from W and Z boson decays, photons from π^0/η decays, and the azimuthally symmetric energy distribution of minimum bias events. Lastly, we present an outlook on the expected Run II performance in the next years.

1. Introduction

The CMS ECAL is a high-resolution, hermetic, and homogeneous electromagnetic calorimeter made of 75,848 scintillating lead tungstate crystals divided among a barrel ($|\eta| < 1.48$) and two endcaps ($1.48 < |\eta| < 3.0$) [1]. These crystals are characterized by fast light emission (80% of light emitted in 25 ns), short radiation length ($X_0 = 0.89$ cm) and small Moliere radius ($R_M = 2.10$ cm). The light emitted by these crystals is detected with avalanche photodiodes (APDs) in the barrel and vaccuum photo-triodes (VPTs) in the endcaps. The signal readout is performed with two avalanche photodiodes (APDs) per crystal in the barrel, and one vacuum phototriode (VPT) in the endcaps. These characteristics, translated into precise energy and timing resolutions, are an invaluable tool for the CMS physics program.

Completing the CMS electromagnetic calorimeter system is a preshower detector (ES), based on lead absorbers equipped with silicon strip sensors. It is installed in front of the ECAL endcaps, covering the region $1.65 < |\eta| < 2.6$. The fine granularity of the ES strips (2 mm wide) can resolve the signals of high-energy photons from the decays of neutral pions into two photons, when the separation angle between the photons is small, and can determine precisely the position of the electromagnetic deposits.

IOP Conf. Series: Journal of Physics: Conf. Series **928** (2017) 012005 doi:10.1088/1742-6596/928/1/012005

During the first year of Run II data taking at 13 TeV, the LHC provided a challenging environment, with one bunch crossing every 25 ns and an average of 10 interactions per crossing (pile up). This is expected to be even more challenging in 2016, with up to 40 pile up interactions. In the 2015 data taking period, the CMS ECAL operated with more than 98% of its channels active, and was responsible for less than 7% of CMS downtime during physics runs.

2. Energy Reconstruction with ECAL

The photon and electron energy reconstruction based on ECAL energy deposits is based on the formula: $E_{e,\gamma} = [\sum_i (S_i(t) \times c_i \times A_i) \times G(\eta) + E_{ES}] \times F_{e,\gamma}$, with A_i, c_i and $S_i(t)$ as, respectively, the per individual channel amplitude, intercalibration constant and light monitoring constant, $G(\eta)$ is the ADC to GeV absolute scale, E_{ES} is the energy deposit in the preshower, and $F_{e,\gamma}$ are the cluster corrections (different for photons and electrons). These terms will be detailed in the next sections.

2.1. Online Reconstruction

The online reconstruction of ECAL deposits starts with the amplification and digitization of the signal from the photodetectors attached to the crystals. This is performed by a multi-gain preamplifier (MGPA), and a 12 bit ADC running at 40 MHz. Ten consecutive samples are recorded and used to perform the pulse reconstruction and amplitude extraction.

The time spacing between two consecutive samples from the ECAL readout electronics is 25 ns, which is the same time spacing between two colliding bunches in the LHC. This implies that, during the readout of one pulse, another scintillation process might start in the same crystal, compromising the in-time amplitude reconstruction. To mitigate this effect, also called out-of-time (OOT) pile up, a new online pulse reconstruction method (multifit) was developed to replace the Run I method [2].

In the multifit method, the pulse shape is reconstructed based on a fit to the time samples, minimizing $\chi^2 = \sum_{i=1}^{10} \left(\sum_j^M A_j p_{ij} - S_i\right)^2 / \sigma^2_{S_i}$. The samples ($S_i$) are fitted with one in-time pulse shape template, plus up to 9 out-of-time templates (p_{ij}) times their respective amplitudes (A_j). σ_{S_i} is noise generated by electronics associated with the crystal readout chain. The OOT templates have the same shape as the in time one, but are shifted in time by multiples of 1 bunch crossing (1 BX = 25 ns), within a range of -5 to +4 BX around the in time signal (BX = 0). The pulse shapes have been measured in early 2015, in special runs in which the LHC delivered isolated bunches (no OOT pile up).

It has been observed in both data and simulation that, with the multifit method, OOT pile up reconstruction is negligible. The energy resolution improvement, with respect to the Run I amplitude reconstruction method, is substantial especially for low E_T photons and electrons, given the larger contribution of deposits from pile up to the total energy.

2.2. Light Monitoring

Time dependent corrections must be applied to the reconstructed amplitude due to changes in detector response with radiation exposure. These changes in response are due to decreases in crystal transparency and variations in VPT response in endcaps.

The changes in the crystal transparency is due to ionizing radiation creating color centers in the lead tungstate. While the scintillation process remains intact, the amount of light detected by the photodetectors decreases. This effect is partially mitigated through thermal annealing, causing the transparency to increase in the absence of radiation.

A light monitoring system is used to monitor the overall changes in response in the ECAL [3]. It consists of a system of lasers (operating at 447 nm, close to the wavelength of peak emission for lead tungstate) that injects light in each ECAL crystal, which is then read by the standard ECAL readout. The difference between input and read laser amplitudes are then used to calculate correction factors.

The history of response change measurements is summarized in Figure 1. The changes are up to 6% in the barrel and reach up to 30% at $|\eta| \approx 2.5$, the limit of the tracker acceptance. For high $|\eta|$ regions, changes are up to 70%. The recovery of the crystal response during the long shutdown period is visible. The response was not fully recovered, however, particularly in the region closest to the beam pipe. The monitoring corrections are validated by comparing isolated electron energy as measured by ECAL (E) and momentum as measured by the CMS Tracker (p), before and after light monitoring corrections. It is seen that the measured corrections bring stability to energy measurements with ECAL.

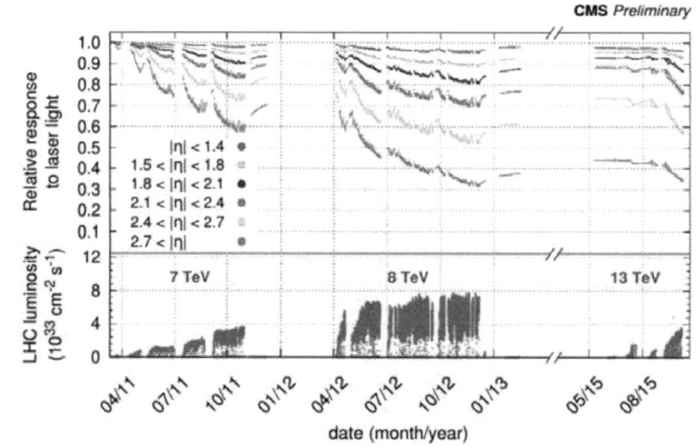

Figure 1. History of channel response changes as measured by the light monitoring system.

2.3. Intercalibration

A relative calibration procedure in all ECAL channels is performed to ensure uniformity across the detector. Different and independent methods are used to calculate intercalibration constants (ICs), which are then combined to achieve the desired precision of $< 0.5\%$. The final 2015 version of the ICs have been calculated with the full 2.6 fb^{-1} dataset recorded by CMS with B=3.8 T. The following methods are the same as in Run I [4].

The ϕ-symmetry method is based on the expected uniformity of the energy flux along ϕ rings (region with fixed η). The ICs are calculated to correct non-uniformities in this flux. This method was used in 2015 to translate the latest ICs, calculated with the full 2012 dataset, to the 2015 detector conditions. This was done by scaling the 2012 ICs by the ratio between 2015 and 2012 ϕ-symmetry ICs. The π^0/η method consists of measuring the invariant mass of these resonances' decays to two photons and maximizing their resolutions by varying the ICs iteratively. The E/p method employs the same logic as the light monitoring validation method, comparing isolated electron energy and momentum. An iterative method is used to minimize the spread of the E/p distribution. The combined intercalibration was obtained from the mean of the individual ICs at a fixed value of η, weighted by their respective precisions. The residual miscalibration of an intercalibration method, which is related to the final method precision, is calculated as the spread of the difference between the method's ICs and the other methods' ICs at a fixed value of η. The combination of ICs achieves the desired goal of less than 0.5% precision.

2.4. η Scale and Absolute Calibration

$Z \rightarrow ee$ events are used both to set the η scale and the absolute calibration [4]. The first is developed to ensure that different η regions have the same relative response, while the second (done separately for barrel and endcaps) sets the absolute energy scale.

A dedicated calibration was performed with 0 T data to account for differences in shower shapes in the absence of magnetic field. For example, in 0 T there is no bremsstrahlung radiation outside the main electron cluster deposit, improving the reconstructed energy resolution.

In addition, the calibration was validated with high energy photons and electrons. The validation was performed by comparing data and Monte Carlo simulations for high energy electrons from $Z \rightarrow ee$. The calibration was found to be stable to 0.5% (0.7%) for electrons up to $p_T = 150$ GeV in the barrel (endcap). Possible saturation effects were corrected for with a multivariate technique, but those effects were found to be $< 2\%$ for photons arriving from resonance masses less than 1.4 TeV.

2.5. Cluster Corrections

The amount of material in front of ECAL, up to $2X_0$ in the barrel outer regions, produces a high rate of bremsstrahlung radiation from electrons and a high probability of photon conversions. To mitigate this effect, a clustering algorithm is used to recombine energy deposits that come from those processes. The cluster energy is corrected via a multivariate technique, separately for photons [5] and electrons [6]. It also aims to correct other effects, such as in time pile up.

3. ECAL Performance with Run II data

The ECAL energy resolution is measured using $Z \rightarrow ee$ events, from an unbinned fit with a Breit-Wigner function convoluted with a Gaussian as signal model. Degradation effects come from the amount of material in front of ECAL and cracks between modules. The relative energy resolution (σ_E/E) achieved using the latest 2015 calibration constants is less than 2% for low bremsstrahlung electrons in the barrel.

4. Summary

The CMS electromagnetic calorimeter has demonstrated excellent performance with Run II data in 2015. The light monitoring system has been working with high stability during Run II, providing response change corrections every ≈ 48 hours. With 2.6 fb^{-1} of data recorded by CMS with a 3.8 T magnetic field, we have achieved our goal of less than 0.5% residual miscalibration in the central barrel region. After absolute calibration, the final energy resolution obtained by ECAL with the 2015 dataset is less than 2% in the central barrel region.

References

[1] [CMS Collaboration], CMS: The electromagnetic calorimeter. Technical design report, CERN-LHCC-97-33, CMS-TDR-4.
[2] P. Adzic et al., Reconstruction of the signal amplitude of the CMS electromagnetic calorimeter, Eur. Phys. J. C 46S1, 23 (2006)
[3] M. Anfreville et al., Laser monitoring system for the CMS lead tungstate 511 crystal calorimeter, Nucl. Instrum. Meth. A 594, 292 (2008).
[4] [CMS Collaboration], Energy Calibration and Resolution of the CMS Electromagnetic Calorimeter in pp Collisions at sqrt(s) = 7 TeV, JINST 8, P09009 (2013)
[5] [CMS Collaboration], "Performance of Photon Reconstruction and Identification with the CMS Detector in Proton-Proton Collisions at sqrt(s) = 8 TeV", JINST 10 (2015) no.08, P08010
[6] [CMS Collaboration], "Performance of Electron Reconstruction and Selection with the CMS Detector in Proton-Proton Collisions at sqrt(s) = 8 TeV", JINST 10 (2015) no.06, P06005
[7] [CMS Collaboration], "First results on Higgs to $\gamma\gamma$ at 13 TeV", CMS-PAS-HIG-15-005

Study of TileCal Scintillator Irradiation using the Minimum Bias Integrators

Cora Fischer on behalf of the ATLAS Collaboration

Institut de Física d'Altes Energies, Universitat Autònoma de Barcelona

E-mail: cora.fischer@cern.ch

Abstract. The Tile Calorimeter (TileCal) is the central hadronic calorimeter of the ATLAS experiment at the LHC. It provides precise energy measurements of hadrons, jets, taus and missing transverse energy. The monitoring and calibration of the calorimeter response at each stage of the signal development is done by a movable ^{137}Cs radioactive source, a laser calibration system and a charge injection system. Moreover, during LHC data taking, an integrator-based readout provides the signals coming from inelastic proton-proton collisions at predominantly low momentum transfer (minimum bias events) and allows monitoring of the instantaneous ATLAS luminosity as well as the response of calorimeter cells. The integrator currents have been used to detect and quantify the effect of TileCal scintillator irradiation using the data taken in 2012 and 2015 that correspond to about 22 fb^{-1} and 4 fb^{-1} of integrated luminosity, respectively. Finally, the response variation for an irradiated cell has been studied combining the information from three calibration systems (cesium, laser and minimum bias). The result of the irradiation on the TileCal response will be reported.

1. Introduction

The ATLAS experiment [1] is one of two general purpose detectors at the Large Hadron Collider (LHC) at CERN. The TileCal is described in detail in Ref. [2] and consists of steel plates and plastic scinitillator material which are arranged in alternating layers. Particles are gradually stopped by the steel plates and photons are produced when the charged particles traverse the scintillator material. This scintillation light is transported by fibers to photomultiplier tubes (PMTs) equipped with an electronic readout. Fig. 1 shows the geometry of the TileCal along the beam axis. It consists of a central Long Barrel (LB) and two Extended Barrel (EB) partitions. The LB covers a region of $0.0 < |\eta| < 1.0$ in pseudorapidity, the EB covers $0.75 < |\eta| < 1.65$. The so-called A-side covers positive, the C-side negative pseudorapidities. Both LB and EB are made up of three layers: the A-layer is the one closest to the beam axis followed by the B(C)-layer and the layer farthest away is the D-layer. The cells marked in yellow are the E-cells. They cover the so-called gap/crack region with $1.0 < |\eta| < 1.6$. They are partly closer to the beam axis than the A-layer cells.

2. TileCal Calibration System

At each level of the signal development inside the TileCal there is a dedicated calibration system to monitor the behaviour of the different detector components. The cesium system tests the whole detector readout, from scintillators to fibers and PMTs. The laser system is capable of measuring gain variations of the PMTs. The charge injection system monitors the front-end

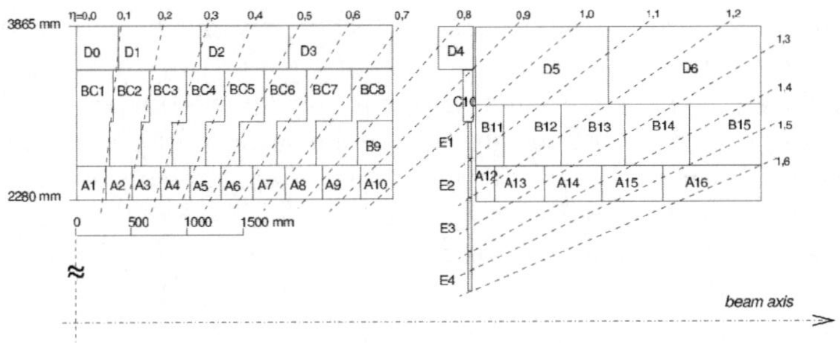

Figure 1. Longitudinal view along beam axis of the ATLAS TileCal cell geometry [3]. One quarter of the longitudinal plane is shown with Long Barrel and Extended Barrel for positive pseudorapidity. The different layers and the corresponding coverage in η of TileCal cells are shown.

electronics. The sketch in Fig. 2 visualizes the different calibration systems along with the paths that signals from various sources take. Fast readout is used for physics analyses of collision data. The integrator readout is the core of the minimum bias (MB) system. It records slow currents from the calorimeter. Here we focus on the MB system. The integrator currents are recorded by all (~10 000) channels of the TileCal. The data recorded includes all data used for physics analyses and is dominated by minimum bias events. Each collision run is devided into so-called lumiblocks (LBs) where a LB is the time interval with a fixed duration where the instantaneous luminosity is about constant. Integrator currents are recorded for each LB unit. The currents are averaged over about 20–25 integrator measurements per LB.

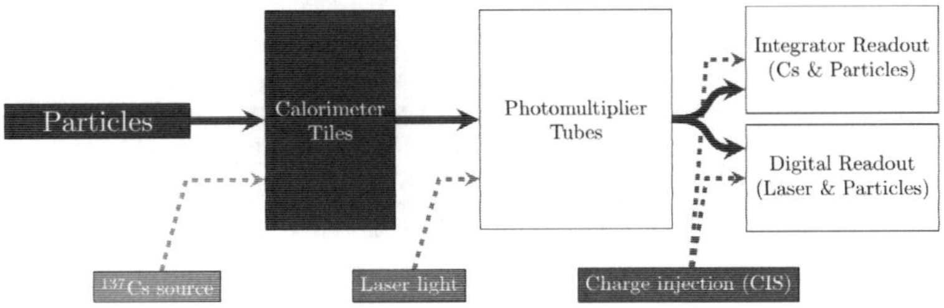

Figure 2. Scheme of the TileCal signal path [4]. Different calibration systems monitor various components of the TileCal at each stage of signal development inside the detector.

3. Luminosity and Integrator Currents

The MB system is used to monitor the instantaneous luminosity. The integrator currents measured during collisions are linearly dependent on the instantaneous luminosity. An example to illustrate this is shown in Fig. 3(a). For runs that have been recorded during the whole 2015 data taking period the current of a specific channel (D5, EBC) has been plotted against the corresponding instantaneous luminosity as measured by the dedicated LUCID detector [5] of ATLAS. The data points have been fitted by a linear function, the lower panel shows the ratio of data over fitted values. A good linear description of the data points can be observed. The cell D5 is located in the EB at $0.9 < |\eta| < 1.1$, protected from irradiation by two other TileCal layers. This cell shows a stable behaviour in time. The fitted slope corresponds to the luminosity coefficient: current/instantaneous luminosity. This quantity is constant in time and therefore allows the monitoring of the instantaneous luminosity via the MB system. One can retrieve these coefficients by dividing the average current over the average instantaneous luminosity. The coefficients obtained with this method are plotted against η in Fig. 3(b) for a run in 2015 . The behaviour of the luminosity coefficients is shown for all 3 layers of the TileCal and the E-cells. The latter ones measure the highest currents due to their position close to the beam axis, with less shielding material in front. Apart from the E-cells, cells A13 are the most irradiated ones and provide the highest currents.

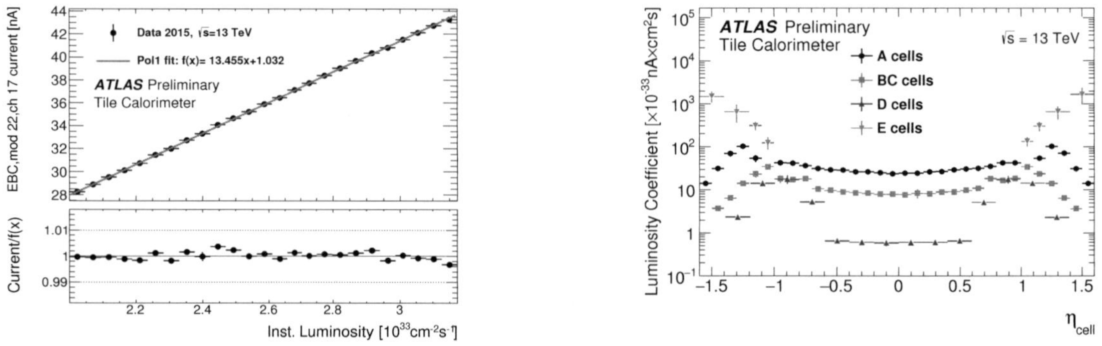

(a) (b)

Figure 3. (a): Current vs instantaneous luminosity for a given channel from cell D5, see Fig. 1. The data has been fitted with a linear function; the fitted parameters are displayed. (b): Luminosity coefficient vs η for all TileCal cells as measured in run 276262 during 2015 data taking [6].

4. Detector Response Variation and Irradiation Studies

The mentioned cell A13 is used to study irradiation effects. Like the cesium system, the MB system is capable of monitoring the detector response in time. This is possible by using a reference cell which is known to be stable in time and is used for luminosity measurements. The reference cell chosen for these studies is the afore metioned cell D5 which is protected from high irradiation. Monitoring the evolution of the currents measured by cell A13 over the currents measured by cell D5 w. r. t. a reference start run gives the detector response variation. The MB measurement is expected to give the same results as the cesium measurement as they both test all TileCal readout components. On the other hand, the laser system measures the PMT gain drifts in time. Any difference in the measurements from Cesium/MB to the laser measurement will therefore point to an effect of scintillator irradiation. The comparison of the three systems for the response variation of A13 has been performed in 2012 and 2015 data periods. Fig. 4 shows the response behaviour of A13 in 2012 and 2015 data taking periods. The 2012 measurement is described in [7]. In both plots the laser points show the PMT gain drift, where drops can be observed during collision phase and a rise during machine development phases where no particle collisions take place. Cesium and MB systems measure in addition any effect of irradiation of the scintillators when the response variation is higher than the PMT gain drift. This can be observed for the 2012 data period. The irradiation effect of the scintillator can be quantified at 2% level at maximum. This irradiation effect will not recover during machine stops. During the 2015 data period no further irradiation is observed. It should be noted that the total integrated luminosity L in 2015 was ∼5 times lower than the one in 2012. The center of mass energy $\sqrt{s}$ was 8 TeV in 2012 and 13 TeV in 2015. Both the integrated luminosity and $\sqrt{s}$ have an impact

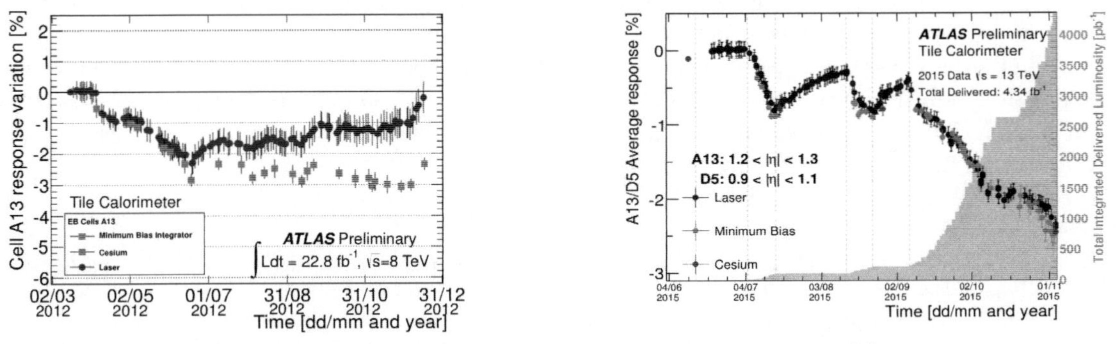

(a) (b)

Figure 4. (a): Response variation of cell A13 in 2012 as measured by laser, cesium and MB system. (b): Response variation of cell A13 w. r. t. cell D5 in 2015 as measured by laser, cesium and MB system [4].

on the overall collected charge, Q. The charge Q is determined with the help of the luminosity coefficient k: $Q(t) = k \cdot \int_{t_0}^{t} L(t)\mathrm{d}t$, t being a given time during collisions and t_0 is a reference start time with $L(t_0)$ =0. The irradiation effect in scintillators will depend on the collected charge and therefore depends on both the $\sqrt{s}$ and the integrated luminosity. Fig. 5 shows the

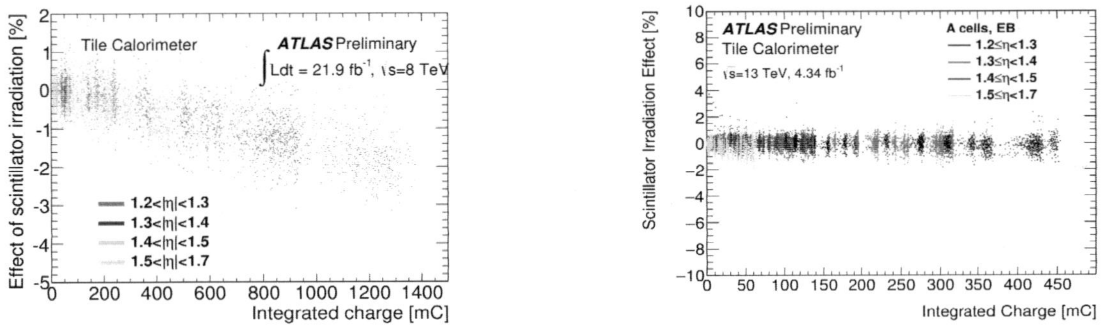

(a) (b)

Figure 5. Irradiation effect in scintillators in the A-layer of the EB vs the collected charge. (a): Data from all channels during 2012 data taking. (b): Data from all channels during 2015 data taking [4].

dependence of the irradiation in cells belonging to the A-layer in the EB on the collected charge for 2012 and 2015 data separately. Also, no irradiation is seen for all the A-layer EB cells in 2015, as Fig. 5(b) illustrates. This differs from 2012 results, when up to 2% loss in the light yield was reported in A13 cells (Fig. 5(a)). It is important to note, that in 2012 the collected charge in the A-layer cells of the EBs goes up to ~1300 mC while it goes up to ~450 mC at the end of 2015 data taking.

5. Conclusion

The minimum bias system of the ATLAS TileCal has been introduced as a powerful tool to monitor both instantaneous luminosity and detector response behaviour in time. With the help

IOP Conf. Series: Journal of Physics: Conf. Series **928** (2017) 012006 doi:10.1088/1742-6596/928/1/012006

of the MB system it is possible to identify irradiation effects on the TileCal scintillators by comparing with measurements from the laser system. A total irradiation effect of at most 2% for the highest exposed cell has been observed for 2012 and 2015 data taking periods. The scintillator irradiation stems from 2012 data taking where collected charges are about three times higher than in 2015, where no further irradiation effect is observed. These studies will continue with the next data taking period in 2016.

References

[1] ATLAS Collaboration 2008 *Journal of Instrumentation* **3** S08003 `http://stacks.iop.org/1748-0221/3/i=08/a=S08003`
[2] ATLAS Collaboration 2010 *Eur. Phys. J.* **C70** 1193–1236 (*Preprint* `1007.5423`)
[3] ATLAS Collaboration `https://twiki.cern.ch/twiki/bin/view/AtlasPublic/ApprovedPlotsTile`
[4] ATLAS Collaboration `https://twiki.cern.ch/twiki/bin/view/AtlasPublic/ApprovedPlotsTileCalibration`
[5] Jenni P and Nessi M 2004 ATLAS Forward Detectors for Luminosity Measurement and Monitoring Tech. Rep. CERN-LHCC-2004-010. LHCC-I-014 CERN Geneva `http://cds.cern.ch/record/721908`
[6] ATLAS Collaboration `https://twiki.cern.ch/twiki/bin/view/AtlasPublic/TileCaloPublicResults`
[7] Fracchia S 2014 Irradiation effect on the response of the scintillators in the ATLAS Tile Calorimeter Tech. Rep. ATL-TILECAL-PROC-2014-013 CERN Geneva `https://cds.cern.ch/record/1710463`

IOP Conf. Series: Journal of Physics: Conf. Series **928** (2017) 012007 doi:10.1088/1742-6596/928/1/012007

The role of the CMS electron and photon trigger in the study of the Higgs boson and high mass searches

Thomas Strebler[1]

Laboratoire Leprince-Ringuet, Ecole Polytechnique / CNRS-IN2P3, Palaiseau, France

E-mail: thomas.strebler@llr.in2p3.fr

Abstract. The Compact Muon Solenoid (CMS) experiment implements a sophisticated two-level triggering system composed of the Level-1, instrumented by custom-design hardware boards, and a software High-Level-Trigger. A new Level-1 trigger architecture with improved performance is now being used to maintain the thresholds used in LHC Run I for the more challenging conditions experienced during Run II. We present the performance of the upgraded CMS electron and photon trigger in the context of Higgs boson decays into final states with photons and electrons. The calorimeter trigger system plays a central role in achieving the ambitious physics program of Run II. The upgraded trigger uses the full granularity of the calorimeters to optimally reconstruct and calibrate the electromagnetic trigger objects. It also implements combinations of calorimeter objects to provide new Level-1 quantities such as invariant mass. Optimized software selection techniques have been developed and advanced algorithms to mitigate the impact of event pileup have been implemented. The selection techniques used to trigger efficiently on these benchmark analyses will be presented, along with the strategies employed to guarantee efficient triggering for new resonances and other new physics signals involving electron/photon final states.

1. Introduction

In LHC experiments, a trigger system is mandatory to select events on the fly and reduce the 40 MHz collision rate to a level sustainable by the data acquisition system, typically on the order of 1 kHz. The development of an efficient triggering system represents a real challenge for the LHC Run II in CMS, as its physics program includes a wide variety of processes, from precision physics at the electroweak scale to the search for new physics at very high energy. Due to the high hadronic activity in proton collisions, leptons and photons provide ideal candidates to trigger on. For the trigger system of the CMS experiment, a two-layer implementation has been adopted:

- **The Level-1 trigger** bases its decision directly on coarse granularity detector readout. Its implementation is hardware-based and its operation is synchronous. Its decision time is 4 μs and the L1 accept rate is kept below 100 kHz.
- **The High-Level Trigger (HLT)** benefits from full detector readout and full granularity. Its implementation is software-based and very close to the offline reconstruction. Its average decision time is about 200 ms and it produces on average 1 kHz of physics data.

[1] On behalf of the CMS Collaboration

The electron and photon reconstruction, offline as online, is based on information from the electromagnetic calorimeter (ECAL) and from the tracker. The CMS ECAL [1] is made of 75848 lead tungstate crystals, arranged between the barrel and two endcaps. Lead tungstate has the advantages of being relatively dense, containing 80% of the energy of an electron or photon within a single crystal, while also having a fast scintillation time. The CMS tracker is made of silicon pixels and strips and is used to reconstruct the tracks from charged particles. For the LHC Run II, the offline electron and photon reconstruction has been fully integrated within the CMS Global Event Description (also called Particle-Flow). This algorithm optimally combines the information produced by each of the CMS subdetectors, which are used to produce different blocks. For the electrons and photons, those blocks correspond to tracks and superclusters built from the energy deposits in the ECAL. Those individual blocks are then linked together to reconstruct individual particles, which leads to a final unique event interpretation. However, as this algorithm is time-consuming, it is not possible to use it directly for the trigger decision. Alternative selection techniques have therefore been developed to be used for the online selection of electrons and photons.

2. The Level-1 electron-photon trigger in CMS

The CMS Level-1 trigger has undergone a major change of architecture to cope with the challenges arising from the increase in luminosity and pileup. The Level-1 electron-photon (L1 EG) trigger has been completely renewed and improved thanks to this upgrade. In the Level-1 trigger, only the calorimetric information is used to reconstruct electron and photon candidates, which are built from clusters of trigger towers (corresponding to 5x5 sets of ECAL crystals). A new dynamic clustering has been developed, which better collects potential Bremsstrahlung radiation along the ϕ direction. The better granularity of the new system also leads to a large improvement in position resolution. Together, these improvements will allow us to develop efficient invariant mass triggers at Level-1. The improvement in terms of triggering efficiency with respect to the algorithm used in Run I is presented in Fig. 1, where the efficiency has been measured from 2012 data using a tag-and-probe method [3].

To reduce as much as possible the level of background and thus the rate of the L1 EG trigger, different features are used:

- the fine-grain, corresponding to the energy distribution within the seed trigger tower of the L1 EG candidates, which is required to be compatible with an electromagnetic object
- the ratio between the energy deposit in the hadronic and electromagnetic calorimeter
- the shape of the clusters, for which the large ones more compatible with jets are rejected or trimmed

In addition, a possible isolation requirement can be used, based on the calorimetric energy deposit around the candidates. The isolation threshold is tuned to keep a constant efficiency as a function of pile-up, using an event-by-event estimate of the level of pile-up. All these configurable features lead to a sizable reduction of the rate with respect to the L1 EG algorithm used during Run I.

3. The High Level Trigger selection for electrons and photons

For LHC Run II, the electron and photon reconstruction at HLT has been made closer to the offline Particle-Flow reconstruction. In particular, thanks to a new superclustering algorithm, combined with regression-based energy corrections, the HLT energy resolution has been significantly improved. The use of the Gaussian Sum Filter algorithm [4] in the track reconstruction, which better accommodates deviations due to Bremsstrahlung radiation than the previous Kalman Filter [5], also improved the momentum resolution for electrons. The tracks also provide additional parameters to identify electrons, when combined with calorimetric

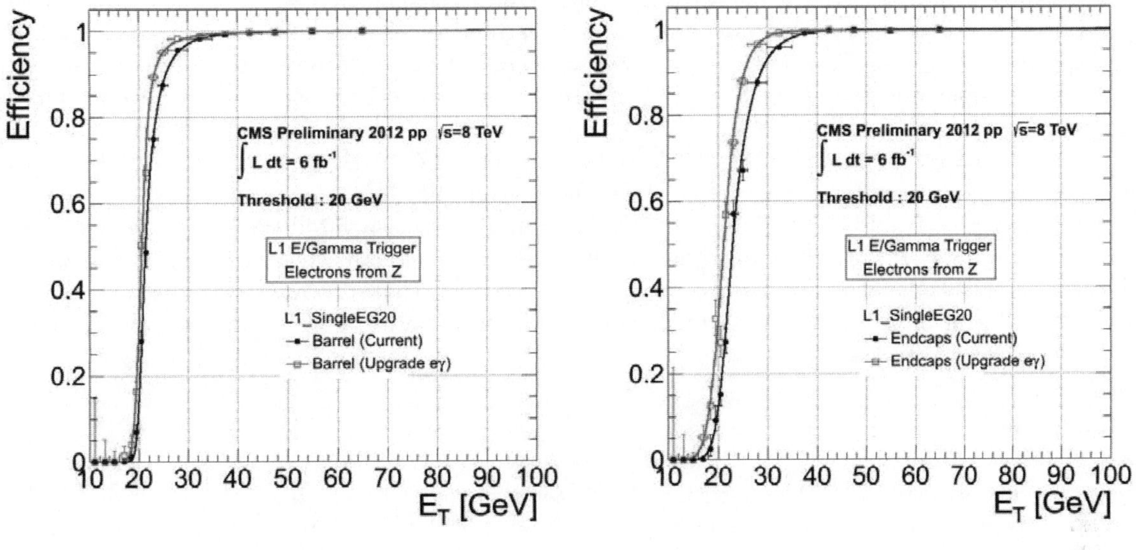

Figure 1. Level-1 e/gamma trigger efficiency as a function of the electron supercluster transverse energy (E_T) for electrons in the barrel (left) and in the endcaps (right) in 2012 data. The performance of the Run II upgraded trigger (red) are compared with those of the Run I system (black) [6].

information: the distance between the track and the calorimetric supercluster in $\Delta\eta$ and $\Delta\phi$ together with the difference between the inverse of the energy $1/E$, measured with the ECAL, and the inverse of the momentum $1/p$, measured with the tracker, are used to reject fake electrons from the QCD background.

Electrons and photons can be further discriminated from hadronic processes using isolation criteria. The Run II HLT isolation has also been modified with respect to Run I to use the same input blocks (calorimeter clusters and tracks) as the offline Particle-Flow algorithm, which resulted in a sizable improvement of the selection efficiency for a given level of background contamination. This new isolation also makes use of pile-up subtraction techniques based on the average energy density measured in the event, which results in a selection efficiency that is relatively robust with respect to the pile-up conditions.

4. Impact of the electron and photon trigger on physics analyses

Data taken with the electron and photon trigger are used in a wide range of CMS analyses. This is for instance the case of diphoton analyses, like $H \to \gamma\gamma$ [7] or the search for high-mass diphoton resonances [8], which rely on double photon triggers. As the p_T spectrum of the photons is relatively hard due to the high mass of the resonances, the acceptance of those analyses is not directly limited by the trigger thresholds. However, the expected number of signal events being relatively low, the triggers used must have efficiencies consistent with 1 for high p_T photons, which is indeed the case.

The electron triggers are also used for $H \to ZZ^* \to 4\ell$ studies [9]. Although this final state provides a clean signature in the hadronic environment of proton-proton collisions, the number of expected signal events with four reconstructed leptons is still relatively low. Moreover, the p_T of the trailing lepton can be very low, because of the presence of an intermediary off-shell Z^* boson. To accommodate for that effect, the offline selections must be as low as $p_T > 7$ GeV for electrons. On the other hand, triggering on those low p_T leptons is very challenging because of the high rates at low trigger thresholds. This issue can be mitigated by requiring the presence of multiple electrons in the event, which enables us to reduce the trigger thresholds with respect to single electron triggers. To maximize the trigger acceptance, a combination of different single, double and triple electrons (together with muon) triggers are then used, leading to a trigger efficiency above 99%.

Finally, it is worth mentioning that the electron and photon triggers are not only used to target Higgs decay modes with electrons or photons in the final state but also for $H \to b\bar{b}$ searches. Indeed, because of the large QCD background, only associated production modes (like the VH [10] or ttH [11] production) have a sensitivity high enough to be observed at the LHC. The decay of the additional vector boson or top quarks can then produce electrons, which can be used to trigger on those events more efficiently than with hadronic triggers, since the electron trigger thresholds are lower.

Conclusion

Due to increases in luminosity and pile-up, maintaining highly efficient triggering on electrons and photons with low thresholds poses a significant challenge during LHC Run II. Maintaining a high trigger efficiency is a critical issue for many physics analyses, like the study of the properties of the Higgs boson and searches for new resonances. The L1 EG algorithm has largely benefited from the new Level-1 trigger system used in CMS, while the selection of electrons and photons in the High-Level Trigger has been improved thanks to techniques closer to the offline reconstruction. It has been possible for CMS to maintain a wide physics program for Run II thanks to the continuous improvements of the electron and photon trigger, which has already contributed to a large number of physics analyses at 13 TeV.

Acknowledgments

The work of the author of this paper has been partly funded by the P2IO LabEx (ANR-10-LABX-0038) in the framework "Investissements d'Avenir" (ANR-11-IDEX-0003-01) managed by the French National Research Agency (ANR).

References

[1] CMS Collaboration, *Performance and Operation of the CMS Electromagnetic Calorimeter*, JINST 5 T03010 (2010)

[2] V. Veszpremi, *Operation and performance of the CMS tracker*, JINST 9 C03005 (2014)

[3] CMS Collaboration, *Measuring Electron Efficiencies at CMS with Early Data*, CMS-PAS-EGM-07-001 (2008)

[4] W. Adam et al., *Reconstruction of Electrons with the Gaussian-Sum Filter in the CMS Tracker at the LHC*, CMS NOTE 2005/001 (2005)

[5] R. Frühwirth *Application of Kalman filtering to track and vertex fitting*, Nucl. Instrum. and Methods A 262 (1987)

[6] CMS Collaboration, *L1 calorimeter trigger upgrade: electron and photon performance*, CMS-DP-2015-003 (2015)

[7] CMS Collaboration, *First results on Higgs to $\gamma\gamma$ at 13 TeV*, CMS-PAS-HIG-15-005 (2016)

[8] CMS Collaboration, *Search for new physics in high mass diphoton events in 3.3* fb^{-1} *of proton-proton collisions at $\sqrt{s}$ =13 TeV and combined interpretation of searches at 8 TeV and 13 TeV*, CMS-PAS-EXO-16-018 (2016)

[9] CMS Collaboration, *Studies of Higgs boson production in the four-lepton final state at $\sqrt{s}$ =13 TeV*, CMS-PAS-HIG-15-004 (2016)

[10] CMS Collaboration, *Search for the standard model Higgs boson produced in association with a W or a Z boson and decaying to bottom quarks*, Phys. Rev. D 89 (2014)

[11] CMS Collaboration, *Search for $t\bar{t}H$ production in the $H \to b\bar{b}$ decay channel with $\sqrt{s}$ =13 TeV pp collisions at the CMS experiment*, Phys. Rev. D 89 (2014)

IOP Conf. Series: Journal of Physics: Conf. Series **928** (2017) 012008 doi:10.1088/1742-6596/928/1/012008

The performance of the CASTOR calorimeter during LHC Run 2

Merijn H. F. van de Klundert on behalf of the CMS collaboration
University of Antwerp, Campus Groenenborger, Groenenborgerlaan 171, 2020 Antwerpen, BE
E-mail: merijn.h.f.van.de.klundert@cern.ch

Abstract. CASTOR is an electromagnetic and hadronic tungsten-quartz sampling Cerenkov calorimeter located at the Compact Muon Solenoid experiment at the Large Hadron Collider. The detector has pseudorapidity borders at -5.2 and -6.6. An overview is presented on the various aspects of CASTOR's performance and their relations during LHC Run 2. The equalisation of CASTOR's channels is performed using beam-halo muons. Thereafter, CASTOR's pedestal spectrum is studied. It is shown that noise estimates which are extracted using a fit, give on average a 10% lower threshold than statistical estimates. Gain correction factors, which are needed for the intercalibration, are obtained using a statistical, in-situ applicable method. The results of this method are shown to be reasonably consistent with laboratory measurements. Penultimately the absolute calibration is discussed, with emphasis on the relation between the scale uncertainty and CASTOR's alignment. It is shown that the alignment's contribution to the systematic uncertainty is decreased by over 50% in LHC Run 2 w.r.t. LHC Run 1. Finally generalisations of the conclusions to other subsystems and future improvements are discussed.

1. Introduction

At collider experiments, forward calorimeters contribute to numerous physics measurements. By their kinematical acceptance, they allow for dedicated QCD studies. But forward calorimeters also complement the central detectors. At the LHC experiments the hermeticity of the calorimetric systems has enabled measuring the missing transverse energy, allowing searches for new physics.

Various aspects of calorimetry, for example the design and hardware, the electronics, and the detector performance provide a unique set of challenges at forward rapidities. A quantitative, well understood estimate of the performance is vital for being able to interpret the measurements of the calorimeter.

In this paper we discuss the performance of the CASTOR calorimeter, which is installed at the Compact Muon Solenoid (CMS) experiment [1] at the Large Hadron Collider (LHC). CASTOR is a unique detector since it is the most forward calorimeter deployed at the LHC (except for Zero-Degree calorimeters). It has pseudorapidity (η) acceptance [-6.6,-5.2]. Many conventional techniques (for example to establish the inter and absolute calibration) have still not been applied successfully at CASTOR. This paper aims to describe not only the methods to assess the performance, but also to emphasise the relationships between the various aspects of CASTOR's performance. The key numbers obtained with the design and improvements w.r.t. LHC Run 1 are stated throughout the text.

2. The design of CASTOR

To measure at the aspired very forward η, CASTOR (which weighs approx. 2 tonnes) is installed at only 1 cm from the beampipe, at 14.4 m from the interaction point. Moreover, when the CMS magnet is turned on, CASTOR becomes displaced from its initial position by approx. 3 mm. These factors make the installation a very delicate task. Constant monitoring of the position sensors during the first magnet cycle is required. Given a fixed initial position, the movements in subsequent cycles have proven to be well predictable though.

CASTOR consists of 16 transversal sectors, 14 longitudinal modules, and has no η segmentation [2]. A channel consists of tungsten plates interleaved with quartz; CASTOR is a Cerenkov calorimeter. The Cerenkov photons are guided to fine-mesh PMTs (Hamamatsu R5505 and R7494), where the signal is collected and amplified. The design was motivated by the relative fast response and radiation hardness. CASTOR is equipped with LED fibers, which can illuminate the PMTs. This has proven to be a very useful tool for various commissioning purposes. The relative energy resolution achieved for pions is $18.3 \oplus 187/\sqrt{E}$ [2].

3. Intercalibrating CASTOR

3.1. Equalising channels with beam-halo muons

The LHC beam causes beam-halo muons, which traverse the CASTOR sectors longitudinally. The muons were collected during the circulating beam periods of the LHC (a period during which circulating beams are present in the LHC but no collisions take place). One can infer that a MIP (Minimally Ionizing Particle) causes on average approx. one photoelectron per channel.

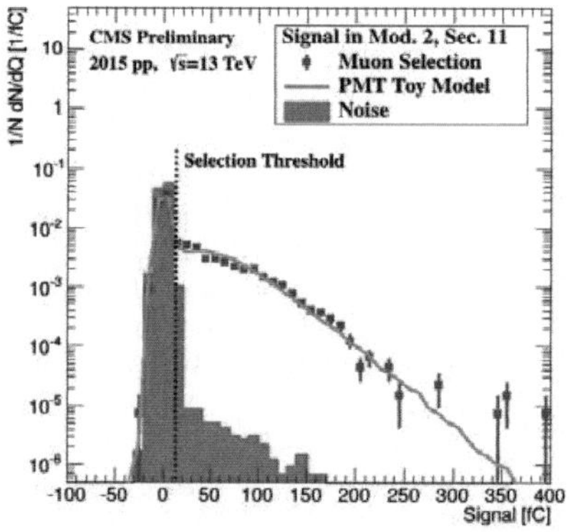

Figure 1. Signal spectrum for a typical CASTOR channel after an offline isolated muon event selection. The overlaid noise distribution is measured from non-colliding bunch data. The model line corresponds to a mesh-type PMT with an average number of photoelectrons $\langle N_{p.e.} \rangle$ of 0.5. The selection threshold used to identify channels above noise is shown as vertical line. Plot from [3]

To maximally distinguish the minute muon signals from the electronic noise background, the muons are collected at the maximal gain settings (1800V, corresponding to an amplification factor of approx. 1M). For most PMT-related sources of noise, the signal-to-noise ratio on contrary is flat as function of the gain on the range considered in this paper. Muons are collected online by demanding exclusive activity in one sector. The offline selection further requires minimally three channels in three different longitudinal sections of the sector to be above a channels-specific threshold. In fig. 1 the signal and noise spectrum after the offline selection are displayed for a particular channel. The good description between the PMT model and data indicates MIPs are indeed mainly measured, as opposed to showering muons. The black dotted vertical line indicates the channel-specific noise threshold. To obtain an event sample which is both statistically relevant and sufficiently pure, it is indispensable to pick this threshold correctly.

3.2. Determining CASTOR's noise threshold per channel

To obtain a precise estimate of the baseline and noise in CASTOR's channels, the pedestal signal spectrum was studied at the maximal gain (see fig. 2 for a typical CASTOR channel). Three different sources of noise are identified:

(i) The Gaussian, fitted peak at the left hand side, corresponding to $\mathcal{O}(0.1)$ GeV, is due to electronic noise. The signal is gain independent

(ii) The shoulder around 100 fC (corresponding to $\mathcal{O}(1)$ GeV) is due to thermal photoelectrons. The signal depends strongly on the gain

(iii) The rare discharges (maximally at $\mathcal{O}(100)$ GeV) in the tails are likely due to ion feedback. These also depend strongly on the gain

The channel thresholds, for distinguishing the muon signals from noise, were determined with the parameters of the Gaussian fits to the electronic noise (see the red line in fig. 2). The width of the Gaussian is on average 10% lower than the statistical width (which was taken as noise estimate instead during LHC Run 1). This is mainly due to the rare, high energetic deposits, to which the fit is insensitive, but which significantly affect the statistical average. Actually, comparing the statistical and fitted width has proven to be a useful method of identifying noisy channels.

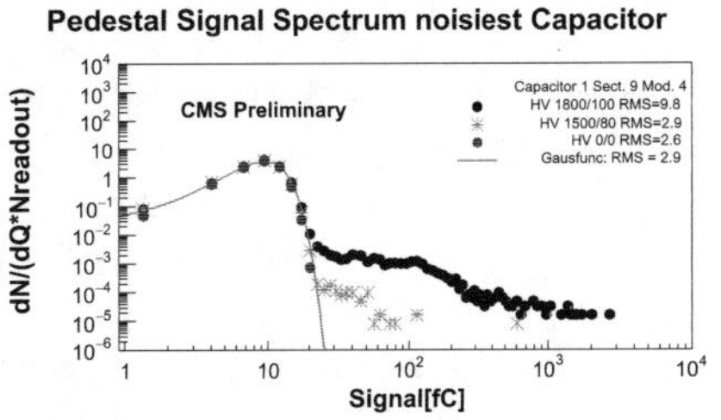

Figure 2. Pedestal charge spectrum for the noisiest capacitor of a typical CASTOR channel for various cathode/last dynode Voltage settings. Plot from [3]

3.3. Results on statistical gain correction factors

The intercalibration constants were determined at the maximal gain settings (corresponding to 1800 V). Physics events, however, were recorded with a gain recipe optimised to the dynamic range. Gain correction factors were therefore needed to scale the intercalibration constants from the maximal gain to the physics gains.

At CASTOR two methods are available to achieve this:

(i) Dedicated laboratory dark-box measurements were performed on CASTOR's PMTs. From the parameterised gains the correction factors can be determined

(ii) A statistical method can be performed by analysing a pedestal and LED spectrum per channel

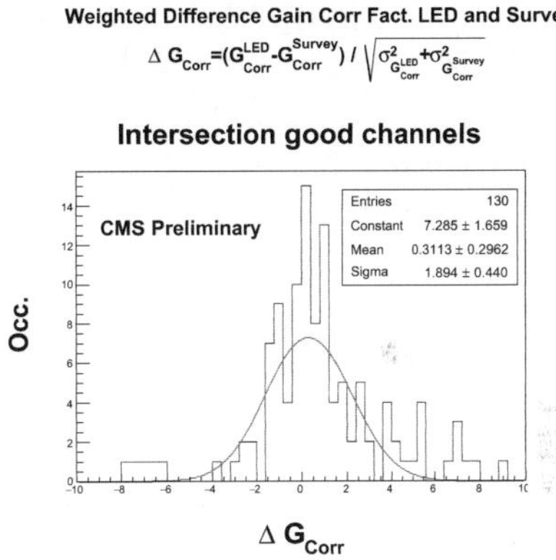

Figure 3. Pull distribution with Gauss fit of gain correction factors determined by a dark box PMT survey (in laboratory) and a statistical analysis of an LED and pedestal spectrum (in situ). Plot from [3]

For the latter, one first needs to obtain a corrected signal S_c by subtracting the pedestal signal S_p from the LED signal S_t. S_c equals the electronic gain $G_{el.}$ times the number of photoelectrons $N_{p.e.}$: $S_c = S_t - S_p = G_{el.} \cdot N_{p.e.}$ The relative width of S_c consists of a contribution of the gain fluctuations and the Poissonian fluctuations of the conversion of photons to photoelectrons and is given by: $\frac{\sigma_{S_c}}{S_c} = \frac{\sigma_{N_{p.e}}}{N_{p.e}} \oplus \frac{\sigma_{G_e}}{G_e}$. The first contribution is $\frac{1}{\sqrt{N_{p.e}}}$. The second contribution $\approx \frac{1}{\sqrt{G_{el.}}}$. The number of photoelectrons is a much smaller number than the electronic gain, therefore the width in corrected signal can be approximated as: $\sigma_{S_c} = G_e \cdot \sqrt{N_{p.e}}$ and thus $G_e = \frac{\sigma_{S_c}^2}{S_c}$.

In fig. 3 the weighted difference between the gain correction factors obtained by both methods is displayed. The statistical properties of the fit to the distribution reveal reasonable consistency between the methods.

Overall, collecting muons at the maximal gain instead of the physics gain recipe significantly improved the collection efficiency, but did not yield smaller uncertainties on the final intercalibration constants. The latter were found to be on average 16%.

IOP Conf. Series: Journal of Physics: Conf. Series **928** (2017) 012008 doi:10.1088/1742-6596/928/1/012008

4. CASTOR's absolute calibration and alignment

A fully data-driven calibration has still not been successfully performed at CASTOR. Generally, conventional methods (for example using the invariant mass of a resonance or applying a jet-balance) have proven nontrivial at CASTOR, due to its peripheral nature and lack of η segmentation. CASTOR is currently calibrated by extrapolating the energy measurement (on particle level) of a nearby subsystem to CASTOR's acceptance and equating this to CASTOR's total energy measurement [4]. Regrettably, the uncertainty due to the model-dependent extrapolation factors is large and the overall uncertainty of this calibration is 15%.

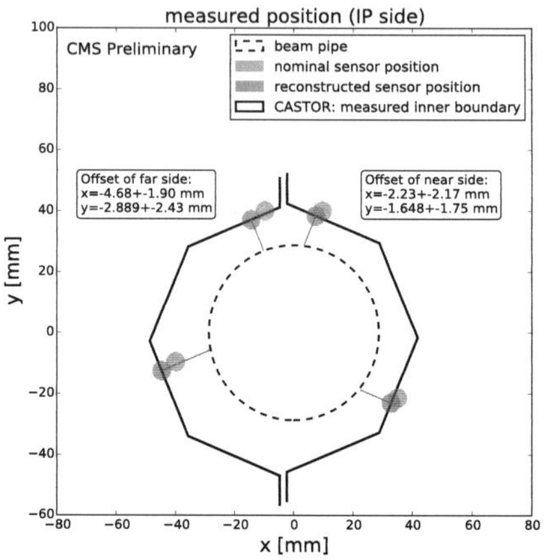

Figure 4. Illustration of the global fit result of the CASTOR position for data taking during LHC Run 2 pp collisions (B=0T). Plot from [3]

The uncertainty on CASTOR's position contributes also to the overall uncertainty, and for an energy-flow measurement can literally be interpreted as a scale-uncertainty. CASTOR is, among others, equipped with infrared sensors which measure the distance w.r.t. the beampipe. CASTOR, the position sensors, and the beampipe have been schematically depicted in fig. 4.

A better calibration during LHC Run 2 of the infrared sensors w.r.t. a curved object reduced the alignment's contribution to the systematic error by over 50% w.r.t. LHC Run 1; the alignment's final contribution to the uncertainty is 2%.

5. Conclusion

We reviewed the main aspects of CASTOR's performance during LHC Run 2. Collecting beam-halo muons at the maximal gains significantly increased the collection efficiency, though no improvement was found on the intercalibration constants' final uncertainty. As a consequence, various noise contributions were also maximally amplified. Therefore, the noise per channel was estimated with a fit.Two methods on obtaining gain correction factors were compared. The results were found to be reasonably consistent. Improving the calibration of CASTOR's infrared sensors reduced the alignment's contribution to the systematic uncertainty by over 50%.

Naturally, the LHC provides a unique operating environment and not all lessons learned apply at other forward calorimeters. Three generally applicable conclusions can be drawn though:

(i) Especially at high gains the noise threshold may need to be determined with a fit
(ii) The statistical method on obtaining gain correction factors, which has a broad, polyvalent range of applicability, yields reasonably consistent results
(iii) A reliable, autonomous alignment system is indispensable for very forward calorimeters

The uncertainty on the statistical gain correction factors may be further improved by parameterising the dependence of the gain on the high voltage. Also the fake signals' contribution from non-electronic noise may be diminished by using CASTOR's longitudinal segmentation.

References

[1] S. Chatrchyan *et al.* [CMS Collaboration], JINST **3** (2008) S08004. doi:10.1088/1748-0221/3/08/S08004
[2] P. Gottlicher *et al.* [CMS CASTOR Collaboration], Nucl. Instrum. Meth. A **623** (2010) 225.
[3] [CMS Collaboration], Results on CASTOR Performance during LHC Run 2. CMS-DP-2016-006, CERN, Geneva, 2016.
[4] [CMS Collaboration], Measurement of the pseudorapidity and centrality dependence of the very forward energy density in PbPb collisions at $\sqrt{s}$=2.76 TeV. CMS-PAS-HIN-12-006, CERN, Geneva, 2012.

IOP Conf. Series: Journal of Physics: Conf. Series **928** (2017) 012009 doi:10.1088/1742-6596/928/1/012009

The NA62 Hadron Calorimeter

Riccardo Aliberti

Johannes Gutenberg Universitaet - Mainz

E-mail: aliberti@uni-mainz.de

Abstract. NA62 is a fixed target experiment located in the north area of CERN. The ambitious aim of the experiment is the measurement of the branching ratio (Br) of the very rare decay $K^+ \to \pi^+ \nu\bar{\nu}$ within 10% precision using a decay-in-flight technique. The branching ratio of this decay is very well calculated in the standard model as $(9.11 \pm 0.72) \times 10^{11}$ and the measurement of this channel represents one of the most promising fields for the search of new physics beyond the standard model. The presence of just one detectable track in the final state is an enormous challenge for any experiment. The decay is fully reconstructed, which leads to a strong background suppression. Still the detector resolution, combined with the tiny branching ratio of the signal, makes the $K^+ \to \mu^+ \nu$ decay (whose Br is 0.64) a critical source of background. The NA62 detector was therefore designed to perform an excellent π/μ separation using a very efficient particle identification system. A major role is played by the calorimeters that provides a muon rejection factor of the order of 10^5 through the measurement of energy and shape of the hadronic showers. The calorimetric system consists of an electromagnetic calorimeter filled with liquid krypton and a hadron calorimeter. This presentation, after illustrating the HAC structure, reports on the calibration procedure of the detector response and preliminary results of the performance of the hadronic energy reconstruction.

1. Introduction

NA62 is a fixed target experiment located at the North Area of the European Center for Nuclear Research (CERN). The proton beam from the SPS accelerator is used to produce a secondary hadron beam and perform flavor physics studies mainly in the charged kaon sector. NA62 is the successor of the NA48[1] experiment and acquired for the first time data in 2007[2] using the same beam line and apparatus as in NA48. Starting from 2008 the experiment underwent an intense phase of R&D and construction of a almost completely new apparatus. The new detector was optimized for the precise measurement of the $K^+ \to \pi^+ \nu\bar{\nu}$ branching ratio and in 2014/2015 has taken data for the first time.

2. The Search for $K^+ \to \pi^+ \nu\bar{\nu}$

The rare decay $K^+ \to \pi^+ \nu\bar{\nu}$ is a Flavor Changing Neutral Current (FCNC) process dominated by short distance interactions and therefore highly suppressed. The clean theoretical environment allows to have a precise computation within the standard model (SM) to $Br_{SM}(K^+ \to \pi^+ \nu\bar{\nu}) = (9.11 \pm 0.72) \times 10^{-11}$ [6], where the main sources of uncertainty are the CKM matrix parameters. Many theoretical models of physics beyond the SM predict deviations for this branching ratio. The high sensitivity of this process to the existence of new leptons or heavy mesons up to the 100 TeV scale makes it a golden channel in the search of physics beyond the SM. Only one previous measurement exists, the E787 and E949 experiments at Brookhaven National

Laboratories (BNL) obtained $Br_{exp} = \left(17.3^{+11.5}_{-10.5}\right) \times 10^{-11}$ [7][8]. The measurement is based on the observation of 7 candidate events and the uncertainty does not allow to precisely probe the standard model validity.

3. Measurement Strategy and NA62 Apparatus

The aim of the NA62 experiment is to collect around 100 $K^+ \rightarrow \pi^+\nu\bar{\nu}$ signal events until end of 2018 in order to measure the branching ratio with 10% accuracy[9][10]. The experiment plans to collect in total 10^{13} kaon decays using the 400 GeV/c proton beam from the CERN SPS accelerator. Assuming a 10% acceptance the background signal ratio should be kept below 20% to achieve the design sensitivity. This implies a background suppression at the level of 10^{12} [11] . The tiny branching ratio combined with the weak signature of the signal makes the rejection of the main K^+ decay modes very challenging. The keys to achieve the foreseen sensitivity are: the rejection of beam induced background, a precise measurement of kinematics, an efficient particle identification and a hermetic photon detection.

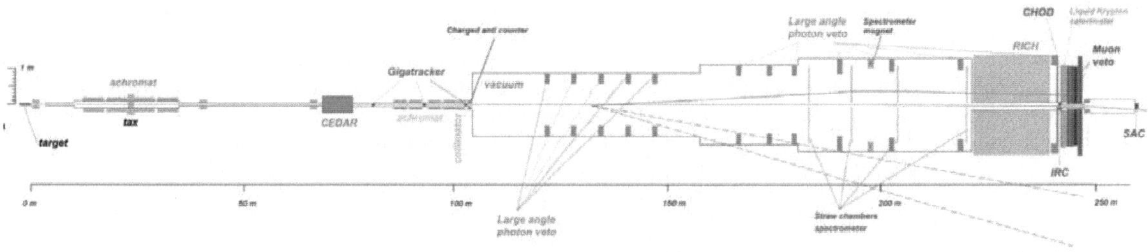

Figure 1. The NA62 apparatus

Beam induced background The 400 GeV protons beam from SPS impinge on a Beryllium target resulting in a secondary hadron beam which reaches the experiment with a central momentum of (75±0.8) GeV/c. The kaons represent 6% of the total 750 MHz particle rate. The remaining components are protons (10%) and pions (74%). A Cherenkov detector filled with Nitrogen (Cedar) provides the kaon identification with 100 ps resolution. The precise timing from the Cedar allows to reject $\pi^+ \rightarrow \mu^+\nu$ that could mimic the signal. A series of scintillator arrays (CHANTI) around the beam line are used to veto the interaction of particles inside the beam spectrometer.

Measurement of the Decay Kinematics The suppression of background from the reconstruction of kinematic variables is required to be at the level of 10^{-5} on the main kaon decays. Precise measurements of both K^+ and π^+ directions and momenta with a sub-percent resolution are performed by two spectrometer systems. The beam spectrometer Gigatracker (GTK), placed right after the Cedar detector, is composed by three, 300 μm thick, silicon pixel stations. The GTK faces the full beam rate and for this reason the timing performances are crucial, the single hit time resolution is at the level of 200 ps. A magnetic spectrometer is used to track the kaon decay products. It is composed of 4 stations of straw tube chambers resulting in a relative momentum resolution (σ_P/p) better than 0.3%. The combined data from the GTK and STRAW detectors allow a reconstructed vertex resolution at the mm level. The last station of the GTK define the start of the fiducial decay region, which ends at the first station of the STRAW tube spectrometer.

Particle Identification The kaon decay $K^+ \rightarrow \mu^+\nu$ with a branching ratio of 63.6%[12] represents the main background source for the $K^+ \rightarrow \pi^+\nu\bar{\nu}$ measurement. In order to keep this background under control a muon suppression factor of 10^8 is required. The NA62 detector has been designed in order to provide a very efficient Particle Identification (PID). The system includes a Ring Cherenkov detector (RICH), 17 m long, filled with Neon and a system of calorimeters which allows to disentangle $\pi/\mu/e$ by studying the energy share between the electromagnetic calorimeter (LKr) and the hadron calorimeter (HAC). Following a 80 cm thick iron wall a fast muon veto (MUV3) provides an online veto trigger with a 300 *ps* time resolution.

Photon Veto A hermetic coverage of the experiment against photons is fundamental to veto all the kaon decays containing a π^o in the final state ($Br \sim 30\%$[12]) with an inefficiency at the level of 10^{-8} . The NA62 experiment is equipped with three different type of detectors devoted to the photon detection from 0 up to 50 mrad: 12 stations of Large Angle Veto (LAV), an electromagnetic calorimeter (LKr) and two Small Angle Vetoes (IRC and SAC) for photons down to 0 mrad.

4. The Hadron Calorimeter (HAC)

The Hadron Calorimeter is a sampling calorimeter using iron as absorber material. It counts in total almost 8 interaction lengths, presenting therefore minimal energy losses even for pions of momenta beyond 100 GeV/c, divided in two independent modules. The front module was newly built for the NA62 experiment with a fine transverse segmentation in order to better disentangle hadronic and electromagnetic shower components. The back module is the front module of the old NA48 hadron calorimeter.

Each of the two modules present 24 iron planes interleaved with scintillators in alternated horizontal and vertical orientation. The strips covering the same area are read out by the same photomultiplier. In the front most of the strips span the full detector width. Wave length shifting fibers are used to readout the scintillators on both ends. The transverse segmentation of the front module is $6cm$.

Each scintillator strip in the back module covers half of the detector width and they are therefore readout on one side only. The light collection in the back module is performed via light guides and the transverse segmentation is 6 cm.

5. HAC Calibration

The performances required for the hadron calorimeter can only be achieved after a precise calibration of the detector response. The calibration procedure consists of four different steps: photomultiplier gain equalization, correction for attenuation inside the scintillators, estimation of the energy scale from muon pulses, and determination of pion energy corrections.

Gains Equalization and Attenuation in Scintillators The adjustment of the gain is performed channel-by-channel by a scan of the supply voltage during the data taking with a muon-based trigger. The single channel response to muon pulses is evaluated as a function of the supply voltage. A major role in the energy resolution of the HAC is played by the correction for the impact point of the particle. To precisely determine the scintillator contribution dedicated runs with only muon tracks are performed. The tracks reconstructed in the STRAW spectrometer are extrapolated to and matched with the HAC. The distribution of the charge collected as a function of the position is then parameterized with a double exponential function to take in account reflections inside the scintillators. Figure 2 shows the effect of the impact point corrections on charge collected in muon events for the front (left) and the back module (right).

Energy Scale The absolute energy scale for the calorimeter is determined independently for the two modules by fitting the collected charge distribution with a Landau function convoluted with a gaussian. The landau distribution describes the energy deposition by minimum-ionizing particles, while the gaussian is used to parameterized the smearing effect from the resolution.

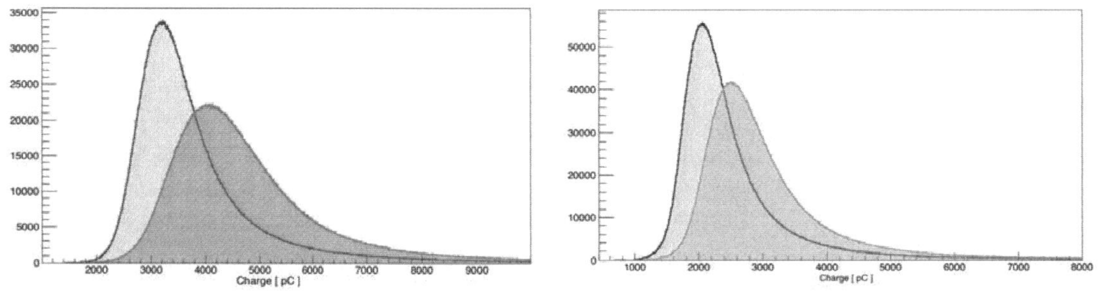

Figure 2. Charge collected by the front (left) and back (right) module when crossed by a muon. The charge distributions are shown before (solid) and after (dashed) the impact point correction.

The energy scale is derived by dividing the peak of energy deposit from Monte Carlo simulation by the most probable value of the fitted landau distribution. The energy scale factors where estimated as 317 keV/pC for the front and 450 keV/pC for the back module.

Pion Calibration Front and back module of the HAC have slightly different structure which results in a different response to impinging pions. In particular the sampling fraction of the front module is almost twice bigger than for the back one. The consequence of the different structure is that the two modules have to be calibrated separately. The calibration is performed by analyzing dedicated MC simulations of $K^+ \rightarrow \pi^+ \nu \bar{\nu}$ without the LKr and only one of the two HAC modules present. The ratio between the energy deposited in the calorimeter modules and the pion momentum is shown as a function of the momentum in figure 3.

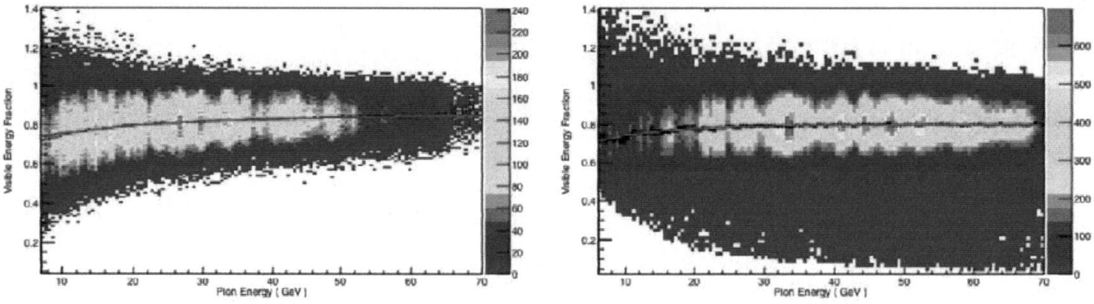

Figure 3. Fraction of visible energy in the front (left) and back (right) HAC module for pions as a function of the momentum.

Showers from hadrons have a development which can vary between two extremes: a full electromagnetic and a full hadronic development. To estimate the contribution from the two components, a weighting parameter (W) was defined as the ratio between the sum of the energy collected in each channel squared and the total energy collected in the module squared: $W = \sum_i E_i^2 / E_{Module}^2$. A small weight indicates a dominance of the hadronic component while a weight parameter near to 1 an electromagnetic shower.

The scale factor for the visible pion energy takes into account the mean contribution by the hadronic and electromagnetic component and is applied dependently to the weight parameter following a Fermi-Dirac function as shown in equation 1.

$$E_{\text{HAC Module Corrected}} = E_{\text{HAC Module}} + E_{\text{Invisible HAC Module}} \times (1 + F(W)) \qquad (1)$$

The calibration is checked in the data by applying the corrections in a step by step basis and verifying the improvement in terms of resolution and linearity of the response. On left of figure 4 it is possible to notice the improvement of the HAC response from before any correction (red line), after applying the visible energy correction (in blue) and after the modulation of the invisible energy correction by the weighting parameter (in green). The resolution was estimated as:

$$\frac{\sigma_E}{E} = (0.115 \pm 0.003) \oplus \frac{0.38 \pm 0.05}{\sqrt{E}} \oplus \frac{1.37 \pm 0.17}{E} \qquad (2)$$

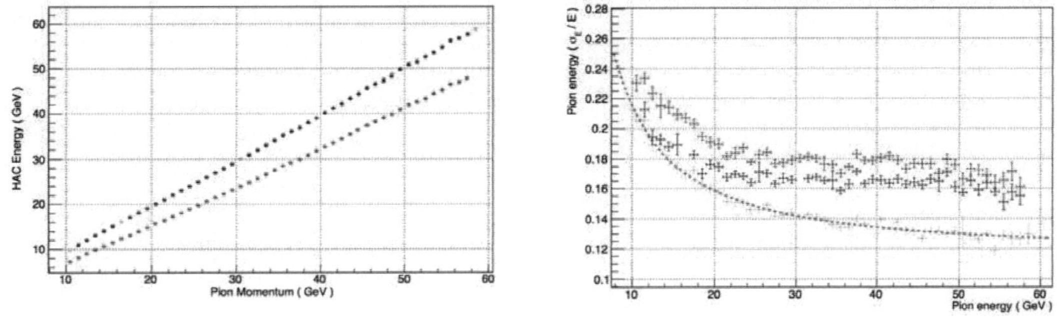

Figure 4. Linearity of the energy response (left) and energy resolution (right). In red before any correction, blue after visible energy scaling and green after the weighting correction. The red dotted line shows the fit to the energy resolution (see formula 2)

Acknowledgments
Thanks to the NA62 Collaboration for the support to this work

References

[1] V.Fanti et al. (NA48 Collaboration), *Nucl. Instrum. Methods* A **574**, 433 (2007).
[2] C. Lazzeroni et al. (NA62 Collaboration), *Phys. Lett.* B **719**, 326 (2013).
[3] M. Gell-Mann and F. Zachariasen, *Phys. Rev.* D **124**, 953 (1961) .
[4] T. Husek and S. Leupold, *Eur. Phys. J* C **75**, 586 (2015).
[5] T. Husek et al., *Phys. Rev.* D **92**, 054027 (2015).
[6] A.J. Buras, D. Buttazzo, J. Girrbach-Noe and R. Knegjens, *JHEP* **1511**, 33 (2005).
[7] S. Adler et al. (E949 and E787 Collaborations), *Phys. Rev.* D **77**, 052003 (2008).
[8] S. Adler et al. (E949 and E787 Collaborations), *Phys. Rev.* D **79**, 092004 (2009).
[9] G. Anelli et al., CERN-SPSC-2005-013; SPSC-P-326.
[10] NA62 Technical Design Document, NA62-10-07; https://cdsweb.cern.ch/record/14049857.
[11] G. Ruggiero (NA62 Collaboration), *PoS KAON* **13** 032 (2013).
[12] C. Patrignani et al.(Particle Data Group), Chin. Phys. C, 40, 100001 (2016).

IOP Conf. Series: Journal of Physics: Conf. Series **928** (2017) 012010 doi:10.1088/1742-6596/928/1/012010

Geometric alignment of the SND detector

A.A.Korol[1,2] **and N.A.Melnikova**[1,*]

[1] Budker Institute of Nuclear Physics, 11 akademika Lavrentieva prospect, Novosibirsk, 630090, Russia

[2] Novosibirsk State University, 2 Pirogova street, Novosibirsk, 630090, Russia

E-mail: *N.A.Melnikova@inp.nsk.su

Abstract. We present the design, implementation and validation of the software procedure used to perform geometric calibration of the electromagnetic calorimeter with respect to the tracking system of the SND detector (BINP, Novosibirsk). This procedure is based on the mathematical model describing the relative calorimeter position by means of a set of parameters. The parameter values are determined by minimizing a χ^2 function using the difference between directions reconstructed in these two subdetectors for $e^+e^- \to e^+e^-$ data events.

1. Introduction

The SND detector [1, 2, 3] is a nonmagnetic detector used at the e^+e^- collider VEPP-2000 [4] (BINP) for the hadronic cross-section measurement experiments in the center of mass energy range $0.3 \div 2.0$ GeV. The SND consists of several subsystems (Fig.1) including the spherical electromagnetic calorimeter (EMC), the cylindrical tracking system (TS), the threshold Cherenkov counters and the muon detector.

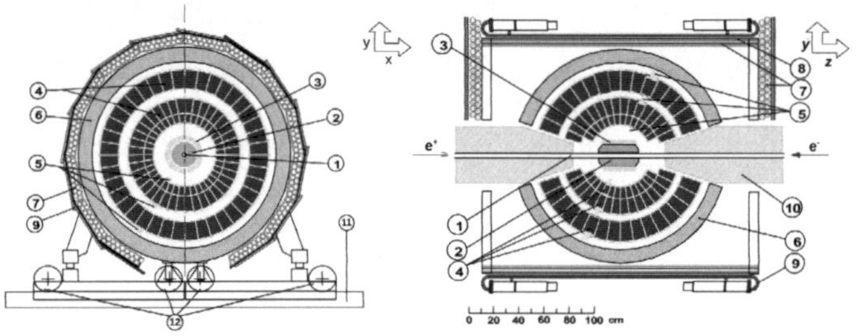

Figure 1. The SND scheme: 1 — vacuum pipe, 2 — tracking system (TS), 3 — threshold Cherenkov counter, 4–5 — electromagnetic calorimeter (NaI (Tl)) (EMC), 6 — iron absorber, 7–9 — muon detector, 10 — focusing solenoids, 11 – rails, 12 – wheels.

The EMC is composed of 1632 NaI(Tl) crystals arranged in 3 spherical layers. The EMC spherical shape covers polar angle range of $18° - 162°$ and provides uniform coverage of solid angle $0.95 \cdot 4\pi$. The EMC crystals are rigidly fixed to two supporting aluminum spheres: the first one is for the first two layers of crystals while the second one is for the third layer. The EMC is divided into 2 hemispheres which can be moved in the horizontal direction orthogonal to the detector axis using rails and wheels (Fig.1: labels 11 and 12). It's used to access the TS and the interior sphere of the SND.

The EMC and the TS provide main information for event reconstruction: the TS measures parameters of charge particle tracks ($z_0; d_0; \varphi; \theta$) while the EMC measures the energy and the angular position of the electromagnetic showers ($\varphi_{EMC}; \theta_{EMC}$). For the events occurred inside the SND to be effectively reconstructed it is important to know precise positions of the EMC sensitive crystals relative to the TS. However the data demonstrates difference between particle angles reconstructed with the TS and with the EMC. We attribute this difference to misalignment of the EMC relative to the TS. This misalignment can result in kinematic discrepancy inside the events containing both charged and neutral particles (like $e^+e^- \to \pi^+\pi^-\pi^0\pi^0$). That is why we perform the software alignment procedure that reveals the true EMC position relative to the TS. We utilize e^+e^- scattering events because electrons and positrons have reliable angle measurements in both the TS and the EMC.

2. Mathematical model

The right handed coordinate system fixed with the TS is used as the reference coordinate system. Its center is at the TS center, the Z axis is directed along the TS horizontal axis, the Y axis is directed vertically upwards and the X axis is directed towards the collider ring center. We consider the EMC halves as rigid bodies.

To describe the EMC position relative to the TS two sets of parameters are used. The first set describes the rotation and the shift of the EMC as a whole: α is the angle of common EMC rotation around the TS Z axis ($0° < \alpha < 360°$); β_x and β_y describes the EMC tilt with respect to the TS Z axis; dx, dy, dz are the EMC shifts in corresponding directions relative to the TS. The second set of parameters describes rotations and shifts of the hemispheres relative to the EMC: μ is the separation half-angle of the EMC hemispheres ($0° < \mu < 90°$); τ is the separation direction angle from the vertical direction, ($-180° < \tau < 180°$); $dx_{rel}, dy_{rel}, dz_{rel}$ are the hemisphere relative shifts; β_{rel} is the angle of the relative rotation around the X axis, ($0° < \beta_{rel} < 360°$).

Subtracting angles of an aligned EMC point from a misaligned EMC point we build the model functions which describe the difference between reconstructed angles for $\sin(\varphi_{TS} - \varphi_{EMC})$ and $\theta_{TS} - \theta_{EMC}$. Here φ_{EMC}, θ_{EMC} are azimuth and polar angles reconstructed in the EMC while φ_{TS} and θ_{TS} represent the expectation for the EMC cluster direction from the TS track and determined as the polar coordinates of the point where the track crosses a sphere with the radius R given by: $R = R_0 + X_0 \cdot (x_{max} - \frac{x_{pre}}{\sin\theta})$. Here R_0 is the EMC inner radius, $x_{max} = \log\frac{E}{E_c} - 0.5$ is the maximum of the longitudinal electromagnetic shower distribution [5] , x_{pre} is an estimation of the sensitive material thickness between the interaction point and the EMC, X_0 is the radiation length for NaI, E_c is its critical energy. Unfortunately, the maximum of the energy deposition in the EMC is located in the first two layers, making impossible to estimate the misalignment of the third layer.

To retrieve the parameter values a χ^2 function is minimized. This function is constructed from the squared differences between the measured angle residuals and the expected ones, parametrized with the model functions, with the differences being scaled by a measurement error squared and summed over 160 spatial bins ($16\,\varphi_{EMC} \times 10\,\theta_{EMC}$). Each error includes both a statistical and systematic component. Systematic uncertainty is attributed to the non-uniformity in the EMC spatial reconstruction and shortcomings of the model. This is expected to be the same for polar and azimuth direction and estimated from the data to be $\sim 0.2^{\circ}$. Minimization is performed using the TMinuit class from the ROOT framework [6]. Typical results of the function fit are illustrated in Fig. 2

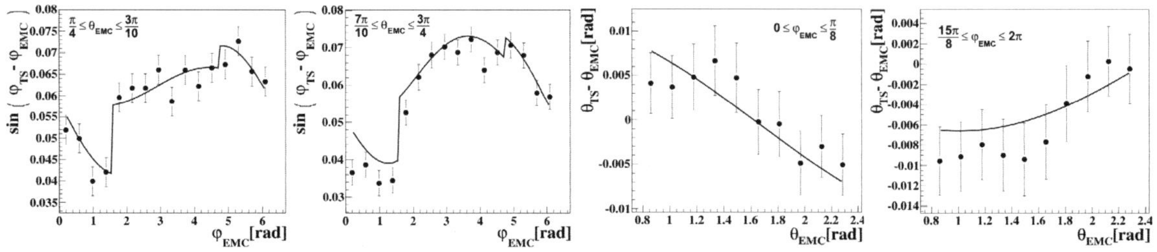

Figure 2. Results of the function fit for one energy point, $E_{beam} = 612.5$ MeV. Here φ_{TS}, θ_{TS} - are azimuth and polar TS angles, φ_{EMC}, θ_{EMC} are azimuth and polar EMC angles

3. Calibration procedure and results

The geometric calibration procedure includes several steps starting from the $e^+e^- \to e^+e^-$ event selection (2 collinear tracks, energy between 0.8 and 1.1 of E_{beam} for each particle).

The next step is to minimize the χ^2 function using the 2D histogram constructed using the selected events. After that the values of the alignment parameters are stored in the conditions data base and can be applied during reconstruction and simulation.

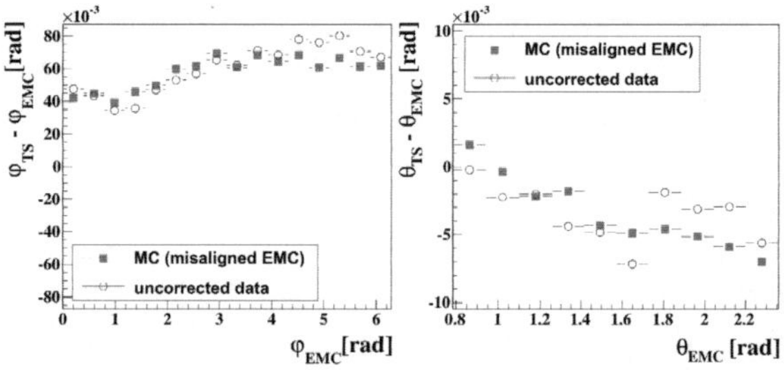

Figure 3. Comparison of angles reconstructed in the EMC and in the TS for the $e^+e^- \to e^+e^-$ events for data (blue unfilled circle) and realistic MC simulation (red filled square), $E_{beam} =$ 612.5 MeV. Here φ_{TS}, (θ_{TS}) - an azimuth (polar) TS angle, φ_{EMC}, (θ_{EMC}) - an azimuth (polar) EMC angle.

The calibration algorithm was validated using MC simulation where the EMC is described misaligned according to the parameters. Comparison of this MC simulation with the data demonstrates a good agreement. (Fig.3).

The procedure described in this section was applied to $\sim$ 2000 runs, recorded in the data taking period of 2010-2011 years, with integrated luminosity integral $25.3\,pb^{-1}$. Fig. 4 shows evolution of the parameter values during this period. The value of α was large but stable during this period while the value of dx changed slightly several times due to disassembling and reassembling the detector.The rest of the parameters were quite small and stable.

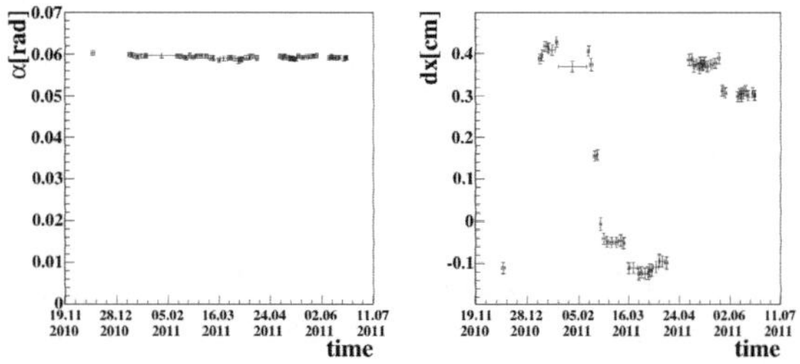

Figure 4. Evolution of the parameter values for α(global rotation around the Z TS axis), dx(global shift along the X TS axis).

Some results of the applied calibration for the $e^+e^- \to e^+e^-$events and the $e^+e^- \to 2\gamma$ annihilation events are shown in Fig.5 and Fig.6.

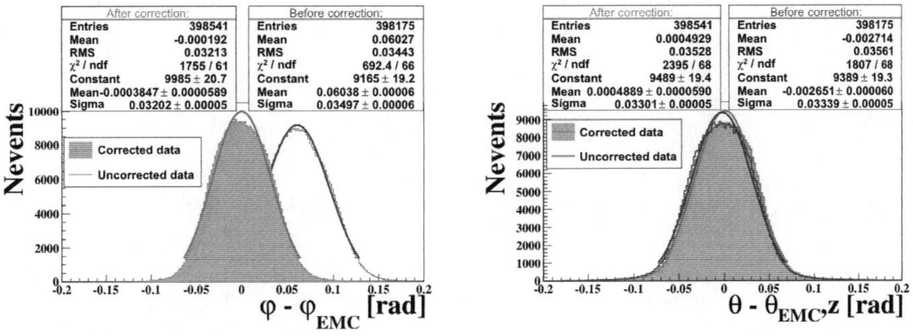

Figure 5. Results of the geometric correction for $e^+e^- \to e^+e^-$ data, $E_{beam} = 612.5$ MeV: difference between angles reconstructed in the EMC and in the TS before (blue unfilled histograms) and after (red filled histograms) the correction. Here $\varphi, (\theta)$ - an azimuth (polar) TS angle, φ_{EMC} - an azimuth EMC angle, $\theta_{EMC,z}$ - a polar EMC angle with correction for the interaction point shift along the TS Z axis.

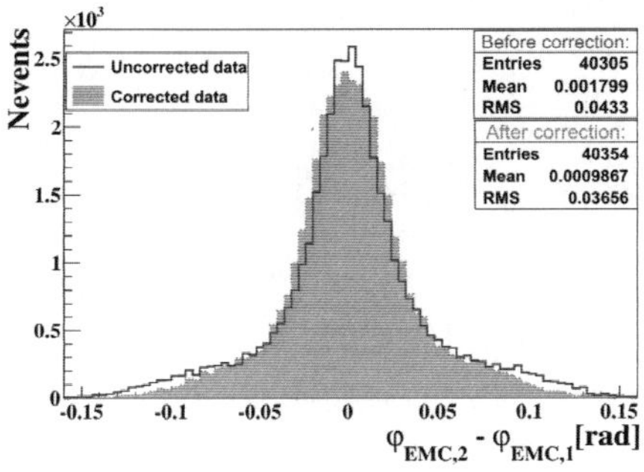

Figure 6. Results of the geometric correction for $e^+e^- \to 2\gamma$ data, $E_{beam} = 612.5$ MeV: difference between azimuth EMC angles reconstructed for two photons (φ_{EMC_1},φ_{EMC_2}) before (blue unfilled histogram) and after (red filled histogram).

4. Conclusions

We have developed and implemented the geometric calibration procedure for the SND detector electromagnetic calorimeter. The algorithm is validated using MC simulation. It is shown that most of the alignment parameter values stay stable between disassembling and reassembling the detector.

The procedure allows us to reduce the polar and azimuth angles discrepancy between the TS and the EMC measurements. Azimuth angle common bias of $\sim 3°$ and irregularity of $\sim 1°$ are removed. We see improvement in the EMC angular resolution due to this correction by 15.6% as well.

The results of geometric calibration are used in the actual experimental data analysis and more realistic MC simulation.

Acknowledgments

This work was supported by the Russian Science Foundation [grant number 14-50-00080].

References

[1] Achasov M N *et al.* 2000 *Nucl. Instrum. Meth.* **A449** 125–139 (*Preprint* hep-ex/9909015)
[2] Abramov G N *et al.* 2001 *eConf* **C010430** T10 [,122(2001)] (*Preprint* hep-ex/0105093)
[3] Aulchenko V M *et al.* 2009 *Nucl. Instrum. Meth.* **A598** 102–104
[4] Khazin B I (CMD-3, SND) 2010 *Nucl. Instrum. Meth.* **A623** 353–355
[5] Olive K A *et al.* (Particle Data Group) 2014 *Chin. Phys.* **C38** 090001
[6] Brun R and Rademakers F 1997 *Nucl. Instrum. Meth.* **A389** 81–86

IOP Conf. Series: Journal of Physics: Conf. Series **928** (2017) 012011 doi:10.1088/1742-6596/928/1/012011

Calorimetry of the CMD-3 detector

R R Akhmetshin[1,2], A V Anisenkov[1,2], V M Aulchenko[1,2], N S Bashtovoy[1], D A Epifanov[1,2], L B Epshteyn[1,2,3], A L Erofeev[1,2], A A Grebenuk[1,2], D N Grigoriev[1,2,3], V F Kazanin[1,2], O A Kovalenko[1,2], A N Kozyrev[1,2,3], A E Kuzmenko[1,2], A S Kuzmin[1,2], I B Logashenko[1,2], K Yu Mikhailov[1,2], V S Okhapkin[1], G P Razuvaev[1,2], A A Ruban[1], V E Shebalin[1,2], B A Shwartz[1,2], V M Titov[1], A A Talyshev[1,2] and Yu V Yudin[1,2]

[1] Budker Institute of Nuclear Physics, SB RAS, Novosibirsk, 630090, Russia
[2] Novosibirsk State University, Novosibirsk, 630090, Russia
[3] Novosibirsk State Technical University, Novosibirsk, 630092, Russia

E-mail: akhmetshin@inp.nsk.su

Abstract. The CMD-3 detector has been collecting data since 2010 at the e^+e^- collider VEPP-2000 in the Budker Institute of Nuclear Physics. VEPP-2000 uses the novel round beam technique and provides high luminosity in a wide c.m.energy range from 0.32 to 2 GeV. The physics goal of the CMD-3 experiment is a study of the e^+e^- annihilation into hadrons. CMD-3 is a general-purpose detector, which provides high efficiency for both charged and neutral particles. The electromagnetic calorimeter consists of the barrel calorimeter based on liquid xenon and CsI crystals, and the endcap calorimeter based on BGO crystals. The main parameters of the calorimeters are presented.

1. Introduction

VEPP-2000 [1, 2] is an e^+e^- collider at the Budker Institute of Nuclear Physics (BINP) in Novosibirsk, Russia. It operates in a center-of-mass (c. m.) energy range from 0.32 up to 2 GeV. The design luminosity of the collider at 2 GeV c. m. energy is 10^{32} cm^{-2}s^{-1}. In order to reach this luminosity in a single-bunch mode a novel round beam technique developed at BINP is used. The beam energy is monitored with a precision of $\Delta E/E = 6{\cdot}10^{-5}$ using the Compton backscattering process [3, 4]. There are two interaction points at the collider and two detectors are mounted there: Spherical Neutral Detector (SND) [5, 6] and Cryogenic Magnetic Detector (CMD-3) [7]. Data taking started in 2010. The integrated luminosity collected by each detector during three years of operation is about 60 pb^{-1}. In 2013 the collider operation was stopped for an upgrade of the collider and detectors, and it is planned to resume experiments in 2016.

The physical program [8, 9, 10] includes a measurement of the $e^+e^- \to$ hadrons cross sections, a study of the properties of known and a search for new vector mesons, a measurement of $n\bar{n}$ and $p\bar{p}$ cross sections near threshold and searches for exotic hadrons. These tasks require a detector with high detection efficiency for multiparticle events and good energy and angular resolution for both charged particles and photons.

CMD-3 is a general-purpose detector. Its layout is presented in figure 1. The electromagnetic calorimeter is one of the most important systems of the CMD-3 detector. Its main goals are

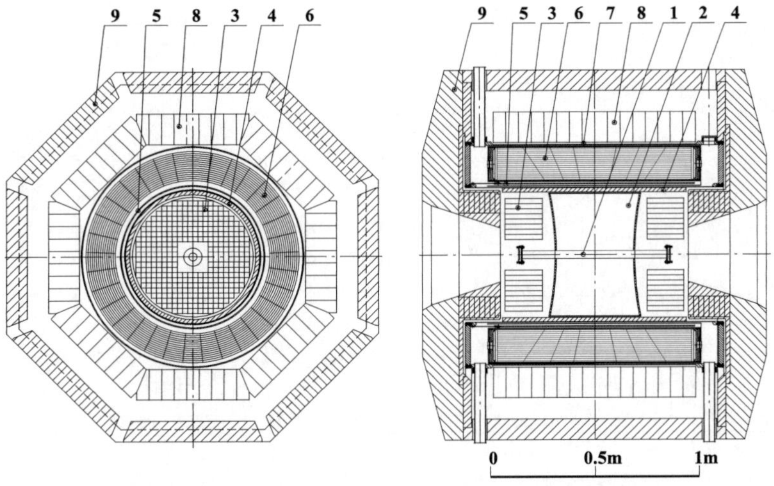

Figure 1. The CMD-3 detector layout.
1 – beam pipe,
2 – drift chamber,
3 – BGO endcap calorimeter,
4 – Z-chamber,
5 – superconducting solenoid,
6 – liquid xenon calorimeter,
7 – time-of-flight system,
8 – CsI calorimeter,
9 – yoke.

measurements of the energy and coordinate of photons, separation of electrons from hadrons and generation of signals for the neutral trigger. The calorimeter of the detector consists of barrel and endcap parts. The barrel calorimeter combines the inner liquid xenon calorimeter (LXe) [11] and the outer one based on CsI crystals [12]. The endcap calorimeter [13] is based on the BGO crystals. The total solid angle coverage of the whole CMD-3 calorimeter is equal to $0.94\times4\pi$.

2. General description

The inner LXe calorimeter is a set of 14 ionization chambers with 7 cylindrical cathodes and 8 anodes divided by a 10.2 mm gaps between them. The calorimeter is placed in the same vacuum vessel with a superconducting solenoid to reduce passive material in front of the calorimeter. The electrodes are made of 0.5 mm thin G-10 plates foiled with copper. The conductive surface of anode cylinders is divided into 264 rectangular pads (8 along the Z-axis and 33 in the r-φ plane) forming so-called "towers" oriented at the beam interaction point. Pads within one tower are electrically connected. An average tower size is $8\times10\times15$ cm^3. The signals from the towers are used to measure the deposited energy. Cathode cylinders are divided into 2112 strips to provide a precise coordinate measurement together with the measurement of specific energy losses. Each side of the cathode cylinder contains about 150 strips. The strips on the opposite sides of the cylinder are mutually perpendicular. One signal strip consists of four connected 2 mm width strips. Such a semitransparent electrode structure provides charge induction on both sides of a cathode electrode. That allows one to determine both coordinates of the point of photon conversion using the information from one gap only. The design of the LXe calorimeter is described in detail in [11, 14].

Since the LXe calorimeter is rather thin (5.4 X_0), it is surrounded by the CsI calorimeter to improve energy resolution. The CsI calorimeter consists of 1152 $6\times6\times15$ cm^3 Na- or Tl-doped CsI crystals assembled in 8 octants. Each octant consists of 9 rows (modules) of crystals. In each octant modules on both sides have special truncated shape in order to avoid gaps between octants. Each module is assembled of 16 counters. The length of crystals corresponds to the thickness of 8.1 X_0. The total sensitive material thickness of the barrel calorimeter for a normal incident particle is equal to 13.5 X_0. The thickness of the passive material in front of the barrel calorimeter is 0.35 X_0 and passive material between LXe and CsI parts of the barrel calorimeter is 0.25 X_0. The design of the CsI calorimeter is described in detail in [12].

To increase solid angle coverage, the CMD-3 is equipped by the endcap calorimeter. It consists of 680 BGO crystals of $2.5\times2.5\times15$ cm^3 arranged in two identical arrays. The endcap part of the calorimeter covers polar angles from 17° to 50° and from 130° to 163°. The length

IOP Conf. Series: Journal of Physics: Conf. Series **928** (2017) 012011 doi:10.1088/1742-6596/928/1/012011

of crystals corresponds to the thickness of 13.4 X_0. The design of this calorimeter is described in detail in [15].

3. Calibrations and resolutions

Several procedures of the calorimeter calibration are used. Calibration with a pulse generator provides a measurement of the pedestal, electronic gain and electronic noise of each channel of the calorimeters. For crystal calorimeters we use cosmic ray particles in special calibration runs. Since the standard CMD-3 trigger does not suppress all cosmic ray events, the experimental data sample contains such events, they are also used for the calibration. This type of calibration is used for both endcap and combined barrel calorimeters. For the LXe calorimeter we use calibration based on e^+e^- elastic scattering events. And for the endcap calorimeter the energy corrections are made using two-photon annihilation process, $e^+e^- \to \gamma\gamma$, to obtain absolute energy calibration. The calibration procedures for the combined barrel calorimeter is described in [16] and for the endcap calorimeter in [13].

Two-photon annihilation events were used to obtain the energy and spatial resolutions of the calorimeters. The results are shown in figures 2 and 3.

The energy resolution for the barrel calorimeter can be parameterized as $\sigma_E/E = \frac{0.034}{\sqrt{E[GeV]}} \oplus 0.020$ and for the endcap one as $\sigma_E/E = \frac{0.024}{\sqrt{E[GeV]}} \oplus 0.023$.

To obtain the spatial resolution of the calorimeters, the distribution of azimuth angle acollinearity $\Delta\varphi = \pi - |\varphi_1 - \varphi_2|$ was used, where φ_1 and φ_2 are azimuthal angles of photons. This distribution was approximated by a Gaussian function, the angular resolution of the calorimeter σ_φ was defined as $\sigma_\varphi = \sigma_{fit}/\sqrt{2}$, where σ_{fit} is the standard deviation of the Gaussian function. The coordinates of most (95%) photons detected in the barrel calorimeter are measured using data from LXe strips with the angular precision of about 0.005 rad, which slightly depends on the photon energy. In this case the angular resolution can be fitted as $\sigma_\varphi[mrad] = 3.70 + \frac{0.33}{0.25+E[GeV]}$. For about 5% of events the conversion point is not reconstructed by strip data and the photon coordinates are determined as a center of gravity of the cluster. For such a case the correction function for the photon coordinates was determined from the simulation. The angular resolution in this case can be fitted as $\sigma_\varphi[mrad] = 37.0 + \frac{3.6}{0.1+E[GeV]}$.

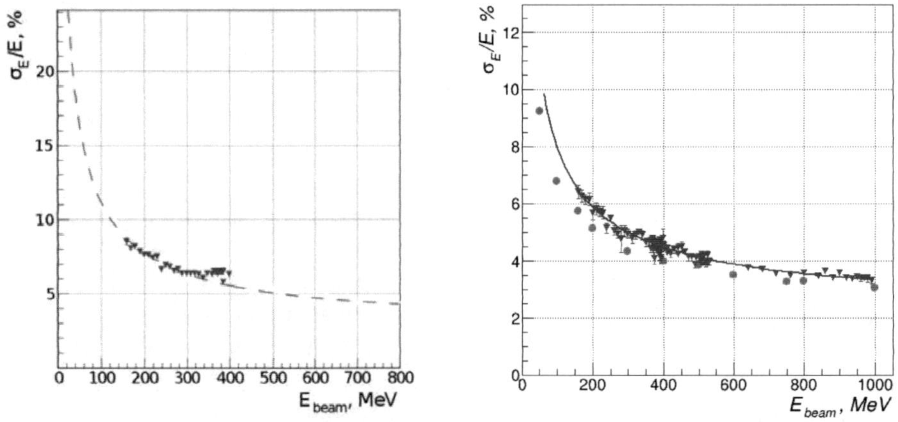

Figure 2. Energy resolution of the barrel (left) and endcap (right) calorimeters for photons: ▼ — experimental data, —— — fit of experimental data, ● — MC simulation, —— — fit of MC simulation.

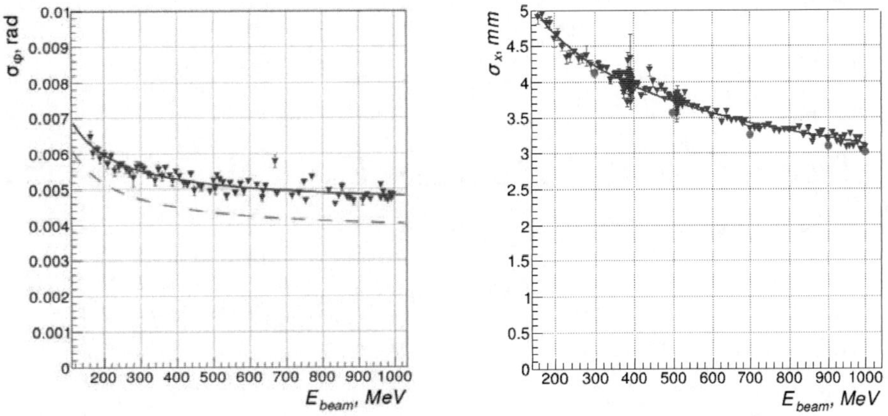

Figure 3. Spatial resolution of the barrel (left) and endcap (right) calorimeters for photons: ▼ — experimental data, ——— — fit of experimental data, ● — MC simulation, - - - - — fit of MC simulation.

The spatial resolution of the endcap calorimeter was calculated as $\sigma_x = \sigma_\varphi \cdot Z_0 \cdot \tan\vartheta/\sqrt{2}$, where Z_0 is the distance from the interaction point to the front plane of the endcap and ϑ is a polar angle of the photon. The resolution can be fitted as $\sigma_x[mm] = \frac{3.120}{\sqrt[4]{E[GeV]}} \oplus 0.197$.

4. Conclusion

The calorimeters were installed into the CMD-3 detector and have been exploited in the experimental data taking since 2010. The calibration procedures of the calorimeters have been developed and used during all three physical seasons. The photon energy reconstruction procedures have been developed and applied. The energy and spatial resolutions at 1 GeV have been determined to be 4.5% and 2 mm for the barrel calorimeter and 3.5% and 3 mm for the endcap calorimeter, respectively.

Acknowledgments

This work has been supported by the Russian Science Foundation (project N 14-50-00080).

References

[1] Shatunov Y M *et al.* 2000 *Conf. Proc.* C **0006262** pp 439–441
[2] Berkaev D *et al.* 2012 *Nucl. Phys. Proc. Suppl.* **225-227** pp 303–308
[3] Abakumova E V *et al.* 2014 *Nucl. Instrum. Meth.* A **744** 35–40
[4] Abakumova E V, Achasov M N, Krasnov A A, Muchnoi N Y and Pyata E E 2015 JINST **10** no.09, T09001
[5] Achasov M N *et al.* 2009 *Nucl. Instrum. Meth.* A **598** 31–32
[6] Achasov M N *et al.* 2015 JINST **10** no.06, T06002
[7] Khazin B 2008 *Nucl. Phys. Proc. Suppl.* **181-182** pp 376–380
[8] Eidelman S 2006 *Nucl. Phys. Proc. Suppl.* **162** pp 323–326
[9] Akhmetshin R R *et al.* [CMD-3 Collaboration] 2015 Phys. Lett. B **740** 273–277
[10] Shemyakin D N *et al.* 2016 Phys. Lett. B **756** 153–160
[11] Anisenkov A V *et al.* 2014 *JINST* **9** C08024
[12] Aulchenko V M *et al.* 2015 *JINST* **10** no.10, P10006
[13] Akhmetshin R R, Grigoriev D N, Kazanin V F, Kuzmenko A E and Yudin Yu V 2014 *JINST* **9** no.10, C10002
[14] Anisyonkov A V *et al.* 2009 *Nucl. Instrum. Meth.* A **598** 266–267
[15] Akhmetshin R R, Grigoriev D N, Kazanin V F, Tsaregorodtsev S M and Yudin Yu V 2009 *Phys. Atom. Nucl.* **72** 477–481 [Yad. Fiz. **72** (2009) 512]
[16] Shebalin V E *et al.* 2014 *JINST* **9** no.10, C10013

IOP Conf. Series: Journal of Physics: Conf. Series **928** (2017) 012012 doi:10.1088/1742-6596/928/1/012012

Status of the Top and Bottom Counting Detectors for the ISS-CREAM Experiment

J.M. Park on behalf of ISS-CREAM Collaboration

Dept. of Physics, Kyungpook National University, Daegu, 41566, Korea

E-mail: jeongmin1024@gmail.com

Abstract. It is important to measure the cosmic ray spectra to study the origin, acceleration and propagation mechanisms of high-energy cosmic rays. A payload of the Cosmic Ray Energetics And Mass experiment is scheduled to be launched in 2017 to the International Space Station for measuring cosmic ray elemental spectra at energies beyond the reach of balloon instruments. Top Counting Detector and Bottom Counting Detector (T/BCD) as a two-dimensional detector are to separate electrons from protons for electron/gamma-ray physics. The T/BCD each consists of a plastic scintillator read out by 20 by 20 photodiodes and is placed before and after the Calorimeter, respectively. Energy and hit information of the T/BCD can distinguish shower profiles of electrons and protons, which show narrower and shorter showers from electrons at a given energy. The T/BCD performance has been studied with the Silicon Charge Detector and the calorimeter by using a GEANT3 + FLUKA 3.21 simulation package. By comparing the number of hits and shower width distributions between electrons and protons, we have studied optimal parameters for the e/p separation.

1. Introduction

Since cosmic rays were discovered in 1912 by Victor Hess [1], we have tried to understand the cosmic rays. But the origin, acceleration and propagation mechanisms of high-energy cosmic rays have remained unexplained. The flux of electrons plus positrons are measured by various experiments such as the ATIC (Advanced Thin Ionization Calorimeter) [2], Fermi-LAT (Fermi Large Area Telescope) [3], AMS (Alpha Magnetic Spectrometer) -02 [4], etc. The cosmic ray electron flux spectrum shows a significant excess above 20 GeV in results of the ATIC, Fermi-LAT and etc. However, the result of AMS-02 shows no significant structure at the same energy region [4]. In order to measure the cosmic ray electron spectrum clearly, it is important to separate electrons from protons.

The CREAM (Cosmic Ray Energetics And Mass) experiment was a balloon born experiment flown 6 times for which has been upgraded for the ISS (International Space Station) as ISS-CREAM. The ISS-CREAM experiment plans to measure the cosmic rays energy spectra from 10^{11} to $>10^{15}$ eV [5], and is scheduled to launch in 2017 to the ISS. The ISS-CREAM consists of the Silicon Charge Detector, a carbon target, the calorimeter (CAL), the T/BCD and a Boronated Scintillator Detector (BSD). A protoype BSD was test at CERN in 2012 and demonstrated e/p discrimination of 100 with 80% acceptance for 350 GeV pions/125 GeV electrons [6].

The T/BCD are designed for studying electron and γ-ray physics [7, 8]. The T/BCD also provide a redundant trigger to the high energy CAL trigger and a low energy electron trigger.

2. Top and Bottom Counting Detectors

The T/BCD consist of a plastic scintillator coupled with a PD array. Dimension of the plastic scintillator of the TCD is 500 × 500 × 5 mm^3, and that of the BCD is 600 × 600 × 10 mm^3. Dimension of the PD is 23 mm × 23 mm × 650 μm, and total 400 PDs are attached to the plastic scintillator of the TCD and BCD, respectively [7]. Figure 1 shows the assembled PD array and assembled plastic scintillator.

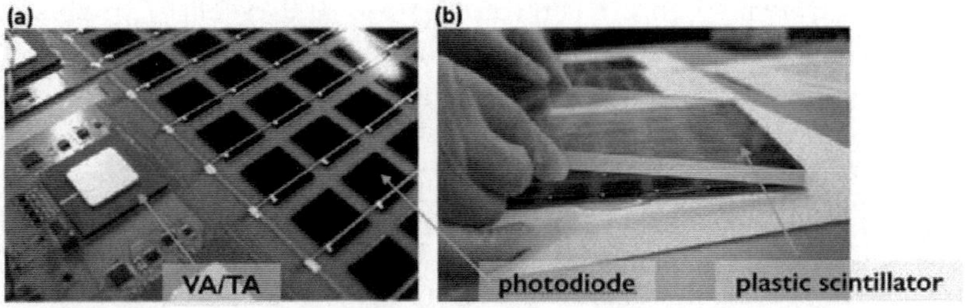

Figure 1. (a) Photodiode array and VA/TA chips, and (b) plastic scintillator and photodiode array being assembled.

Figure 2 shows the photos of the TCD and BCD before assembling Al enclosure cover. The outer sides of the T/BCD are the electronics parts, and the center is the detector part. The electronics parts consist of 17 VA/TA chips, 4 high voltage circuits and 2 FPGA (Field-Programmable Gate Array) chips, respectively [7]. The TCD is located between the CAL and measures the transverse shower profile and back scattering particles. The BCD is located below the CAL and measures the transverse and longitudinal shower profiles. The electromagnetic shower is narrower and shorter than the hadronic shower. Thus the electrons are separated from protons by using these T/BCD information [7, 8].

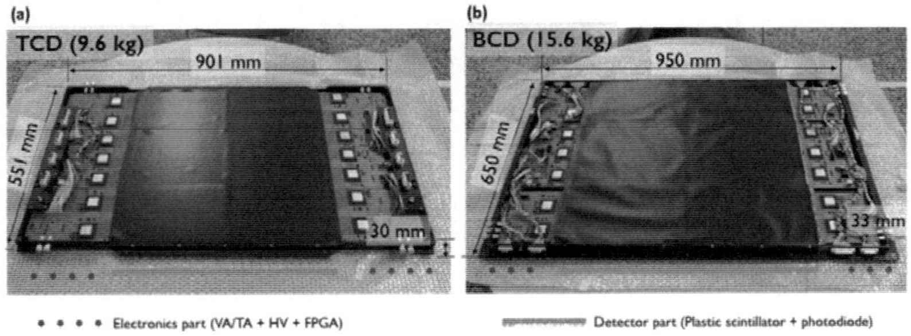

Figure 2. Photos of (a) TCD and (b) BCD before assembling Al enclosure cover.

3. Environmental Tests

Since the ISS-CREAM will be installed on the ISS, the T/BCD are constructed under critical requirements of mechanical structure [9].

The vibration test and the thermal vacuum test in the detector level are performed at the Keimyung University and the KARI (Korea Aerospace Research Institute) in Korea. The environmental tests in the payload level, such as EMI (Electro-Magnetic Interference) & EMC (Electro-Magnetic Compatibility) tests, vibration test and thermal vacuum test, are performed at the NASA Goddard Space Flight Center. The T/BCD were operated normally at these tests. The pedestal RMS values were < 15 ADC (~ 2 fC), and the gain values were ~ 330 DAC/ADC on average with the flight electronics.

4. A preliminary Study of e(γ)/p Separation by a Monte Carlo Simulation

Electron and proton events are generated by using a GEANT3 simulation package [10]. In order to study the proton rejection power, we simulate a 2.3×10^7 proton events with 150 GeV $\sim$ 15 TeV and 5.7×10^4 electron events with 150 GeV $\sim$ 10 TeV [11]. In this study, 300 GeV electron and protons which are expected to be simulated to the 300 GeV electron are used. The electronics noise level was smeared by using the previous CREAM experiment results assuming a similar performance to the CREAM [12]. The CAL trigger and TCD 2 MIP (Minimum Ionizing Particle) trigger were used for event selection. The e/p separation was studied by comparing the hits and RMS distribution of electrons and protons in the T/BCD [9]. Figure 3(a) shows the hit distributions in the T/BCD, and a two dimensional cut is applied for e/p separation. Figure 3(b) and (c) show the shower width distribution in the TCD and f-factor distribution in BCD, respectively. The RMS and f-factor are defined as

$$RMS^2 = E_{TCD_i} \times \{(x_i - x_c)^2 + (y_i - y_c)^2\},$$
$$f - factor = RMS^2 \times E_{BCD}/E_{CAL},$$

where x_i, y_i are the center coordinates of the sensor where energy E_{TCD} deposited, x_c, y_c are the coordinates of the energy center in the TCD, and E_{TCD_i} is the energy deposited in the ith sensor of the TCD. Also E_{BCD} is the energy deposition in the BCD and E_{CAL} is the total energy deposition in the CAL.

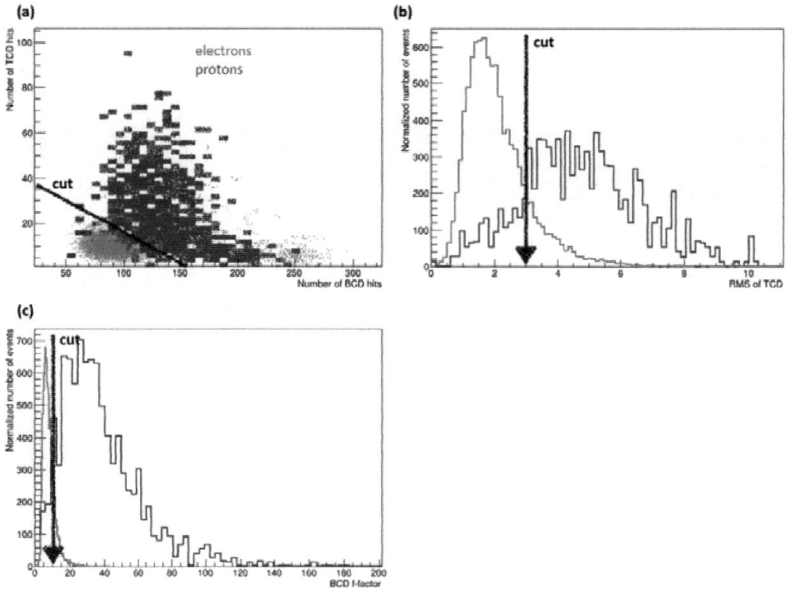

Figure 3. (a) TCD number of hits *vs.* BCD number of hits, (b) TCD shower width distribution, and (c) BCD f-factor distribution for electrons and protons.

The proton rejection power is defined as

$$\textit{Proton rejection power} = \textit{electron efficiency} \; / \; \textit{proton efficiency} \; [13].$$

In this study, the electron efficiency and proton rejection power are found to be approximately 60% and 4.7×10^4, respectively.

5. Conclusion

The T/BCD are one of the ISS-CREAM instruments, and the ISS-CREAM is being planned to be launched in 2017. The T/BCD are developed for e/γ physics, providing a redundant trigger for CAL and providing a low energy electron trigger. Environment tests are performed at the NASA Goddard Space Flight center, and T/BCD behaved normally during these tests. The ISS-CREAM is currently being prepared for the launch at the NASA Kennedy Space Center. The capability of the T/BCD to separate electrons from protons has been studied preliminarily with a GEANT3 simulation. The proton rejection power is $\sim 4.7 \times 10^4$ and electron efficiency is $\sim 60\%$ at 300 GeV electrons.

References

[1] V.F. Hess, Über Beobachtungen der durchdringenden Strahlung bei sieben Freiballonfahrten, Phys. Zeit 13, 1084 (1912).
[2] J.Chang et al., An excess of cosmic ray electrons at energies of 300-800 GeV, Nature 456, 362 (2008).
[3] M. Ackermann et al., Fermi LAT observations of cosmic-ray electrons from 7 GeV to 1 TeV, Phys. Rev. D 82, 092004 (2010).
[4] M. Aguilar et al., Precision Measurement of the ($e^+ + e^-$) Flux in Primary Cosmic Rays from 0.5 GeV to 1 TeV with the Alpha Magnetic Spectrometer of the International Space Station, Phys. Rev. Lett. 113, 221102 (2014).
[5] E.S. Seo et al., Cosmic Ray Energetics And Mass for the International Space Station (ISS-CREAM), Adv. Space. Res. 53, 1451 (2014).
[6] A. Tyler, "Exploring the cosmic-ray spectrum with the CREAM experiment", PHD Thesis, Penn State University 2013.
[7] Y.S. Hwang et al., Construction and testing of a Top Counting Detector and a Bottom Counting Detector for the Cosmic Ray Energetics And Mass experiment on the International Space Station, JINST 10, P07018 (2015).
[8] H.J. Hyun et al., Performances of photodiode detectors for top and bottom counting detectors of ISS-CREAM experiment, Nucl. Instrum. Meth. A 787, 134 (2015).
[9] J.M. Park et al., Construction and Test of the Top and Bottom Counting Detectors for the ISS-CREAM Experiment, Proc. of the 34th Int. Cosmic Ray Conf., Hague, Netherlands, PoS 583 (2015).
[10] ROOT Data Analysis Framework http://root.cern.ch/.
[11] F. Aharonian et al., Probing the ATIC peak in the cosmic-ray electron spectrum with HESS, Astronomy&Astrophysics 508, 561 (2009).
[12] M.H. Lee et al., The CREAM Calorimeter: Performance in Tests and Flights, AIP conference proceedings 867, 167 (2006).
[13] F. Palma et al., Simulation studies of the expected proton rejection capabilities of CALET, Proc. of the 34th Int. Cosmic Ray Conf., Hague, Netherlands, PoS 1196 (2015).

Acknowledgments

The authors thank the NASA GSFC WFF for project management and engineering support, and the NASA JSC ISS Program Office for the launch support and the ISS accommodation. This work was supported in the U.S. by NASA grants NNX11AC52G, NNX08AC15G, NNX08AC16G and their predecessor grants, as well as by directed RTOP funds to NASA GSFC WFF. It is supported in Korea by the Creative Research Initiatives of MEST/NRF and by National Research Foundation Grants NRF-2014R1A2A2A01002734, NRF-2014R1A1A2006456, NRF-2015R1A2A1A13001843. It is supported in France by IN2P3/CNRS and CNES and in Mexico by DGAPA-UNAM and CONACYT. The authors also thank H.S. Choi, Korea Institute of Industrial Technology, for contributions to the SCD thermal vacuum tests, M. Geske, Penn State, for contributions to the BSD, and M. A. Coplan, University of Maryland, contributions to CAL electronics vacuum tests.

CaloCube: an innovative homogeneous calorimeter for the next-generation space experiments

L. Pacini[abc], O. Adriani[bc], A. Agnesi[de], S. Albergo[fg], L. Auditore[hs], A. Basti[k], E. Berti[bc], G. Bigongiari[jk], L. Bonechi[bc], S. Bonechi[jk], M. Bongi[bc], V. Bonvicini[l], S. Bottai[b], P. Brogi[jk],G. Cappello[g], G. Carotenuto[m], G. Castellini[ab], P.W. Cattaneo[e], M. Chiari[b], N. Daddi[c], R. DAlessandro[bc], S. Detti[b], M. Fasoli[no], N. Finetti[pb], P. Lenzi[b], P. Maestro[jk], P.S. Marrocchesi[jk], M. Miritello[r], N. Mori[bc], G. Orzan[l], M. Olmi[bc], P. Papini[b], M.G. Pellegriti[g], F. Pirzio[de], A. Rappoldi[e], S. Ricciarini[ab], P. Spillantini[c], O. Starodubtsev[b], F. Stolzi[jk], J.E. Suh[jk], A. Sulaj[jk], A. Tiberio[bc], A. Tricomi[fg], A.Trifirò[hi], M. Trimarchi[hi], E. Vannuccini [bc], A. Vedda[no], G. Zampa[l], N. Zampa[l]

[a]IFAC (CNR), via Madonna del Piano 10, I-50019 Sesto Fiorentino (Firenze), Italy
[b]INFN Firenze, via B. Rossi 1, I-50019 Sesto Fiorentino (Firenze), Italy
[c]University of Florence, Department of Physics and Astronomy, via G. Sansone 1, I-50019 Sesto Fiorentino (Firenze), Italy
[d]University of Pavia, Dipartimento di Ingegneria Industriale e dellInformazione, Pavia, Italy
[e]INFN Pavia, via A. Bassi 6, I-27100 Pavia, Italy
[f]University of Catania, Department of Physics and Astronomy, via S. Sofia 64, I-95123 Catania, Italy
[g]INFN Catania, via S. Sofia 64, I-95123 Catania, Italy
[h]University of Messina, Dipartimento di Scienze Matematiche e Informatiche, Scienze Fisiche e Scienze della Terra, sal. Sperone 31, I-98166 Messina, Italy
[i]INFN Messina, c.da Papardo, sal. Sperone, 31 - 98166 S. Agata, Messina
[j]University of Siena, Department of Physical Sciences, Earth and Environment, I-53100 Siena, Italy
[k]INFN Pisa, via F. Buonarroti 2, I-56127 Pisa, Italy
[l]INFN Trieste, via Valerio 2, I-34127 Trieste, Italy
[m]Institute for Polymer, Composites and Biomaterials (CNR), piazzale E. Fermi 1, I-80055 Portici (Napoli), Italy
[n]University of Milano-Bicocca, Department of Materials Science, via Cozzi 55, I-20125 Milano, Italy
[o]INFN Milano-Bicocca, Piazza della Scienza 3, Milano, Italy
[p]University of L'Aquila, Department of Physical and Chemical Sciences, Via Vetoio, Coppito, 67100 L'Aquila, Italy
[q]University of Trieste, Department of Physics, via Valerio 2, I-34127 Trieste, Italy
[r]MATIS (CNR), via S. Sofia 64, I-95123 Catania, Italy
[s]A.O.U. Policlinico "G. Martino", U.O.C. di Medicina Nucleare, Messina, Italy

E-mail: `l.pacini@ifac.cnr.it`

Abstract.
The direct measurement of the cosmic-ray spectrum, up to the knee region, is one of the instrumental challenges for next generation space experiments. The main issue for these measurements is a steeply falling spectrum with increasing energy, so the physics performance of the space calorimeters are primarily determined by their geometrical acceptance and energy

resolution. CaloCube is a three-year R&D project, approved and financed by INFN in 2014, aiming to optimize the design of a space-born calorimeter. The peculiarity of the design of CaloCube is its capability of detecting particles coming from any direction, and not only those on its upper surface. To ensure that the quality of the measurement does not depend on the arrival direction of the particles, the calorimeter will be designed as homogeneous and isotropic as possible. In addition, to achieve a high discrimination power for hadrons and nuclei with respect to electrons, the sensitive elements of the calorimeter need to have a fine 3-D sampling capability. In order to optimize the detector performances with respect to the total mass of the apparatus, which is the most important constraint for a space launch, a comparative study of different scintillating materials has been performed using detailed Monte Carlo simulation based on the FLUKA package. In parallel to simulation studies, a prototype consisting in 14 layers of 3 x 3 CsI(Tl) crystals per layer has been assembled and tested with particle beams. An overview of the obtained results during the first two years of the project will be presented and the future of the detector will be discussed too.

1. Introduction

Precise measurements of the cosmic-ray (CR) spectrum and composition are essential to improve the constraints on the theoretical models of CR propagation and acceleration. Since the differential all-particle flux of CR is very steep, a large acceptance is required to achieve statistically significant measurements of high energy CR flux. Measurements of the CR all-particle spectrum with ground experiments show that it becomes suddenly steeper and the composition progressively heavier in the PeV region. This feature, known as the "knee", is believed to indicate the energetic limit of the galactic accelerators. A precise knowledge of the spectrum and composition in this region would address key items in the field of high-energy astrophysics. Furthermore, measurements of primary and secondary species is important in understanding CR propagation in the Galaxy. On the other hand, electrons and positrons suffer significant energy losses due to radiative emission along their path to Earth, so detailed measurement of the high energy electron spectrum and anisotropy could provide information about nearby CR sources.

Ground experiments achieve very large statistics for high energy CRs thanks to their large acceptance but are affected by large systematic errors due to modelling of the interaction of particles with the atmosphere. Direct observation of CR, by experiments installed on satellites and balloons, provides precise information of single-particle spectra. The main limitations of these experiments are due to the constraints on volume and mass. In order to extend nuclei spectra above 100 PeV, and electron+positron spectra up to 10 TeV, future space experiments need an acceptance of at least 2.5 m^2sr $\times 5$ yr, an energy resolution better than 40% for protons and nuclei and 2% for electrons, a good particle identification and a high dynamic range.

2. The CaloCube baseline and the expected performance

CaloCube [1] proposes an innovative solution for space experiments, based on a deep homogeneous isotropic calorimeter with three dimensional segmentation. The current space experiments (e.g. [2, 3, 4, 5]) accept particles only from the top surface while CaloCube can accept particles from almost all directions: only the bottom surface is unusable due to the mechanical structures and the Earth. The base line for such a detector is a cubic calorimeter consisting of $20\times20\times20$ cubes. The crystal side is 3.6 cm, approximately one Moliere radius. The spacing between crystals is 0.3 cm and the total depth for vertical particle is 39 X_0 (1.9 λ_I). Due to the five-faced detection, the total geometric factor (GF) is very large, 9.55 m^2sr. Scintillation light is detected by two photodiodes (PDs): a large area one and a small area one to achieve a very high dynamic range. The PDs are connected to an automatic double-gain front-end electronics by kapton cables as shown in Fig. 1.

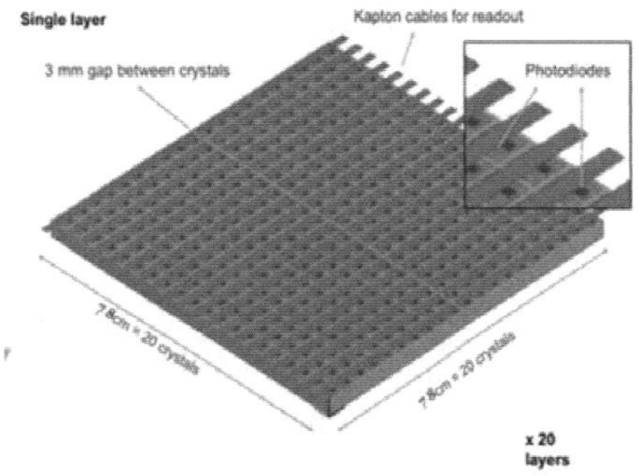

Figure 1. One layer of the CaloCube baseline geometry. The complete calorimeter consists of 20 identical layers.

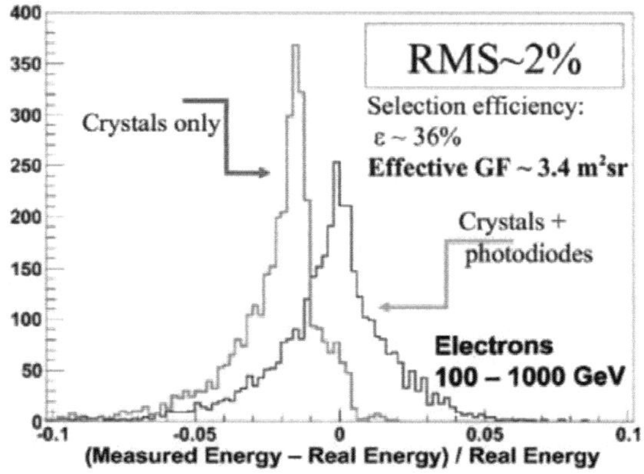

Figure 2. Measured energy in the calorimeter for electrons in the energy range 100 GeV - 1 TeV.

A Monte Carlo simulation based on the FLUKA package have been developed. The CaloCube baseline geometry is implemented and both the scintillation light and the energy deposit in the large area PD due to the direct ionization are converted into photo-electrons. Electrons from 100 GeV to 1 TeV are simulated with an isotropic generation on the top surface of the calorimeter. The distribution of energy measured in the calorimeter is shown in Fig. 2: the red line includes only the energy deposit in crystals and the black line includes the direct ionization on the PDs. The energy resolution is about 2%, and a very large effective GF[1] is achieved, 3.4 m^2sr.

Protons from 1 TeV to 1 PeV have been simulated. In order to get a good energy resolution for hadronic showers an off-line compensation method is required: the measured energy in such thin calorimeter strongly depends on the shower length[2]. As an example, the trend of the measured energy for protons at 1 TeV versus the shower length is shown in Fig. 3: the integral of a Gamma function is used in the fit. Event by event corrections of the longitudinal leakage based on the fitted function improve the energy resolution for protons: about 35% with an effective GF$\sim$ 2.5 m^2sr at 1 TeV. Simulations show that the energy resolution and the effective GF are weakly dependent on the proton energy, only $\sim$ 10% differences in the simulated range.

In order to optimize both the energy resolution and the effective GF, different materials for the scintillating crystal (CsI(Tl), BaF_2, YAP:Yb, BGO, LYSO:Ce) have been tested using MonteCarlo simulations. The geometric parameters have been defined by assuming about 1.6 *tons* of active material, 78% of active-volume fraction, and the size of the single element has been fixed to one Moliere radius. The energy resolution dependence on the GF for the tested materials applying a progressively lower limit on the shower containment are shown in fig. 4. As expected an increase in the shower containment translates into an improvement of the energy resolution while the GF gets worse. LYSO is found to be a promising material to maximize energy resolution and GF.

3. The prototype: beam test results

As a validation of the calorimeter simulations, a small-scale prototype has been assembled [6]. This calorimeter consists of 15 layers of 3 × 3 CsI(Tl) cubic crystals with a side equal to 3.6 cm. The spacing between crystals is 0.4 cm. A large area PD (VTH2090, Excelitas) is used for

[1] The *effective geometric factor* is defined as the geometric factor of five-faced detection multiplied by the event selection efficiency.

[2] The shower length is the distance from the shower starting point and the exit point from the calorimeter

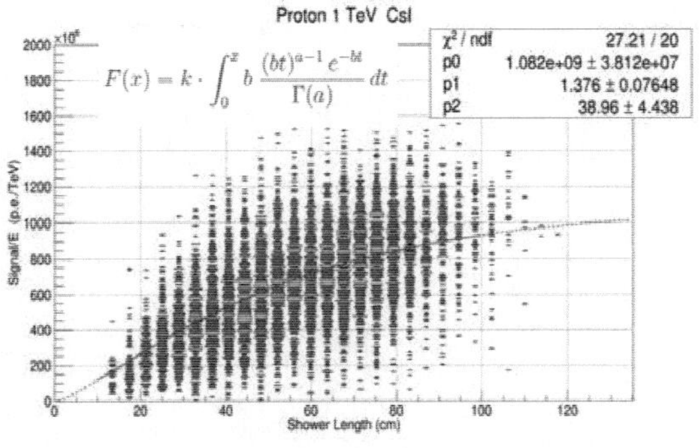

Figure 3. Measured energy in the calorimeter (CsI(Tl)) versus the shower length for protons at 1 TeV. The red curve is the integral of the gamma function used in the fit.

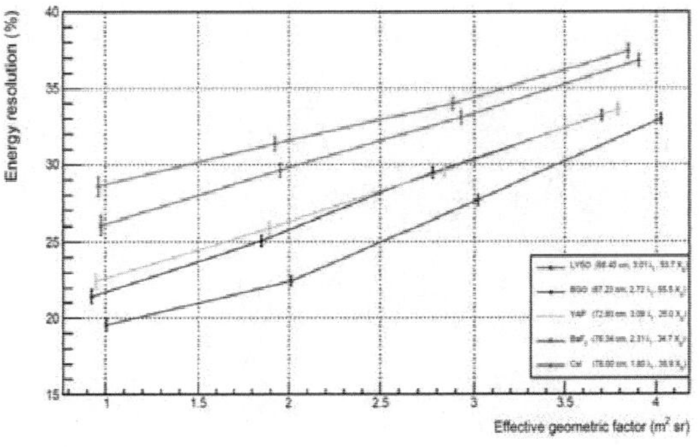

Figure 4. Energy resolution and effective geometric factor for protons at 1 TeV with different scintillating materials. LYSO is one of the best candidates for future hadronic calorimeters.

the read-out of the scintillating crystal. Nine CASIS ASICs developed by the INFN in Trieste [7] are the front-end electronics. The CASIS chip consists in 16 channels with a double-gain charge-amplifier with automatic gain selection circuitry and correlated double sampling filtering: thank to this design high dynamic range, up to 52.6 pC, low noise (ENC $\sim$ 3000 e^-) and low power consumption, 2.8 mW/ per channel, are achieved. Different wrapping materials of the crystals have been studied: laboratory measurements have shown that Vikuiti optimizes the light collection.

Beam tests with ions, from Deuterium to Iron at 12.8 GeV and 30 GeV per nucleon, were performed at CERN-SPS in 2013. The results are presented in [6]: a very good linearity up to ion kinetic energies of 1.6 TeV is shown. A Monte Carlo simulation based on the FLUKA package of the prototype has been developed and a good agreement between beam test data and simulation is found.

An upgraded version of the prototype has been tested with electron and muon beams at CERN-SPS in 2015. During this test, ADAMO [8], a tracking system based on a silicon micro-strip detector, was located in front of the calorimeter. Data with muon beam at 150 GeV and electron beam at energies from 50 GeV to 200 GeV were acquired. The muon dataset has been used for channel-by-channel calibration and for the reconstruction of the geometrical position of both the crystals and the PD active areas thanks to the direct ionization within the silicon, Fig. 5. This information is crucial to understand correctly the performance of the calorimeter for electrons; in a beam test configuration all the incident particles are almost perpendicular to the top face of the calorimeter, therefore an electron shower can intercept all the PD active areas. This difference with respect to the situation with real CR, where the flux is isotropic and the number of PDs directly hit is a stochastic variable, affects the expected energy resolution estimated by the simulation of an isotropic flux (sec. 2). The energy resolution for electrons at 50 GeV without selection on the incident position is $\sim 3\%$, but rejecting events directly interacting in the PDs results in a energy resolution of $\sim 1.5\%$, Fig. 6. A FLUKA based simulation of the beam test configuration has been developed: a good agreement with the beam test data is found and the estimated energy resolution is consistent with the one measured from beam test data for both unselected and the geometrical-selected events.

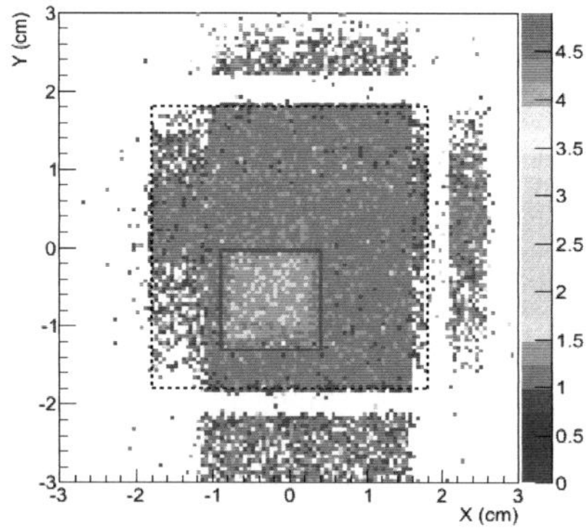

Figure 5. Measured energy for muons at 150 GeV in the central crystal of the central column of the calorimeter versus the position reconstructed by ADAMO. The crystal (black dashed line) and the PD active area (red line) are highlighted.

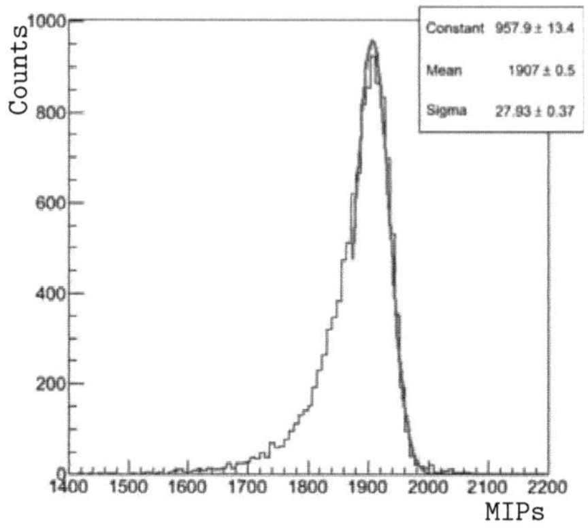

Figure 6. Measured energy for an electron beam at 50 GeV. Only events that do not intercept the PDs are taken into account. A good energy resolution, about 1.5% is achieved.

4. Conclusion

An innovative design for a space calorimeter is presented. Thanks to the five-faced acceptance and to the high granularity a good energy resolution and a very large effective geometric factor for both electrons and protons are achieved. Monte Carlo simulations of different scintillating crystals show that LYSO is one of the best candidates as active material of future space hadronic calorimeters. A small-scale prototype has been tested with ion and electron beams: a good agreement between simulations and data has been obtained.

An upgrade of the prototype is under construction: the new version will consist of 18 layers, each consisting of 6 × 6 CsI(Tl) crystals, and a small area photodiode will be added to each crystal to extend the current dynamic range (some TeV for hadronic showers) to the hundred TeV region.

Acknowledgment

This project has received founding from the European Union's Horizon 2020 research and innovation programme under grant agreement No. 654168.

References

[1] O. Adriani, Development of a 3D cubic crystal calorimeter for space: CaloCube, Proceedings of the 1st Conference on Calorimetry for the High-Energy Frontier (CHEF 2013), 454459
[2] P. Maestro, arXiv:1302.1257 [astro-ph.IM]
[3] V. Gallo et al., PoS(ICRC2015)1199
[4] O. Adriani et al., arXiv:1103.4055 [astro-ph.HE]
[5] M. Aguilar et al.,Phys. Rev. Lett. 114, 171103 Published 30 April 2015
[6] O. Adriani et al., CaloCube - A highly segmented calorimeter for space based experiment, NIM A (2016)
[7] V. Bonvicini, et al., IEEE 5 (2010) NS-57
[8] L. Bonechi et al., Proc. XXIX International Cosmic Ray Conference (ICRC), Pune (India) 2005, session HE2.4, Vol. 9, pag. 283-286.

IOP Conf. Series: Journal of Physics: Conf. Series **928** (2017) 012014 doi:10.1088/1742-6596/928/1/012014

Status and New Results for the sPHENIX Calorimeter Systems

C. Woody, for the sPHENIX Collaboration

Physics Department, Building 510C
Brookhaven National Lab
Upton, NY 11973

E-mail: woody@bnl.gov

Abstract. The PHENIX Experiment at RHIC is planning a major upgrade that involves building an entirely new spectrometer, sPHENIX, that is based around the former BaBar solenoid magnet which will enable a comprehensive study of jets and heavy quarkonia in relativistic heavy ion collisions. It will include two new calorimeter systems, one electromagnetic and one hadronic, that will cover an acceptance of ±1.1 units in pseudorapidity and 2π in azimuth. The hadronic calorimeter will be a steel plate scintillating tile design that is read out with wavelength shifting fibers and silicon photomultipliers. It will be divided into two sections: one (the Inner HCAL) will be situated inside the magnet and the other (the Outer HCAL) will be outside the magnet. The electromagnetic calorimeter will be a SPACAL design consisting of a tungsten powder epoxy matrix absorber with embedded scintillating fibers which are also read out with silicon photomultipliers. The design of sPHENIX and its calorimeter systems has made considerable progress over the past several years and is described in this paper. Prototypes of all three calorimeters were built and tested in the test beam at Fermilab in April of 2016, and the first preliminary results from this test, along with a comparison to Monte Carlo simulations, are also discussed.

1. Introduction

The sPHENIX Collaboration is planning a series of major upgrades to the PHENIX experiment at RHIC that will enable an extensive set of new physics programs over the next decade. This will involve replacing the current PHENIX spectrometers with a new central spectrometer based around the former BaBar solenoid magnet. It will require the construction of two new large calorimeter systems, one electromagnetic and the other hadronic, covering ± 1.1 units in pseudorapidity and 2π in azimuth. They will be used to measure jets and heavy quarkonia in heavy ion collisions and allow a detailed study of the Quark Gluon Plasma near the region of its critical temperature. The hadronic calorimeter will be the first mid rapidity hadronic calorimeter ever used in a RHIC experiment and will provide a much better measurement of jets than has previously been possible at RHIC energies. The requirements for the calorimeters are an energy resolution ~ $100\%/\sqrt{E}$ for the hadronic calorimeter for single particles and ~ $15\%\sqrt{E}$ for the electromagnetic calorimeter, and to provide an electron/hadron separation ~ 100:1 for measuring Upsilons in heavy ion collisions.

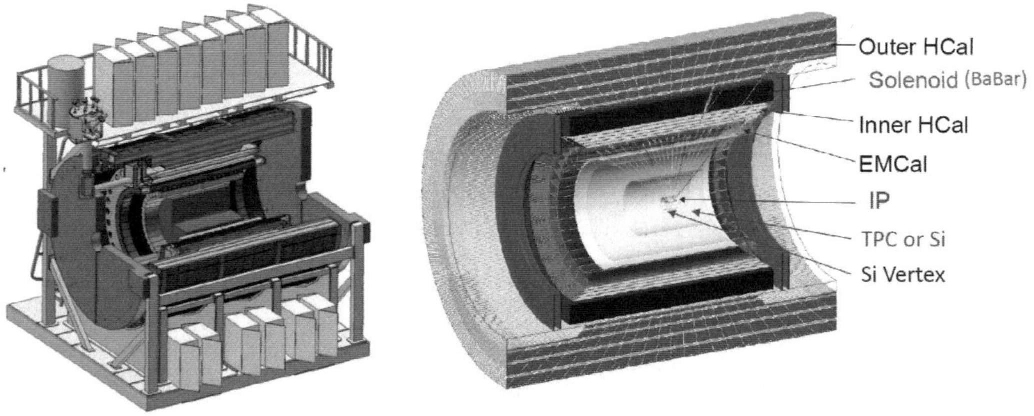

Figure 1. The sPHENIX detector utilizing the former BaBar solenoid magnet, new electromagnetic and hadronic calorimeters, a central TPC tracker and a silicon vertex tracker.

Earlier designs of the sPHENIX detector have been presented at previous conferences [1,2]. This design has now advanced considerably and is shown in Figure 1. The BaBar solenoid magnet has already been acquired from SLAC and is currently undergoing magnetic field testing at BNL. Inside the magnet, there will be a tracking system consisting of a TPC with GEM readout and a MAPS silicon vertex tracker. Figure 2 on the left shows a radial view of the two calorimeter systems. The hadronic calorimeter will be divided into an Inner HCAL located inside the magnet and an Outer HCAL located outside the magnet. It will be a steel plate and scintillating tile design that is read out with wavelength shifting fibers and silicon photomultipliers. The steel plates will be oriented parallel to the beam, serving also as a flux return, and tilted in the azimuthal direction to prevent channelling. The electromagnetic calorimeter will be a compact SPACAL design consisting of a matrix of tungsten powder and epoxy with embedded scintillating fibers that are also read out with silicon photomultipliers. This design was originally developed by the group at UCLA [3,4] and has now been further developed by a commercial company, Tungsten Heavy Powder (THP), and the group at the University of Illinois at Urbana Champaign (UIUC).

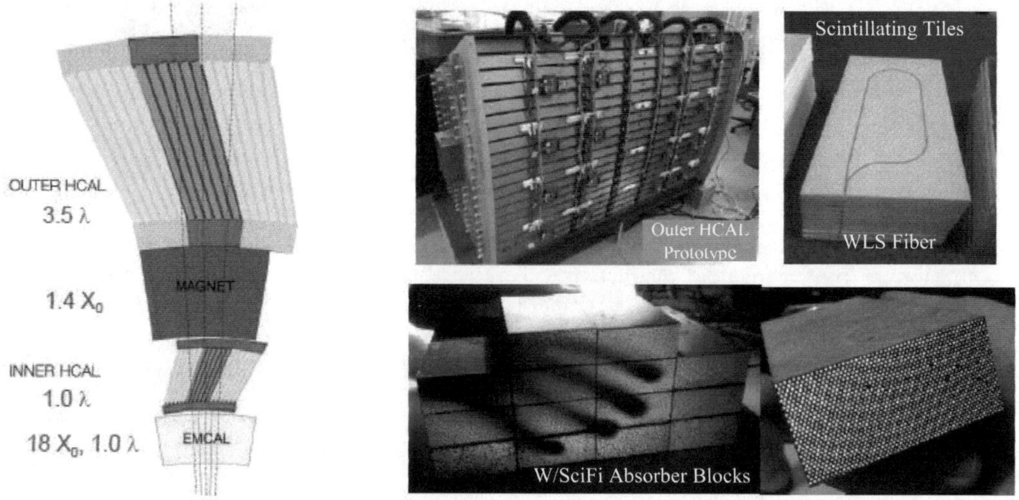

Figure 2. Left: Radial view of the sPHENIX calorimeter systems and BaBar magnet. Right: Prototype Outer HCAL with scintillating tiles (top) and W/SciFi absorber blocks for the prototype EMCAL (bottom).

IOP Conf. Series: Journal of Physics: Conf. Series **928** (2017) 012014 doi:10.1088/1742-6596/928/1/012014

2. Prototype Calorimeter Tests

Prototypes of all three calorimeters were built and tested at Fermilab in April 2016. The Inner and Outer HCAL (IHCAL and OHCAL) prototypes each consisted of a 4x4 array of towers. Each tower contained 5 scintillating tiles, each with its own embedded scintillating fiber that was read out on both ends with a single SiPM. All five SiPMs were summed together to form the readout from a given tower. Figure 2 top right shows the OHCAL prototype with its tilted steel plates and a stack of scintillating tiles that go in between the steel plates with the groove for the wavelength shifting fiber. The SiPMs are coupled to the fiber ends at the edge of the tile.

The EMCAL prototype consisted of an 8x8 array of towers constructed out of 1x2 tower absorber blocks that were tapered in one direction. The bottom right side of Figure 2 shows some of the absorber blocks used in the prototype along with a close up of the fibers within a block. Half of the absorber blocks were manufactured by THP and the other half was made at UIUC.

The calorimeters were measured over an energy range from 1 GeV to 32 GeV with electrons and hadrons, both individually and as a combined system as they will be used in sPHENIX, and for incident particle angles of zero and ± 4.5°. The calorimeters were initially calibrated with minimum ionizing particles to equalize the response of each tower and establish an approximate energy calibration. The HCAL was calibrated in situ with cosmic ray muons and the EMCAL was calibrated using the 120 GeV primary proton beam passing through each tower. For the EMCAL, an improved energy calibration was also developed that used electron showers to equalized the response of each tower.

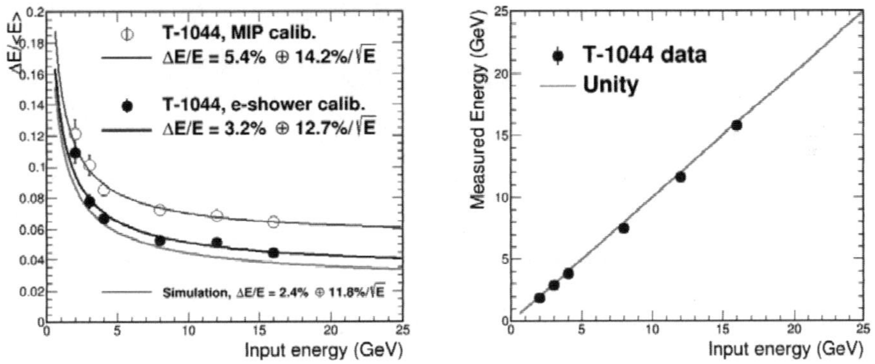

Figure 3. Energy resolution and linearity of the EMCAL prototype. The initial calibration was done using MIPs and a more sophisticated calibration was done using electron showers.

Figure 3 shows preliminary results for the energy resolution and linearity measured for the EMCAL using both the MIP and electron calibration. The energy resolution improved from 14.2%/√E ⊕ 5.4% with the MIP calibration to 12.7%/√E ⊕ 3.2% with the electron calibration. Neither resolution has yet been corrected for the beam momentum spread, which believe is ~ 2% over this energy range. The data also showed good linearity over the energy range measured.

For the combined energy resolution of the EMCAL + HCAL system, a procedure was used to select hadronic showers starting each separate detector which was then used to equalize the energy response of all three calorimeters. Figure 4 shows the energy distributions for particles which start showering in the various parts of the calorimeter (MIP in the EMCAL, MIP in the EMCAL and IHCAL, and all showers irrespective of their starting position). Figure 5 shows preliminary results for the energy resolution and linearity measured for the combined calorimeter system. Also shown on the resolution plot are curves for 95.7%/√E and 70.6%/√E, indicating that the data lie between these two values. A two component fit to the data gives a resolution of ~ 70-80%/√E ⊕ ~ 10-15%.

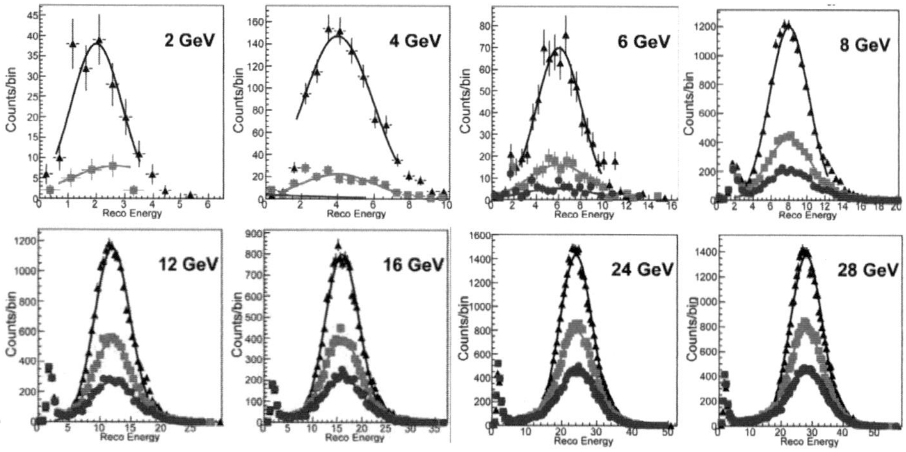

Figure 4. Energy distributions for various energy hadronic showers originating in different parts of the combined calorimeter after applying equalizing weighting factors. Red: MIP in EMCAL, Blue: MIP in EMCAL+IHCAL, Black: All showers irrespective of starting position.

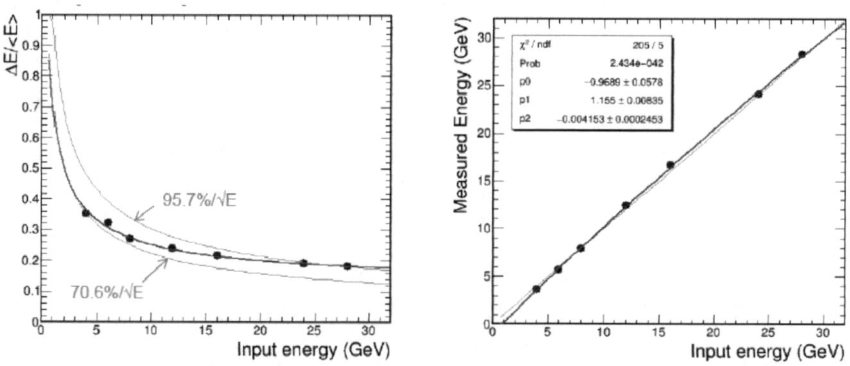

Figure 5. Combined EMCAL + HCAL (Inner + Outer) energy resolution and linearity. Curves are also shown for energy resolutions of 95.7%/√E and 70.6%/√E for comparison.

3. Summary and Conclusions

The sPHENIX experiment at RHIC has developed a detailed design of the entire detector including the new electromagnetic and hadronic calorimeters. Prototypes of these calorimeters were built and tested at Fermilab in April of 2016 and the first preliminary results indicate that both calorimeters, as well as the combined calorimeter system, will meet the design requirements of the sPHENIX experiment. Further analysis of the test beam data is currently under way and we expect new and improved results within the next few months.

4. References

[1] C.Woody and E.Kistenev, Proceedings of the 15th International Conference on Calorimetry in High Energy Physics (CALOR2012), Journal of Physics Conference Series, 404 (2012).
[2] C.Woody and E.Kistenev, Proceedings of the 16th International Conference on Calorimetry in High Energy Physics (CALOR2014), Journal of Physics Conference Series, 587 (2015).
[3] O. Tsai, Proceedings of the 15th International Conference on Calorimetry in High Energy Physics (CALOR2012), Journal of Physics Conference Series, 404 (2012).
[4] O. Tsai, Proceedings of the 16th International Conference on Calorimetry in High Energy Physics (CALOR2014), Journal of Physics Conference Series, 587 (2015).

IOP Conf. Series: Journal of Physics: Conf. Series **928** (2017) 012015 doi:10.1088/1742-6596/928/1/012015

A Very Compact Crystal Shashlik Electromagnetic Calorimeter for Future HEP Experiments

Ren-Yuan Zhu
256-48, HEP, California Institute of Technology, Pasadena, CA 91125, USA

E-mail: zhu@hep.caltech.edu

Abstract. A very compact crystal based shashlik calorimeter is proposed for future HEP experiments in an extreme harsh radiation environment, such as the proposed HL-LHC. Thin crystal plates are used as the sensitive medium to reduce the light path length and thus the radiation damage effects and the calorimeter cost. A design of such a calorimeter uses tungsten as absorber, LYSO crystals as active medium, and liquid scintillator filled quartz capillaries as WLS to transport scintillating light to photodetectors. Initial calorimeter design and performance of prototype modules are presented. Possible optimizations are discussed.

1. Introduction

In high energy physics (HEP) and nuclear physics (NP) experiments, total absorption electromagnetic calorimeters (ECAL) made of inorganic crystals are well known for their superb energy resolution and detection efficiency for photon and electron measurements [1]. The CMS PWO ECAL, for example, played a crucial role in the discovery of the Higgs boson [2]. One crucial issue is crystal's radiation damage in the severe radiation environment at the LHC, which requires precision monitoring to correct variations of crystal transparency [3]. Figure 1 shows up to 70% loss of light output as well as partial recovery during the shutdown periods in the CMS PWO crystals observed in high rapidity when the LHC was running at different energies with a luminosity of up to 7×10^{33} cm^{-2}s^{-1} [4].

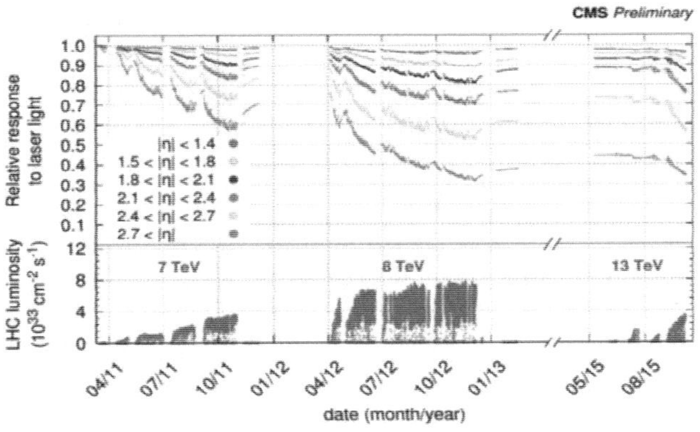

Figure 1. History of monitoring response is shown as a function of rapidity for CMS PWO crystals since LHC starts in 2011.

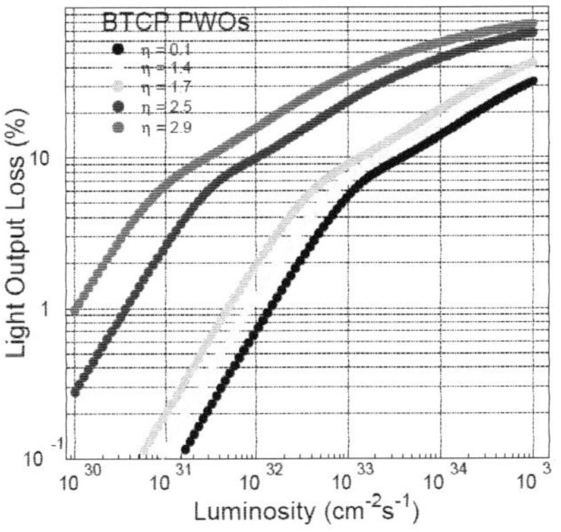

Figure 2. Predicated LO loss as a function of the luminosity and rapidity.

The dose rate dependent damage and recovery in PWO crystals are well understood [5, 6]. Figure 2 shows the expected light output loss as a function of luminosity for PWO crystals at different rapidity, predicted in 2010 before LHC running by using γ-ray induced radiation damage data [7]. It agrees well with Figure 1, indicating that the radiation damage in PWO crystals observed in Figure 1 is mainly caused by ionization dose. Radiation damage caused by charged hadrons was also studied [8]. Because of these damages the CMS endcap ECAL is planned to be replaced by more radiation hard technologies. An LYSO crystal based very compact Shashlik sampling calorimeter was proposed as one option [9].

2. A very Compact Crystal Shashlik Sampling Calorimeter

Figure 3 shows a photo of an LYSO/W Shashlik tower with four Y-11 wavelength shifter (WLS) fibers to transport scintillation light in LYSO to a PMT. This detector concept improves the overall radiation hardness of the calorimeter because of the much reduced light path length in the sensitive medium. An initial design consists of 30 LYSO plates of 14×14×1.5 mm and 29 W plates of 14×14×2.5 mm. Each tower has a depth of 25 X_0 to accommodate electrons and photons up to the TeV range. The sampling fraction was chosen to be 20% to provide a stochastic term at a level of 10% for the energy resolution. Because of the high density of both LYSO and W, the average radiation length (0.51 cm) and Moliere radius (1.3 cm) are much smaller as compared to commonly used heavy crystal scintillators. It thus provides fine granularity to reduce the pile-up effect at the HL-LHC.

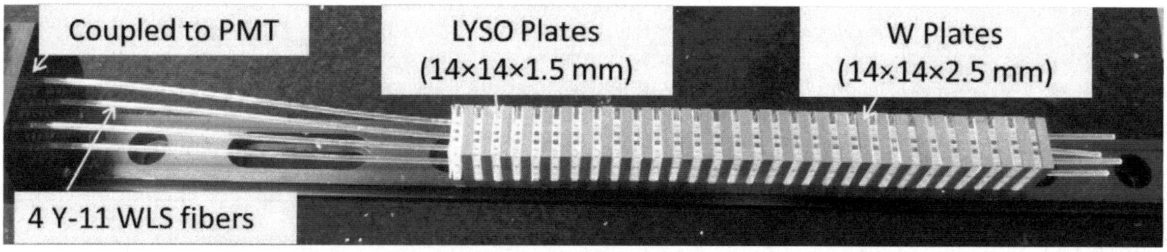

Figure 3. A photo showing an LYSO/W Shashlik tower with four Y-11 WLS fibers inserted.

Beam tests were carried out at Fermilab and CERN for an LYSO/W Shashlik matrix consisting of sixteen towers with Y-11 and SiPM readout. Figure 4 shows an energy resolution of 1.5% observed for 100 GeV electrons. Figure 5 shows an LYSO crystal based Shashlik calorimeter detector concept with four quartz capillaries filled with liquid scintillator serving as WLS and a quartz monitoring leaky fiber at the center [10]. The quartz capillaries allow light transportation in quartz so greatly enhance radiation hardness of the WLS and the entire calorimeter [11].

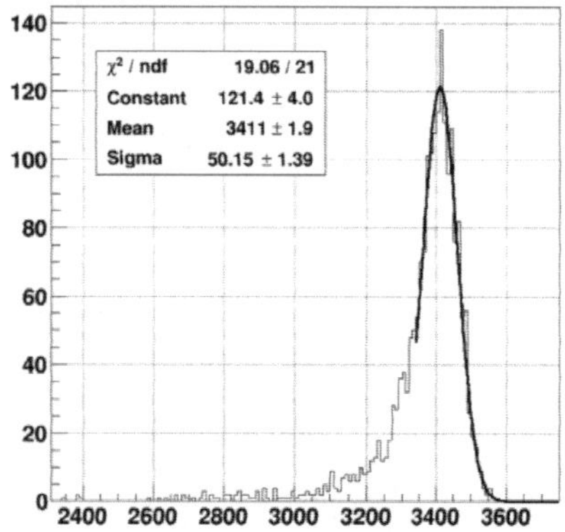

Figure 4. Energy resolution measured by a LYSO/W Shashlik matrix with 16 towers for 100 GeV electrons.

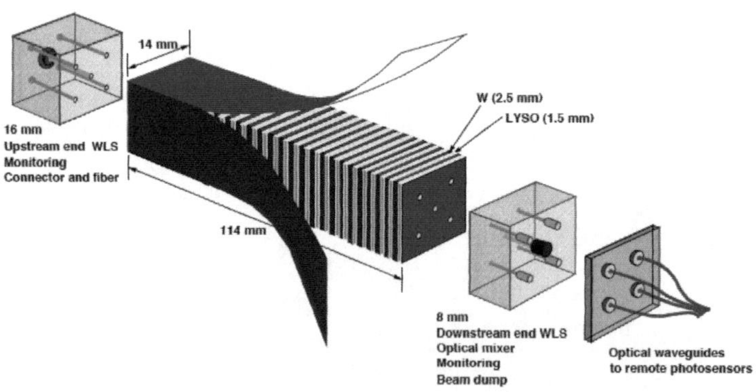

Figure 5. The LYSO/W Shashlik calorimeter concept consisting of 1.5 mm LYSO plates, 2.5 mm tungsten plates, four quartz capillaries as WLS and one quartz leaky fiber for light monitoring.

3. Bright, Fast and Radiation Hard LYSO Crystals and LYSO/W/Quartz Capillary Tower

Because of their high density (7.4 g/cm^3), short radiation length (1.14 cm), fast (40 ns) and bright (4 times BGO) scintillation, cerium doped lutetium oxyorthosilicate (Lu_2SiO_5:Ce, LSO) and lutetium yttrium oxyorthosilicate ($Lu_{2(1-x)}Y_{2x}SiO_5$:Ce, LYSO) crystals have attracted a broad interest in the high energy physics community. Their excellent radiation hardness against γ-rays [12], neutrons [13] and charged hadrons [14] also makes them a preferred material for the HL-LHC.

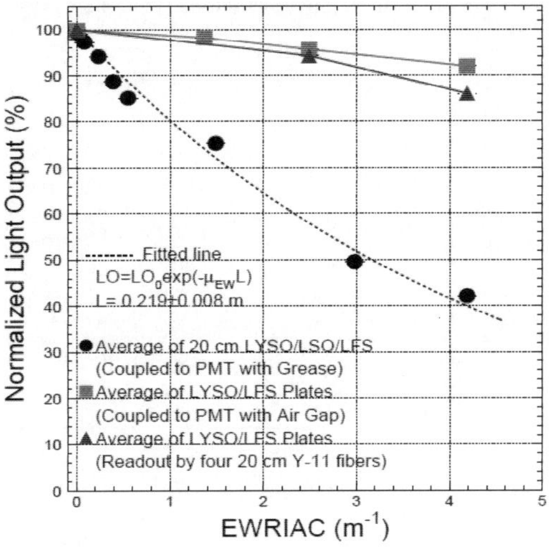

Figure 6. Light output is shown as a function of the induced absorption in LYSO crystals.

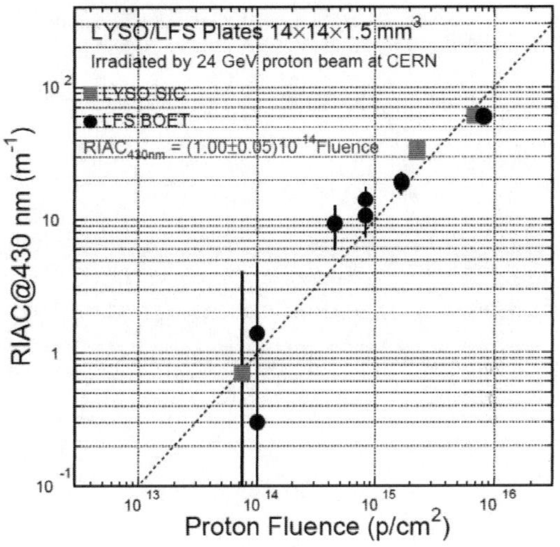

Figure 7. Induced absorption is shown as a function of 24 GeV proton fluence for LYSO crystals.

Figure 6 shows 5% and 50% losses observed for LYSO crystals of 14×14×1.5 mm and 2.5×2.5×20 cm respectively after 150 Mrad, showing clearly the advantage of reduced light path length. Figure 7 shows induced absorption as a function of 24 GeV proton fluence for 14×14×1.5 mm LYSO plates. About 3 m^{-1} is observed after 3 × 10^{14} cm^{-2} which is the maximum expected by the CMS FCAL at the HL-LHC.

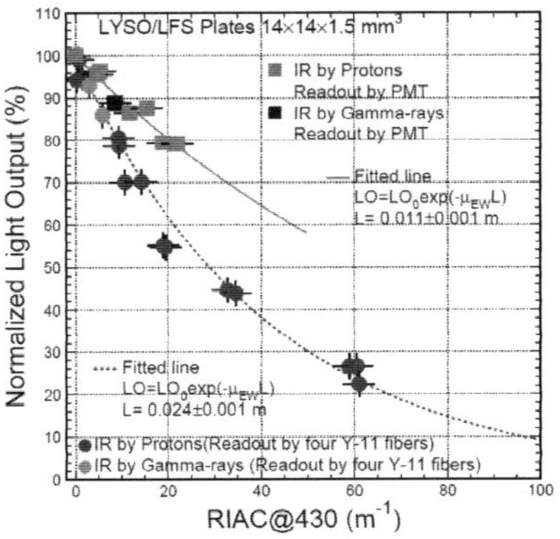

Figure 8. Light output is shown as a function of γ-ray induced absorption in 14 x 14 x 1.5 mm LYSO plates

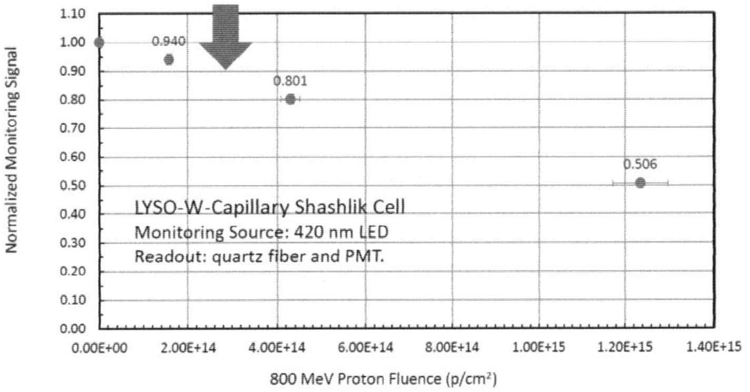

Figure 9. Monitoring response is shown as a function of 800 MeV proton fluence for an LYSO/W/Quartz capillary tower with a red arrow showing the maxim fluence expected by CMS FCAL

Figure 8 shows light output as a function of induced absorption for 14×14×1.5 mm LYSO plates after proton and γ-ray irradiation. Consistent losses at 6% are observed for crystals with 3 m^{-1} absorption by both protons and ɣ-rays. Figure 9 shows monitoring response as a function of 800 MeV proton fluence for an LYSO/W/Quartz capillary tower. About 10% loss is observed after 3 × 10^{14} cm^{-2} proton fluence, indicating that this detector concept provides a very stable calorimeter under severe radiation.

IOP Conf. Series: Journal of Physics: Conf. Series **928** (2017) 012015 doi:10.1088/1742-6596/928/1/012015

4. Optimization

Figure 10 and 11 show light output as a function of integration time for two 25×25×0.4 mm LuAG:Ce ceramic samples before and after 220 Mrad and 3×10^{14} cm^{-2} proton fluence and 220 Mrad respectively. No degradation was observed. Because of low material loss in processing, such a material is promising as a cost-effective active medium for calorimeter of this type.

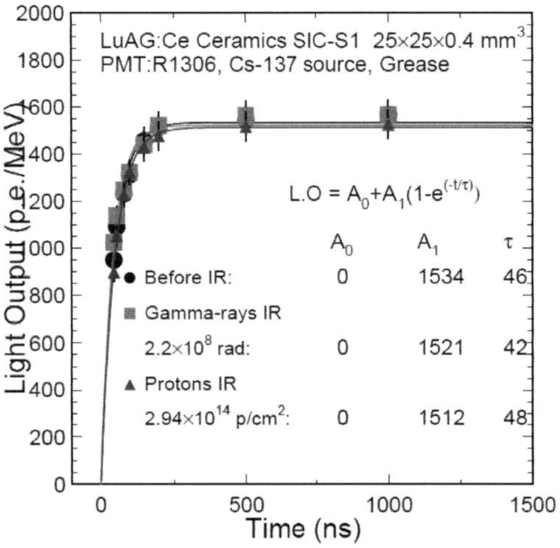

Figure 10. Light output is shown as a function of the integration time for a LuAG:Ce ceramic sample

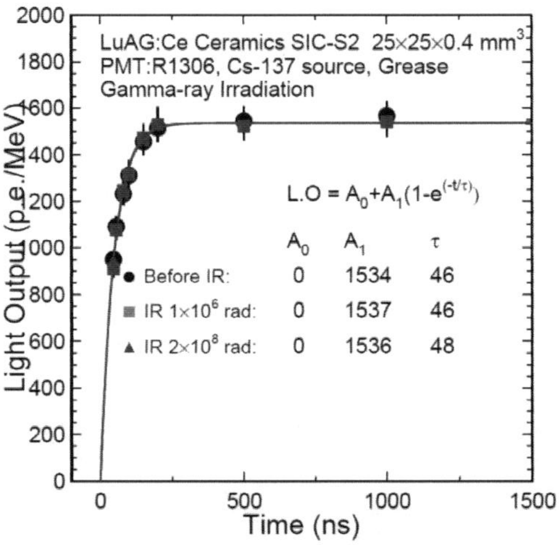

Figure 11 Light output is shown as a function of the integration time for a LuAG:Ce ceramic sample

5. Summary

To preserve precision e/γ measurement and face the challenge of the severe radiation environment expected by future HEP experiments at the energy frontier, a very compact crystal Shashlik ECAL was designed and tested in beam. Bright, fast and radiation hard LSO/LYSO crystals and quartz capillary based WLS provide a solid foundation for a stable and robust Shashlik calorimeter at the proposed HL-LHC. Scintillating ceramics, such as LuAG:Ce, may play an important role for a cost-effective Shashlik calorimeter.

Acknowledgments

This material is based upon work supported by the U.S. Department of Energy, Office of Science, Office of High Energy Physics program under Award Number DE-SC0011925. Much works carried out by the Shashlik collaboration is sincerely acknowledged.

References

[1] Zhu R-Y, 2012, *Physics Procedia* **37**, 372
[2] Chatrchyan S, Khachatryan V, Sirunyan A M, Tumasyan A, Adam W, Aguilo E, *et al.*, 2012, *Physics Letters B,* **716**, 30
[3] Zhu R-Y, 2004, *IEEE Trans. Nucl. Sci.,* **51**, 1560
[4] Lima R T, on behalf of the CMS Collaboration, 2016, in these proceedings
[5] Zhu R-Y, 1998, *Nucl Instrum Meth A,* **413**, 297
[6] Mao R H, Zhang L Y, and Zhu R-Y, 2004, *IEEE Trans. Nucl. Sci.,* **51**, 1777
[7] Zhu R-Y, http://www.hep.caltech.edu/~zhu/talks/ryz_101209_pwo.pdf
[8] Dissertori G, Luckey D, Nessi-Tedaldi F, Pauss F, Quittnat M, Wallny R, *et al.*, 2014, *Nucl Instrum Meth A,* **745**, 1
[9] Zhang L Y, Mao R H, Yang F, and Zhu R-Y, 2014, *IEEE Trans. Nucl. Sci.,* **61**, 483
[10] Yang F, Zhang L Y and Zhu, R-Y, 2016, *IEEE Trans. Nucl. Sci.,* **63**, 605
[11] Baumbaugh B E, Ford K, McKenna M, Ruggiero D and Ruchti R, 2015, in IEEE NSS 2015 CR
[12] Yang F, Zhang L Y and Zhu R-Y, 2016, in these proceedings
[13] Zhang L Y, Mao R H and Zhu R-Y, 2009, IEEE NSS Conference Record, paper N32-4
[14] Yang F, Zhang L Y, Zhu R-Y, Kapustinsky J, Nelson R and Wang Z, 2016, in these proceedings

IOP Conf. Series: Journal of Physics: Conf. Series **928** (2017) 012016 doi:10.1088/1742-6596/928/1/012016

Precision Timing with shower maximum detectors based on pixelated micro-channel plates

A Bornheim[1], A Apresyan[1], A Ronzhin[2], S Xie[1], M Spiropulu[1], J Trevor[1], C Pena[1], F Presutti[1], S Los[2]

[1]Caltech, 1200 E. California Blvd., 91125 Pasadena, CA, USA
[2]FNAL, Kirk Road and Pine Street, 60510 Batavia, IL, USA

E-mail: bornheim@hep.caltech.edu

Abstract. Future calorimeters and shower maximum detectors at high luminosity colliders need to be highly radiation resistant and very fast. One exciting option for such a detector is a calorimeter composed of a secondary emitter as the active element. In this report we outline the study and development of a secondary emission calorimeter prototype using micro-channel plates (MCP) as the active element, which directly amplify the electromagnetic shower signal. We demonstrate the feasibility of using a bare MCP within an inexpensive and robust housing without the need for any photo cathode, which is a key requirement for high radiation tolerance. Test beam measurements of the prototype were performed with 120 GeV primary protons and secondary beams at the Fermilab Test Beam Facility, demonstrating basic calorimetric measurements and precision timing capabilities. Using multiple pixel readout on the MCP, we demonstrate a transverse spatial resolution of 0.8 mm, and time resolution better than 40 ps for electromagnetic showers.

1. Introduction

In this paper we discuss our studies on the usage of micro-channel plates (MCP) as precision timing sensors in high energy physics. MCPs have been suggested as the active element in a sampling calorimeter for a long time [1] where the charged shower particles induce signals in the MCP structure which we shall refer to as secondary emission (SE). MCPs are also commonly used in combination with a photo cathod as MCP-PMTs. They feature excellent timing performance due to the very fast signal rise time of order 100 ps. Commercial MCP-PMTs with photo cathode may also be used to detect relativistic charged particles [2]. The particles create Cherenkov radiation as they pass through the window of the MCP which ensures high detection efficiency. SE induced signals enhance this signal further. We use MCP-PMTs in this fashion as timing reference counters [3]. Our studies with MCPs cover several aspects of precision timing measurements in high energy physics : MCPs as very precise reference timing detectors, as tools to study the timing properties of showers in calorimeters and MCPs as active elements in precision timing calorimeter. Precision timing calorimeters promise to be a very powerful tools for physics at the energy and intensity frontier [4]. The studies presented here focus on the timing performance of MCPs measuring electromagnetic showers. As we have discussed previously, the temporal evolution of electromagnetic showers is very coherent [5], allowing multiple measurements on a single shower, increasing the precision of the combined measurements.

2. Experimental Setup

We use a Photek MCP-PMT 240 as our main sensor both for the reference timing measurements as well as for shower timing measurements. As data acquisition system we use a Domino Ring Sampler (DRS) [6] either in the evaluation board available from PSI or as implemented in the CAEN V1742 digitizer board [7]. We record data with 5 GS/s. The DRS4 evaluation board is rated at an analog bandwidth of 700 MHz with a noise level of 0.35 mV. Our studies of the transverse shower evolution are performed with a Photonis MCP model XP85012 with segmented readout. It has 8 times 8 pixels on a area of about 5 by 5 cm, resulting in a pixel size of about $(6\text{mm})^2$. Further details on the setups and the performance of its individual components can be found in the references [3, 8, 9].

In Fig. 1 we show the experimental setup with two Photek MCP-PMTs as start and stop counter and a Photonis MCP as the test vehicle. Tungsten absorber was used to induce a shower. The beam impact point is defined with a scintillator trigger counter with a lateral size of $(2\text{mm})^2$. Beam was provided from the FNAL M-test facility. A 120 GeV proton beam was used for calibration and alignment purposes. Electron beams of 4, 8, 16 and 32 GeV are used to perform the shower timing measurements. The time of arrival is extracted from the signal

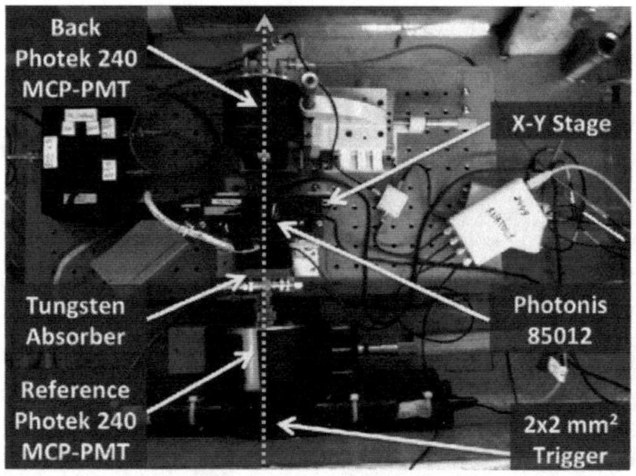

Figure 1. Experimental setup with two Photek MCP-PMTs and a Photonis MCP with segmented readout. Here the setup to test the pixelated readout of the Photonis is shown. For this a motorized stager allows to shift the Photonis without changing the alignment of the trigger with respect to the Photek. The setup testing the shower timing with the more precise Photek is equivalent.

pulses by performing a Gaussian fit to the peak of the reference MCPs spanning 8 to 10 bins, corresponding to 1.6 to 2.0 ns at an equivalent bin width of 0.2 ns at 5 GS/s. The differential time resolution between two Photek MCP-PMT was measured to be 10 ps in response to 120 GeV protons. No additional radiator was used for this measurement. This corresponds to a single device resolution of around 6 ps. The time resolution of the DRS wiht typical MIP signal amplitude from the Photek MCP-PMT was measured to be around 5 ps from split signals feeding the DRS. We conclude that the time resolution of the Photek MCP-PMT when measuring the time of arrival of charged particles in our setup is dominated by the readout system.

Electromagnetic showers are induced by inserting tungsten absorber of varying thickness into the beam. The absorber was placed as closely as possible to the sensor under study to avoid spreading of the shower.

3. Experimental Results

In Fig.: 2 we show the signal amplitude in a Photek MCP at various shower depths for a 32 GeV electron beam. We find the signal amplitude to scale in the expected way with the shower depth, increasing to a depth of about 5 radiation length and then decreasing. This demonstrates that the MCP may serve as a calorimetric device. In Fig.: 3 we show the time resolution between the reference MCP and the MCP behind the tungsten absorber, measuring the shower time of arrival. We find the time resolution to be around 13 ps, only slightly worse than for the bare MCP in response to a 120 GeV proton. The time resolution is largely independent of the shower depth. This confirms our statement, based on earlier studies and simulations, that the

IOP Conf. Series: Journal of Physics: Conf. Series **928** (2017) 012016 doi:10.1088/1742-6596/928/1/012016

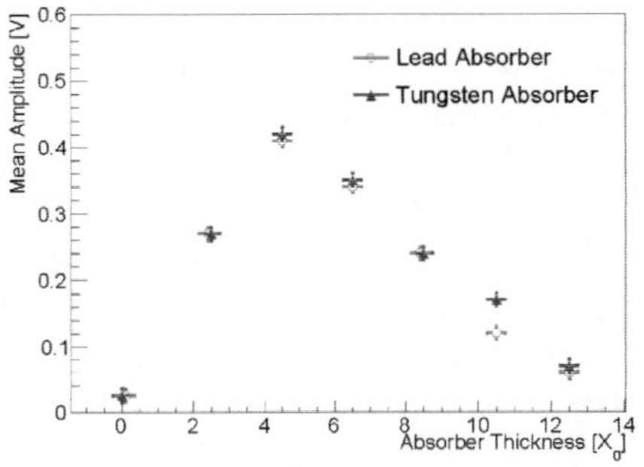

Figure 2. Signal amplitude in a Photek MCP at various shower depth for a 32 GeV electron beam. We find the signal amplitude to scale in the expected way with the shower depth, increasing to a depth of about 5 radiation length and the decreasing. We use tungsten as absorber. With a Moliere radius of 0.93 cm the shower is largely contained transversely in the entrance window of the Photek MCP-PMT which has a diameter of 4 cm.

temporal evolution of electromagnetic showers is very coherent. To further study the feasibility

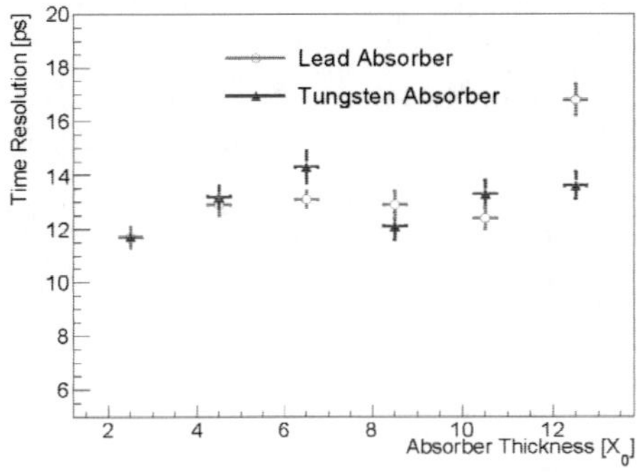

Figure 3. Time resolution between the reference MCP and the MCP behind the tungsten absorber, measuring the shower time of arrival. The time resolution is independent of the depth it is measured at. Note that the signal amplitude varies as a function of the shower depth as shown in Fig.: 2.

to MCPs as an active element in a precision timing calorimeter we perform measurements with a segmented MCP. Transverse segmentation allows to measure the lateral position of a shower. It also allows to perform multiple measurements across the shower which can be exploited to enhance the timing performance. In Fig.: 4 we show the time resolution of the Photonis MCP

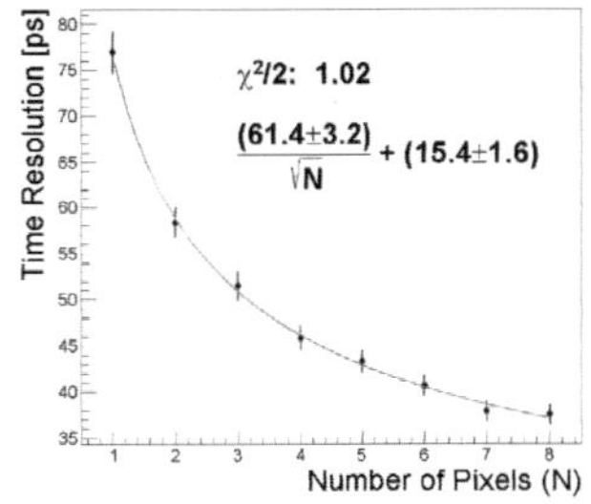

Figure 4. Time resolution for electromagnetic shower measured with the Photonis MCP. Combining the measured time from up to 8 pixels one improves the time resolution down to around 37 ps for 8 pixels. This allows to achieve a very good resolution despite a modest single channel resolution.

as a function of the number of pixels used in a combined timing measurement. Here, a time stamp is reconstructed for each pixel and the results are combined in quadrature. The single pixel timing performance of the Photonis MCP is worse than for the Photek MCP. However combining 8 pixels, the maximum we could read out in this setup, we achieve a timing resolution of around 37 ps. A further improvement seems achievable with combining more pixels. With the information of multiple pixels we also are able to reconstruct the lateral position of the showers. We achieved a spatial resolution of 0.9 mm with the pixel size of the Photonis MCP of 6 mm. In Fig.: 4 we show the time resolution of the Photonis MCP with the photo cathode switched on or off. We find the timing resolution to be independent of the photo cathode bias. This illustrates that the signal in showers is dominated by SE induced signals.

Operating MCPs in SE mode is crucial for their application in a calorimeter, in particular in a high energy proton collider. In such an application the device will be exposed to very significant

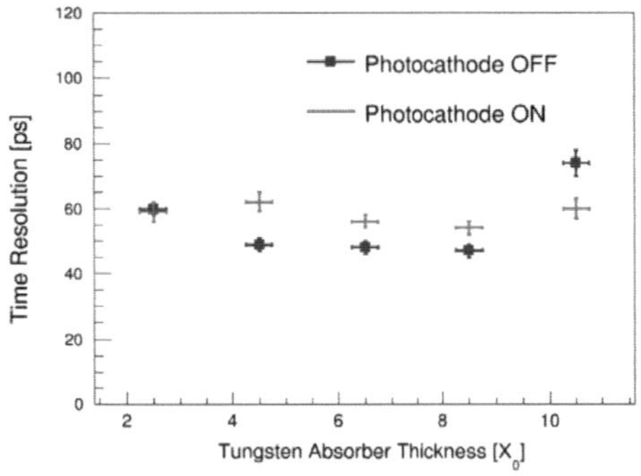

Figure 5. Time resolution per pixel of the Photonis MCP as a function of the shower depth. The data recorded with the photo cathode on is compared with data recorded with photo cahtode off. The time resolution is identical within the uncertainties. This demostrates that MCPs operated in SE mode are capable to provide precise timing in a calorimetric application.

radiation levels which are challenging conditions for a photo cathode. Bare MCPs however may exhibit much higher tolerance to radiation.

4. Summary

We demonstrate that MCPs are well suited to be used as an active elements in a precision timing calorimeter. Timing precision of around 13 ps can be achieved for electromagentic showers. With this we also demonstrate that the temporal evolution of electromagnetic showers is coherent on the level of a few ps. Transverseley segmented MCPs allow to measure the position of showers and utilize multiple measurements on the same shower to improve the timing resolution beyond the single pixel performance.

4.1. Acknowledgments

Operated by Fermi Research Alliance, LLC under Contract no. DE-AC02-07CH11359 with the United States Department of Energy. Supported by funding from California Institute of Technology High Energy Physics under Contract DE-SC0011925 with the United States Department of Energy. We thank the FTBF personnel for very good beam conditions during our test beam time.

References

[1] A. A. Derevshchikov, V. Yu. Khodyrev, V.I. Kryshkin, V.E. Rakhmatov, A. I. Ronzhin, On possibility to make a new type of calorimeter: radiation resistant and fast. Preprint IFVE 90-99, Protvino, Russia, 1990.
[2] A 5 ps TOF-counter with an MCPPMT, K. Inami, , N. Kishimoto, Y. Enari, M. Nagamine, T. Ohshima, NIM-A, doi:10.1016/j.nima.2015.05.029
[3] Development of a new fast shower maximum detector based on microchannel plates photomultipliers (MCP-PMT) as an active element, A. Ronzhin, S. Los, E. Ramberg, M. Spiropulu, A. Apresyan, S. Xie, H. Kim, A. Zatserklyaniy, NIM-A, doi.org/10.1016/j.nima.2014.05.039
[4] LYSO based precision timing calorimeters, A. Bornheim, CALOR2016, these proceedings
[5] Calorimeters for Precision Timing Measurements in High Energy Physics, A. Bornheim, CALOR2014, doi:10.1088/1742-6596/587/1/012057
[6] https://www.psi.ch/drs/
[7] http://www.caen.it/
[8] Study of the timing performance of micro-channel plate photomultiplier for use as an active layer in a shower maximum detector, A. Ronzhin, S. Los, E. Ramberg, A. Apresyan, S. Xie, M. Spiropulu, H. Kim, NIM-A, dx.doi.org/10.1016/j.nima.2015.06.006
[9] Direct tests of micro channel plates as the active element of a new shower maximum detector, A. Ronzhina, S. Los, E. Ramberg, A. Apresyan, S. Xie, M. Spiropulu, H. Kim, NIM-A, dx.doi.org/10.1016/j.nima.2015.05.029
[10] On timing properties of LYSO-based calorimeters, D. Anderson, A. Apresyan, A. Bornheim, J. Duarte, C. Pena, A. Ronzhin, M. Spiropulu, J. Trevor, S. Xie, NIM-A, doi:10.1016/j.nima.2015.04.013, 2015

IOP Conf. Series: Journal of Physics: Conf. Series **928** (2017) 012017 doi:10.1088/1742-6596/928/1/012017

Design, status and test of the Mu2e crystal calorimeter

N. Atanov[1], V. Baranov[1], J. Budagov[1], R. Carosi[5], F. Cervelli[5], F. Colao[2], M. Cordelli[2], G. Corradi[2], E. Dané[2], Y. I. Davydov[1], S. Di Falco[5], S. Donati[5,7], R. Donghia[2], B. Echenard[3], K. Flood[3], S. Giovannella[2], V. Glagolev[1], F. Grancagnolo[8], F. Happacher[2], D. G. Hitlin[3], M. Martini[*,2,4], S. Miscetti[2], T. Miyashita[3], L. Morescalchi[5,6], P. Murat[9], G. M. Piacentino[8], G. Pezzullo[5,7], F. Raffaelli[5], A. Saputi[2], I. Sarra[2], F. Spinella[5], G. Tassielli[8], V. Tereshchenko[1], Z. Usubov[1] and R. Y. Zhu[3].

[1]Joint Institute for Nuclear Research, Dubna, Russia
[2]Laboratori Nazionali di Frascati dell'INFN, Frascati, Italy
[3]California Institute of Technology, Pasadena, United States
[4]Università Guglielmo Marconi, Roma, Italy
[5]INFN Sezione di Pisa, Pisa, Italy
[6]Dipartimento di Fisica dell'Università di Siena, Siena, Italy
[7]Dipartimento di Fisica dell'Università di Pisa, Pisa, Italy
[8]INFN, Sezione di Lecce and Dipartimento di Matematica e Fisica dell'Università del Salento, Lecce, Italy.
[9]Fermi National Laboratory, Batavia, Illinois, USA

E-mail: matteo.martini@lnf.infn.it

Abstract. The Mu2e experiment at Fermilab searches for the charged-lepton flavor violating neutrino-less conversion of a negative muon into an electron in the field of an aluminum nucleus. The dynamics of such a process is well modeled by a two-body decay, resulting in a monoenergetic electron with an energy slightly below the muon rest mass (104.967 MeV). The calorimeter of this experiment plays an important role to provide excellent particle identification capabilities and an online trigger filter while aiding the track reconstruction capabilities. The baseline calorimeter configuration consists of two disks each made with $\sim$ 700 undoped CsI crystals read out by two large area UV-extended Silicon Photomultipliers. These crystals match the requirements for stability of response, high resolution and radiation hardness. In this paper we present the final calorimeter design.

1. Introduction

The Mu2e experiment at Fermilab [1] aims to search for Charged Lepton Flavor Violation (CLFV) in the neutrino-less, coherent conversion of a negative muon into an electron in the Coulomb field of an ^{27}Al nucleus. The $\mu \rightarrow e$ conversion results in a mono energetic electron with an energy equal to the muon rest mass minus the corrections for the nuclear recoil and the binding energy. For ^{27}Al the energy of the mono energetic electron is equal to 104.97 MeV.

The experiment is designed to reach the single event sensitivity of 2.4×10^{-17} in three years [1]. This value represents an improvement of four orders of magnitude over the current best

IOP Conf. Series: Journal of Physics: Conf. Series **928** (2017) 012017 doi:10.1088/1742-6596/928/1/012017

experimental limit set by SINDRUM II experiment [2].

The Standard Model predicted rate for this process is $\mathcal{O}(10^{-52})$ [3], therefore any signal observed by Mu2e would be compelling evidence of new physics.

2. Mu2e electromagnetic calorimeter

The Mu2e detector is designed to be almost background free and it is located inside a large superconducting solenoid with a magnetic field of 1 T in the region of the calorimeter. This detector is located just behind the tracker and complements it by providing: powerful μ/e particle identification, seed for pattern recognition in the tracker and an independent software trigger system. Efficient particle identification requires a time resolution better than 500 ps and an energy resolution $\mathcal{O}(5\%)$.

The calorimeter should be able to operate in an environment where a dose up to 100 krad and a neutron fluency of 10^{12} n/cm^2 are expected. It must also work in a 1 T magnetic field and 10^{-4} Torr vacuum ensuring the redundancy of the component.

Due to physical and geometrical constraints, we decided to adopt a solution with two annuli made of undoped CsI crystals each read out using two Silicon Photomultipliers. Each disk has an internal (external) radius of 374 mm (660 mm) and is filled with $34 \times 34 \times 200$ mm^3 crystals (see Fig.1). The disks are separated of 75 cm. The analog read out electronics is connected to the SiPM while the digital boards are housed in the crates in the top of each disk.

The discussed requirements pushed the experiment to adopt a calorimeter made by undoped CsI crystals optically coupled to 14×20 mm^2 large area UV-extended SiPM.

Figure 1. Structure of the Mu2e calorimeter

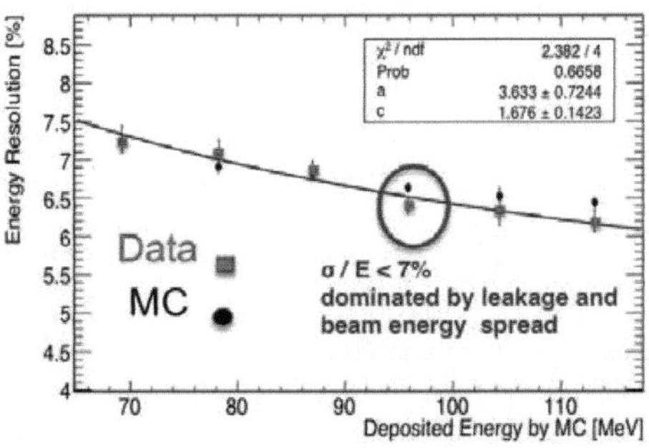

Figure 2. Measured energy resolution for a 3×3 CsI matrix with Hamamatsu TSV 12×12 mm^2 MPPCs using electrons between 80 and 120 MeV.

3. Crystal choice and test

The requirements of the electromagnetic calorimeter imply to use crystals with:

- high light output
- good light response uniformity, LRU, ($>10\%$)
- fast signal with small slow component ($\tau < 40$ ns)
- radiation hard with maximum light output loss below 40%
- small radiation induced readout noise (below 0.6%)

In the Conceptual Design Report [4], the baseline calorimeter choice was LYSO crystals readout with APD and many tests were carried out for this option [5]. A large increase on the Lu_2O_3 salt price in 2013 made this option unaffordable, so that for the TDR [1] we have opted for cheaper crystals such as BaF2 and CsI. After a long R&D program [6], we have finally selected undoped CsI crystals as baseline choice (See Tab.1).

Table 1. Crystals suitable for Mu2e calorimeter.

Crystal	BaF_2	LYSO	CsI	$PbWO_4$
Density $[g/cm^3]$	4.89	7.28	4.51	8.28
Radiation Length [cm]	2.03	1.14	1.86	0.9
Moliere radius [cm]	3.10	2.07	3.57	2.0
dE/dx [MeV/cm]	6.5	10.0	5.56	13.0
Refractive Index at λ_{MAX}	1.50	1.82	1.95	2.20
Peak Luminescence [nm]	220/300	402	310	420
Decay time [ns]	0.9/650	40	26	30/10
Light Yield [% NaI]	4.1/3.6	85	3.6	0.3/0.1
Light Yield Variation with temperature $[\%/^oC]$	0.1/-1.9	-0.2	-1.4	-2.5
Hygroscopicity	None	None	Slight	None

Tests on CsI crystals have been performed for three different vendors: ISMA (Ukraine), SICCAS (China) and Optomaterial (Italy). These crystals have been irradiated up to 900 Gy and to a neutron fluency up to 9×10^{11} n_{1MeV}/cm^2. The ionization dose does not modify LRU while a 20% reduction in light yield has been observed at 900 Gy. Similarly, the neutron flux causes a 15% light yield deterioration. These results are compatible with the requirements of the calorimeter.

A small unroped CsI matrix has been built and tested in Frascati Beam Test Facility using electrons with energy between 80 and 120 MeV. The prototype is a 3×3 matrix and each crystal is read out using an array of sixteen 3×3 mm^2 Hamamatsu TSV MPPCs. During this test we measured a light yield of 30 (20) pe/MeV with (without) optical grease with Tyvek wrapping. The measured time and energy resolution, 110 ps and 7% respectively, perfectly match our initial requirements (Ref. Fig.2).

4. UV-extended SiPM

The requirement of having a small air gap between crystal and photodetector and the request of redundancy in the read out implies the use of custom SiPMs. For the Mu2e experiment we have increased the transverse dimension of the CsI from 30×30 to 34×34 mm^2 in order to accomodate two 2×3-arrays of 6×6 mm^2 UV-extended SiPM. This allows to work with an air-gap while satisfying the pe/MeV requirement with a single SiPM. Presently, we are testing both Hamamatsu MPPCs and FBK UV-extended Silicon Photomultipliers.

The photosensors are packaged using a parallel arrangement of two groups of three cells biased in series. The samples already acquired show a good PDE ($\sim$ 30% at 315 nm) with a gain greater than 10^6 at an over-voltage of 3 V with respect to the breakdown voltage. The series connection produce a signal with a total width of 70 ns. A first array has been assembled with 6 Hamamatsu 6x6 MPPCs and has been optically connected to a CsI Tyvek wrapped crystal measuring a time resolution for 1 MIP ($\sim$ 20 MeV) of 170 ps.

Following calorimeter requirements, one important aspect to be considered for the read out of the crystals is the radiation hardness of the SiPM. In this context, we have performed different tests using Hamamatsu and FBK UV-extended devices. SiPMs irradiated with a dose of up to 20 krad photons source do not show any effect on the leakage current. Different effects are observed with neutrons. When exposing sensors to 14 MeV neutrons with a total flux of 2.2×10^{11} n/cm^2 (corresponding to 2.2 times the experiment lifetime), we observe a too high increase of the leakage current (up to 2.3 mA when using 3×3 mm^2 Hamamatsu cells). To reduce the leakage current to acceptable value, we need to cool down all SiPM to a temperature of 0 degrees. In order to do so, we will use a dedicated cooling station for the calorimeter. The final choice of the coolant and the parameters of the station are currently under study.

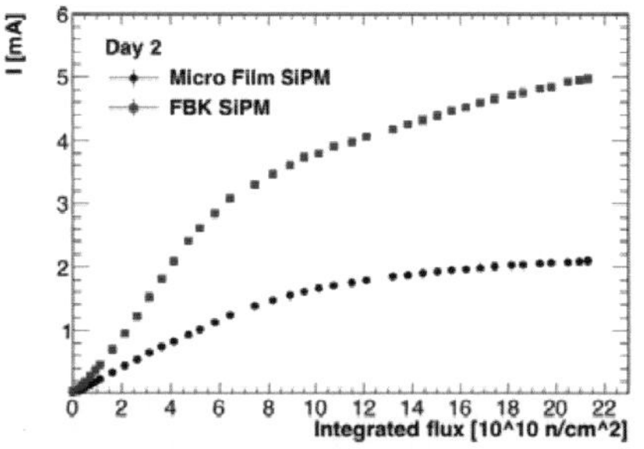

Figure 3. Measured leakage current versus integrated neutron flux.

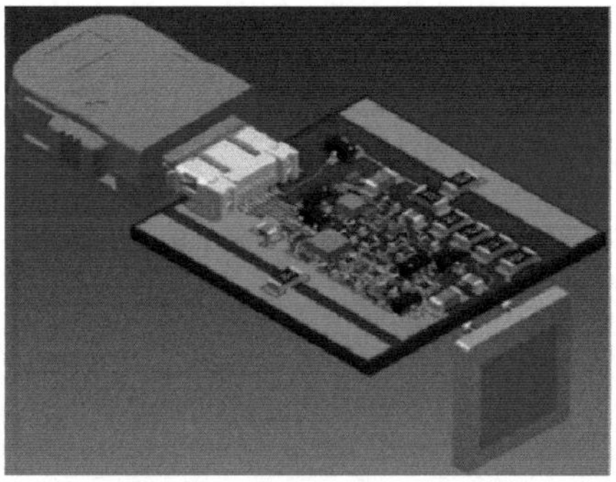

Figure 4. Rendering of the FEE board connected with SiPM.

5. Front End Electronics

Each SiPM is directly connected to a dedicated board (See Fig.4) housing a transimpedence preamplifier with a settable gain $\times 15$ or $\times 30$, 2 V dynamic range and 15 ns rise time. This board provides also a pulse signal for testing the preamplifier and a slow control readout of temperature and leakage current. The digital boards are housed into 11 crates per disk with 20 differential channels per board. These boards host a mezzanine which: receives signals from SIPM, manage HV setting and includes a Waveform Digitizer section that is based on SmartFusion II FPGA with 200 Msps 12 bit ADC.

Acknowledgments

This work was supported by the EU Horizon 2020 Research and Innovation Programme under the Marie Sklodowska-Curie Grant Agreement No. 690835.

References

[1] L. Bartoszek, et al., Mu2e Technical Design Report arXiv:1501.05241.
[2] W. H. Bertl, et al., A Search for muon to electron conversion in muonic gold, Eur.Phys.J. C47 (2006) 337-346.
[3] W. J. Marciano, T. Mori, J. M. Roney, Charged Lepton Flavor Violation Experiments, Ann.Rev.Nucl.Part.Sci. 58 (2008) 315-341.
[4] J. R. Abrams et al, Mu2e conceptual design report arXiv:1211.7019.
[5] N. Atanov *et al.*, Nucl. Inst. Meth. A 812 (2016), 104.
[6] N. Atanov *et al.*, Nucl. Inst. Meth. A 824 (2016), 695.

IOP Conf. Series: Journal of Physics: Conf. Series **928** (2017) 012018 doi:10.1088/1742-6596/928/1/012018

Polarimetry concept based on heavy crystal hadron calorimeter

I Keshelashvili[1], F Müller[1,2] and D Mchedlishvili[1,3], on behalf of JEDI collaboration.

[1]Forschungszentrum Jülich GmbH, Leo-Brandt-Strae, 52428 Jülich, Germany
[2] Physics Institute III B, RWTH Aachen, Otto-Blumenthal-Strasse, 52074 Aachen, Germany
[3]High Energy Phisics Institute of Tbilisi State University, Tbilisi, Georgia

E-mail: i.keshelashvili@fz-juelich.de

Abstract. In the ongoing JEDI (**J**ülich **E**lectric **D**ipole moment **I**nvestigations) project, the essential point will be to measure a tiny beam polarization change over an extended period of time. The particle scarcity in the polarized deuteron or proton beams and the required slow extraction rate puts tough experimental constrains on the polarimetry. For the EDM measurements, a dedicated high precision polarimeter is required. To fulfill specifications, a fast, dense, high resolution (energy and time), and radioactive hard novel crystal scintillating material is required. LYSO crystals are supposed to be used as an ideal scintillating material for this kind of detector. The LYSO crystal PMT and SiPM readout, with a FADC based system is under development. The first proton and deuteron beam test of the prototypes are presented here. In this paper, the new polarimetry concept and preliminary results from first proton and deuteron beam time are presented.

1. Introduction

In the search for an electric dipole moment with a storage ring (srEDM), the polarimeter must be sensitive to very small changes in the vertical component of the polarization during a beam store. Such changes require vast statistics and excellent sensitivity; thus, the polarimeter must operate extremely efficiently. Due to smallness of the expected EDM signal the precision or calibration of the measurement is not the most important feature. Thus, the design of the polarimeter favors efficient targets and the observation of an elastic scattering, along with low q-value reactions, at forward angles where the spin-orbit interactions create a large polarization sensitivity. This requirement leads to the use of calorimetric detectors that can easily select such events with the use of a lower threshold for the total energy measurement. To maintain sufficient sensitivity to vertical component changes, the detectors and their thresholds must be very stable over time and across changes in rate or beam properties (direction and angle). It is not as necessary to restrict the acceptance to a single or a few reaction channels as would be needed for a nuclear physics study. This leads to the emphasis on forward-angle elastic scattering and the use of single component detectors. In ref.[1], the vector analyzing power, the differential cross section, and the figure-of-merit (FOM) of deuteron carbon elastic scattering at 200 MeV and 270 MeV deuteron kinetic energies are published. At both energies, the vector analyzing power is positive and quite large over the forward polar angular range covered by the new concept of the JEDI polarimeter [2], show in fig. 1. The FOM is calculated as a product of the differential cross section and the squared analyzing power. But in the real experiment, the detection efficiency (defined as a ratio between the identified number of elastically scattered deuterons and the incoming deuterons) must be taken into account. In such a case, FOM can be defined as:

$$FOM(E) = \{\sigma(E) \times \varepsilon(E)\} \times A_y^2(E) \tag{1}$$

where the additional parameter $\varepsilon(E)$ is introduced as a detection or in our case elastic reaction selection efficiency. This efficiency is determined by placing a threshold at the lower edge of the observed elastic scattering peak (in the deposited energy spectrum), and it depends on the energy resolution of the crystals and the features of the reaction and background tail below the peak. The detection efficiency itself hase been measured as a first experiment [3, 4], where the LYSO modules were directly exposed to the low intensity ($< 10kHz$) tagged deuteron and proton beam.

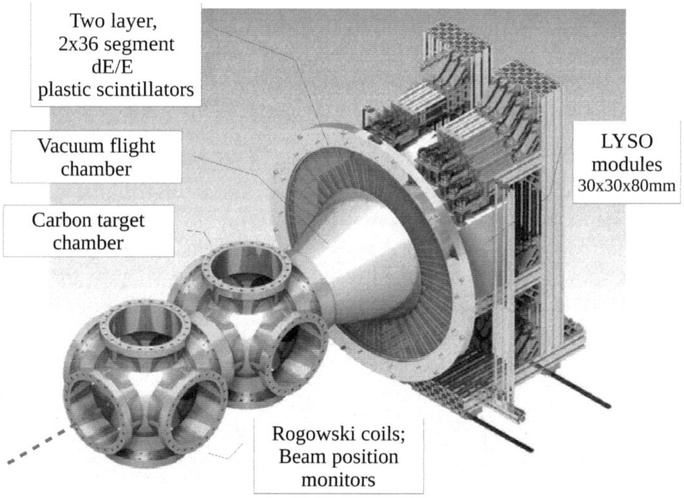

Figure 1. The new JEDI Polarimeter concept is shown. From left to right there are two cross type flanges, one for beam position monitors (BPM) and the second for the carbon targets. In the middle, a vacuum flight chamber. Next, two layers of ϕ sensitive "pizza" shaped, 2x36 plastic scintillators. The last is the LYSO HCAL to totally absorb the energy of the scattered particles.

2. Experiment

The goal of the first step experiment was to investigate characteristics of scintillating material LYSO with deuteron beam. In these tests, five LYSO crystals have been tested (four Saint-Gobain (SG) [5] 2x 30x30x100mm and 2x 15x30x100mm and one EPIC Crystals (EP) [6] 30x30x100mm), forming the four independent calorimeter modules (see fig.: 2). The readout was done using dual channel PMT [8] and SensL 6x6mm SiPMs 4x4 arrays. The optical contact has been made by optical grease and air contact. The air contact was preferred due to unwanted capillary effects in case of the optical grease flowing between crystal and Teflon wrapping. All events are written using Struck FADC [7] and signal forms were analyzed offline. The modules were exposed to direct proton ($100, 150MeV$) and deuteron ($100, 150, 200, 235, 270MeV$) beams mounted to the movable test table fig.:3. Using these crystals, two different module configurations can be assembled: three for low (large θ angles) and one for high count rate (small θ angles) use.

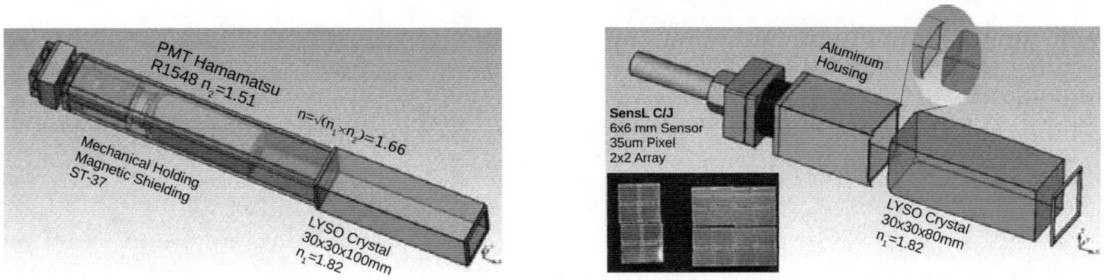

Figure 2. Left: the drawing of a single LYSO module with mechanical holding structure, high voltage passive divider, squared Hamamatsu PMT [8], a light guide, and the LYSO crystal. Right: example of the new prototype module combining a $30 \times 30 \times 80$ mm LYSO crystal with a SiPM [9] readout.

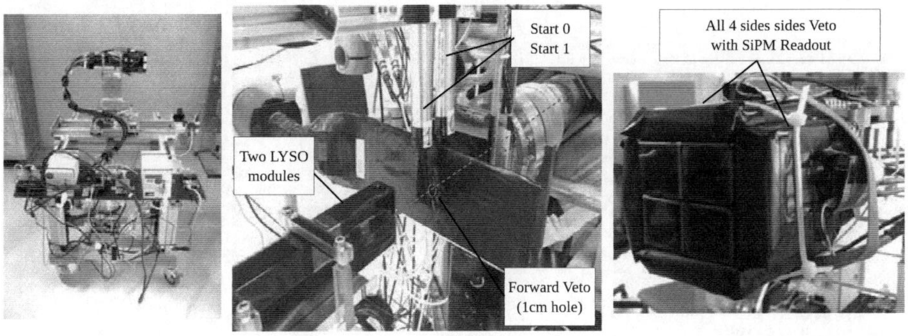

Figure 3. Left: X, Y, Θ test support table; Middle: two LYSO modules in beam with two start, and forward veto counter. Right: the cluster of 4 LYSO modules surrounded with plastic side veto scintillators.

3. Preliminary results and outlook

The main goals of the beam time: LYSO energy resolution, deuteron, proton identification efficiencies and all technical issues have been tested successfully. The energy resolution has been measured as a function of deuteron beam using QDC and FADC. The measurement using QDC is shown in the right plot of the fig.:4. The FADC results can be seen in more details in F. Müller's contribution [4]. The measured deuteron identification efficiency at 270 MeV is estimated as $\sim 70\%$ mainly due to deuteron breakup reaction inside the crystal [2]. Also, the crystal homogeneity scans, the front face scan, and beam incoming with different angles have been explored. All in all, the achieved resolution for the 270 MeV deuteron kinetic energy is 0.5%. The achieved time resolution is well below 300 ps and is good enough for an efficient operation of the polarimeter. The above-mentioned identification efficiency and the dead-time less FADC based data acquisition system are a perfect demonstration of our inorganic scintillator based polarimeter.

The next step is to measure scattered polarized deuterons from different target sizes and materials at the external beamline fig.:3. The main emphasis is to measure elastic and total cross-section ratio dependence on the polar angle at different energies. Also, the possibility to measure asymmetries, during one beam extraction with various targets gives us the opportunity to learn more about target materials. The table construction is made to be very versatile. One can change the polar angles of each arm independently with remote control. The distance between target and crystals can also be modified. The module orientation will also be adjustable. Also, the third static arm with plastic scintillator counter just below the beam line will be used

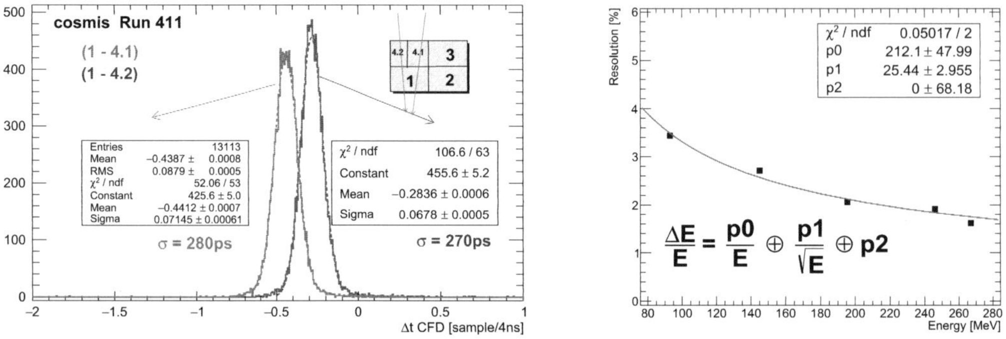

Figure 4. Left: measured time difference for cosmic events between neighbor PMT modules with sampling rate of 250 MS/s. For the time extraction the CFD method have been used. Right: measured LYSO energy resolution using PMT and QDC *vs* incoming beam energies.

Figure 5. Experimental setup to be constructed for the second step experiment aiming to measure elastic to total cross section ratio and asymmetries with different target materials.

for the normalization purposes.

Acknowledgments

Authors wishing to acknowledge assistance by technical staff Nils Giese, Maike Maubach, Tanja Hahnraths von der Gracht and Peter Wüstner. We would like to express our special thanks for participation in the experiment to Krzysztof Nowakowski, Arkadiusz Popczak from the Jagiellonian University, Krakow Poland, Simone Basile, Luca Barion from the University of Ferrara, Italy and Nodar Lomidze, Mirian Tabidze from High Energy Physics Institute of Tbilisi State University, Georgia.

References

[1] Y. Satou et al., Phys. Lett. B **549**, 307 (2002).
[2] Towards EDM Polarimetry, PoS(PSTP2015)026.
[3] JEDI proposal E2: "Towards the EDM polarimetry", (May 2015).
[4] F. Müller, Proceeding of CALOR '16, Korea (2016).
[5] Saint-Gobain, Cerium doped Lutetium PreLude420 crystal.
[6] Epic-Crystal, LYSO Ce scintillator.
[7] Struck, FADC, SIS3316-250-14 16 channel 250 MSPS 14-bit
[8] PMT, Hamamatsu R1548-07.
[9] SiPM, SensL, MicroFC-60035-SMT

IOP Conf. Series: Journal of Physics: Conf. Series **928** (2017) 012019 doi:10.1088/1742-6596/928/1/012019

LYSO crystal testing for an EDM polarimeter

F Müller[1,2], I Keshelashvili[1], and D Mchedlishvili[1,3] for the JEDI collaboration

[1] Forschungszentrum Jülich GmbH, IKP, 52425 Jülich, Germany
[2] Physics Institute III B, RWTH Aachen, 52074 Aachen, Germany
[3] High Energy Phisics Institute of Tbilisi State University, 0186 Tbilisi, Georgia

E-mail: fa.mueller@fz-juelich.de

Abstract. Four detector modules, built from three different LYSO crystals and two different types of light sensors (PMTs and SiPM arrays), have been tested with a deuteron beam from 100 MeV – 270 MeV at the COSY accelerator facility for the srEDM project at the Forschungszentrum Jülich in Germany. The detector modules were arranged in a cluster hand mounted on a positioning table. The deuteron beam was targeted at the center of each individual crystal for data analysis. The signals were digitized using a 14 bit, 250 MS/s flash ADC. Further, the energy spectra were calibrated using the known beam energies from the accelerator. From the calibrated spectra, the energy resolution was calculated. A resolution of 3% for the low energies and down to 1% for the high energy of 270 MeV was achieved. A deuteron reconstruction efficiency of almost 100% for low energies and around 70% for high energies was achieved. The SiPM light sensor showed a very good performance and will be used for the next generation of detector modules.

1. Introduction

As stated in [1, 2] our aim is to build a polarimeter detector consisting of individual LYSO crystal detector modules for deuteron electric dipole moment (EDM) measurements [3]. As a first prototype, four detector modules using different LYSO crystals and different types of light sensing parts have been built and tested. The modules were analyzed in a deuteron beam with four different beam energies of 100, 200, 235 and 270 MeV. All test were done at the COSY accelerator facility at the Research Center Jülich in Germany. The test setup was installed on a external beam line.

2. Experimental Setup

The experimental setup for the LYSO crystal tests mainly consisted of two parts: A cluster of four detector modules consisting of LYSO crystals and light sensing parts, and a positioning table that allowed the simulation of the arrangement of the detector modules in the final polarimeter. Figure 1 shows a schematic overview of the experimental setup. After leaving the beam pipe, the deuterons had to penetrate two 2 mm plastic scintillators before they were stopped in the LYSO crystal. In between these two plastic scintillators (start counters) an additional scintillator with a hole was installed, and acted as a forward veto. The trigger condition of the DAQ system was as follows: A hit in both start counters but no hit in the forward veto. The four detector modules were arranged as a 2x2 cluster. Surrounding this cluster, four 6 mm plastic scintillators were installed and each of it was read out individually by four SiPMs mounted on the chamfered

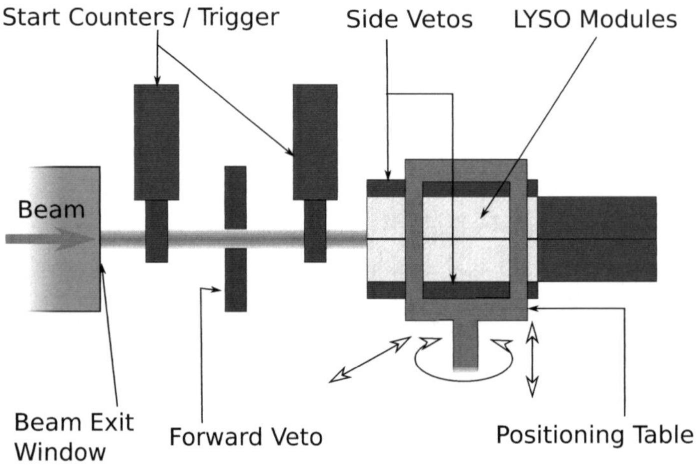

Figure 1. Schematic overview of the experimental setup.

edges. These scintillators were used as side vetos in the offline analysis to exclude events were a particle was escaping the LYSO crystal.

The positioning table was used to move the detector module cluster in the vertical and the horizontal direction as well as rotating it around the vertical axis. This allowed the simulation of different positions of the cluster in the final polarimeter as well as to target the deuteron beam to the center of the individual LYSO crystals. The positioning table was moved using stepper motors and designated drivers that were controlled by a RaspberryPi computer. A web-based user interface was used to set the positions of the cluster remotely as the experiment was not accessible when the beam was on. The position of the cluster was monitored using two webcams. Further, the web-based user interface allowed to set the high-voltage on the PMT based detector modules and switching on and off the side vetos.

The detector modules were built using LYSO crystals from two different manufacturers: Saint-Gobain [4] (Module 2 - 4) and Epic Crystals [5] (Module 1). Two different sizes of LYSO crystals were used: Module 1, 2 and 3 consisted of one LYSO crystal with a dimension of 30x30x100 mm. Module 4 was built of two LYSO crystals, each with a dimension of 15x30x100 mm.

To read out the LYSO crystals, two different types of sensors were utilized. The fist type of light sensing part (Module 1, 2 and 4), was made of a dual channel PMT [7], which was connected with the LYSO crystal by a light guide. This PMT was driven by a specially designed high-voltage divider. All components were enclosed by a steel housing and attached to the LYSO crystal using Kapton strips. These strips were strained by a spring tensioning device attached to the end of the light sensing part housing and were therefore pressing the light guide to the LYSO crystal. Module 4 consisting of two LYSO crystals had two individual light guides from each crystal to a channel of the PMT.

The second type of light sensing part (Module 3) was comprised with four SiPM arrays. Each array was made of 2x2 SiPMs, each with a side length of 6mm [8]. This SiPM array was connected to a passive circuit for the forward voltage supply and enclosed in a 3D printed ABS housing. As the PMT type, the LYSO crystal was connected to the SiPM array by Kapton strips and a spring tensioning device.

IOP Conf. Series: Journal of Physics: Conf. Series **928** (2017) 012019 doi:10.1088/1742-6596/928/1/012019

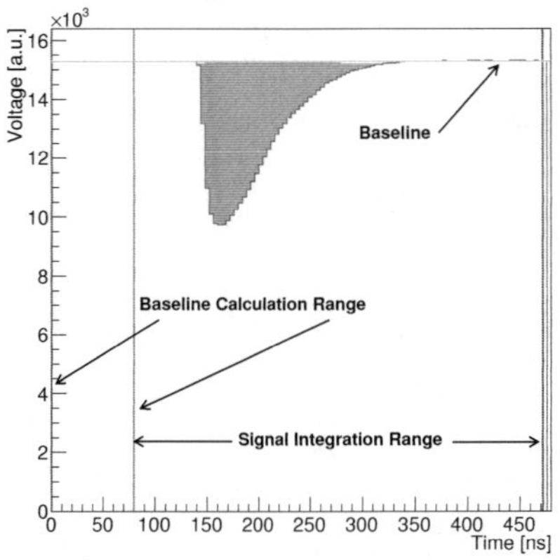

Figure 2. Valid signal shape. The shaded area is proportional to the deposited energy of the deuteron.

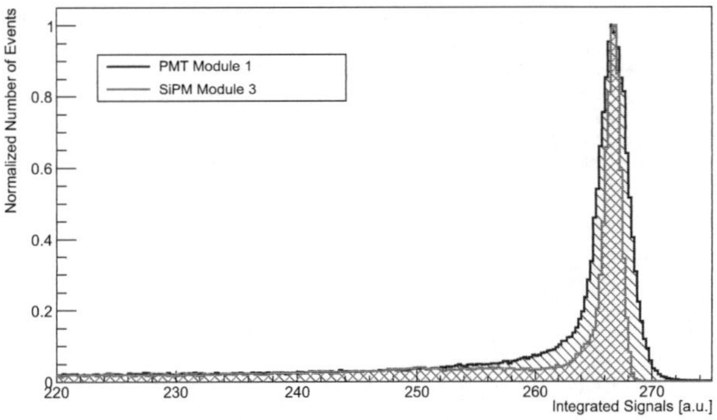

Figure 3. Comparison of the spectra at 270 MeV beam energy from Module 1 (blue) with a PMT and Module 3 (red) with a SiPM as light sensor.

3. Data Analysis

For the data taking, the deuteron beam was directed to the center of each detector module successively. The signals from each module were digitized by a 14 bit, 250 MS/s Flash ADC from Struck [6]. A baseline was calculated from the flat part before the actual signal shape (baseline calculation range, see Figure 2). To exclude misaligned baselines, a χ^2 test was performed on each baseline calculation. The signal shape was integrated (signal integration range, see Figure 2) and the baseline was subtracted from the integrated signal. To avoid events where a pile-up occurred, a peak count algorithm was implemented, and only events with one peak in the signal shape were preserved for the further analysis.

The integrated signals were filled into histograms to create spectra that shows the energy distribution (see Figure 3). A cut on the spectra of the start counters was applied to exclude head-on pile-ups where only one peak was visible in the signal shape. Additionally, a cut on the spectra of the side vetos and the neighboring modules was performed to exclude events in which a break-up reaction took place, or the deuteron was scattered out of the crystal.

To perform an energy calibration for each module, the peak position in the spectrum of the integrated signals was plotted against the beam energy. A GEANT4 simulation was used to correct the beam energy and take into account the energy losses in the start counters and the beam exit widow. This plot was fitted with a second order polynomial.

4. Preliminary Results

Two main results were obtained from the calibrated energy spectra. For each beam energy and module, the peak was fitted using a Gaussian. The energy resolution $R(E)$ was defined as $R(E)$ = FWHM(E) / $< E >$. The resolution for all beam energies and modules is shows in Figure 4 and was fitted using the standard formula for the resolution.

The second result that was obtained is the deuteron reconstruction efficiency. This is a very important quantity as the main task of the final polarimeter will be the identification of scattered deuterons. The deuteron reconstruction efficiency ϵ is defined as follows: $\epsilon = \int$ Signal Range / $\int$ Full Range. The expectation value of the fitted peak $< E >$ was defined as 100%. For the signal, a range from 90% up to $+6\ \sigma$ of the Gaussian was used. The asymmetrical range around the peak was taken due to the fact that the peak cannot be fully described by a Gaussian. The values of ϵ for all modules were plotted *vs.* the beam energy (see Figure 5) and fitted with an exponential decay function.

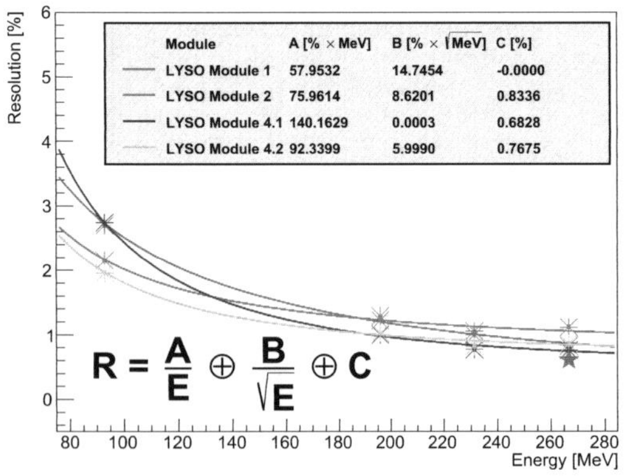

Figure 4. Energy resolution for all LYSO detector modules. The red star at 270 MeV denotes the resolution from the SiPM module 3.

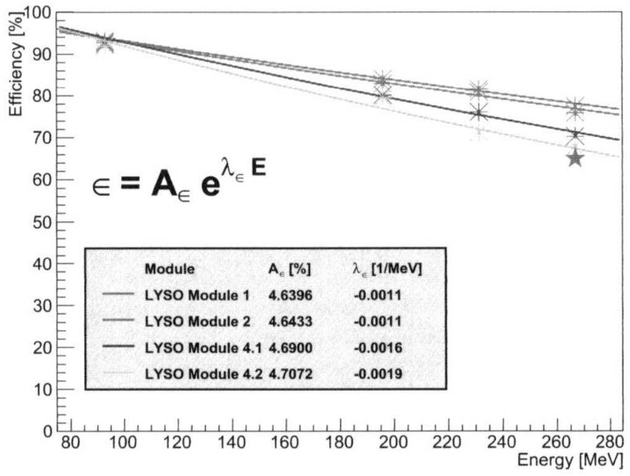

Figure 5. Deuteron reconstruction efficiency for all LYSO detector modules. The red star at 270 MeV denotes the measurement from the SiPM module 3.

5. Conclusion

With the experimental setup described above, we were able to obtain resolutions of 3% for low energies and even around 1% for the energy of interest of 270 MeV. These values are more than satisfactory for the needs of the final polarimeter. The deuteron reconstruction efficiency lies around 100% for low energies and decreases to around 70% (due to break-up reaction in the LYSO crystal) for higher energies, which still is a good value and sufficient for our needs.

Module 3 with a SiPM readout shows very good performance although we were only able to measure at the energy of interest of 270 MeV due to technical difficulties with the accelerator at the end of the beam-time. As the SiPM array was read out without a signal amplifier, a simple circuit consisting of only passive components can be used for it. Besides, a bias voltage of only 30 V was needed for this array. If the final polarimeter would be built of modules using SiPMs as a light sensor, there would be no need for a multichannel high-voltage supply.

In a more sophisticated next experiment, we will further investigate on the SiPM modules and directly measure elastically scattered deuterons from a carbon target [9].

References

[1] I. Keshelashvili, Proceeding of CALOR '16, Korea (2016).
[2] I Keshelashvili, Towards EDM Polarimetry, PoS(PSTP2015)026.
[3] Homepage of the JEDI collaboration.
[4] Saint-Gobain, Cerium doped Lutetium PreLude420 crystal.
[5] Epic-Crystal, LYSO Ce scintillator.
[6] Struck, FADC, SIS3316-250-14, 16 channel 250 MS/s 14-bit.
[7] PMT, Hamamatsu, R1548-07.
[8] SiPM, SensL, MicroFC-60035-SMT.
[9] I. Keshelashvili and B. Lorentz for the JEDI collaboration, Towards the EDM Polarimetry (E2.2).

IOP Conf. Series: Journal of Physics: Conf. Series **928** (2017) 012020 doi:10.1088/1742-6596/928/1/012020

Precision Timing with Silicon Sensors for Use in Calorimetry

A Bornheim[1], A Ronzhin[2], H Kim[3], G Bolla[2], C Pena[1], S Xie[1], A Apresyan[1], S Los[2], M Spiropulu[1], E Ramberg[2]
[1]Caltech, 1200 E. California Blvd., 91125 Pasadena, USA
[2]FNAL, Batavia, IL, USA
[3]University of Chicago, Chicago, IL, USA

E-mail: bornheim@hep.caltech.edu

Abstract. The high luminosity upgrade of the Large Hadron Collider (HL-LHC) at CERN is expected to provide instantaneous luminosities of $5 \times 10^{34} cm^{-2} s^{-1}$. The high luminosities expected at the HL-LHC will be accompanied by a factor of 5 to 10 more pileup compared with LHC conditions in 2015, causing general confusion for particle identification and event reconstruction. Precision timing allows to extend calorimetric measurements into such a high density environment by subtracting the energy deposits from pileup interactions. Calorimeters employing silicon as the active component have recently become a popular choice for the HL-LHC and future collider experiments which face very high radiation environments. We present studies of basic calorimetric and precision timing measurements using a prototype composed of tungsten absorber and silicon sensor as the active medium. We show that for the bulk of electromagnetic showers induced by electrons in the range of 20 GeV to 30 GeV, we can achieve time resolutions better than 25 ps per single pad sensor.

1. Introduction
To meet the challenges posed by current and future high energy colliders, particle physics detectors have to operate in ever increasing particle fluxes and at higher interaction rates. Aside from finer segmentation both for tracking devices as well as for calorimeters, precision timing capabilities are being viewed as a very powerful additional functionality to enhance the physics performance of the detectors. Timing resolutions of order few 10 ps allow to cleanly associate signals to primary interactions. A 10 ps resolution is also equivalent to a spatial resolution of a few mm. This allows tracking of photons with such a precision which is hugely important in hadron collisions which have a significant fraction of the total energy carried by neutral mesons decaying into photons. Precision timing calorimetry with scintillating crystals [1] and multi-channel plates [2] is well established. In this paper we present our measurements using silicon sensors as sensitive element in a precision timing calorimeter. Large scale silicon sensors are considered as an upgrade option for the HL-LHC [3].

2. Experimental Setup
For our measurements, we used a silicon sensor produced by Hamamatsu [4]. The thickness of the silicon was measured to be 325 μm. The transverse size of the sensor is $6 \times 6\,\text{mm}^2$. The negative bias voltage was applied to the p-side of the silicon. We observe that the silicon is

fully depleted above about 120 V. Timing measurements are expected to improve with larger bias voltage as the the carrier velocity increases. Attention was paid to provide good filtering for bias voltage, to reduce ground loop effects, and to minimize inductive loop for the signal readout. The timing characteristics of the signal pulses are dominated primarily by properties of the silicon sensor rather than the details of the circuit. The silicon diode was placed inside a light-tight box of thickness 1.5 cm, which also provides electromagnetic shielding. The box is made of 0.2 mm steel. The bias voltage was supplied to the circuitry by a cable with a balun filter, terminated with an high voltage connector. The silicon diode output signal is read out through an SMA connector electrically grounded to the box. The dark current was measured at several values of the bias voltage. The maximum value of the dark current was less than 1.0 nA at -500 V, which is the largest bias voltage used in the measurements reported in this paper. The signals from the silicon sensor were amplified by two fast, high-bandwidth pre-amplifiers connected in series. The first amplifier is an ORTEC VT120C pre-amplifier, and the second amplifier is a Hamamatsu C5594 amplifier. Using a pulse-generator, we measured the combined gain of the two amplifiers in series as a function of the input signal amplitude and found some degree of non-linearity for typical signals produced by the silicon sensor under study, and we corrected for them. Further details of the experimental setup can be found in [6]. As a reference timing sensor we use a Photek MCP-PMT 240. The data are recorded with a DRS [5] based ring sampler digitizer at at rate of 5 GS/s. Data was recorded at the FNAL M-test facility. Further details on the reference timing system and its performance can be found in [2]. The timing

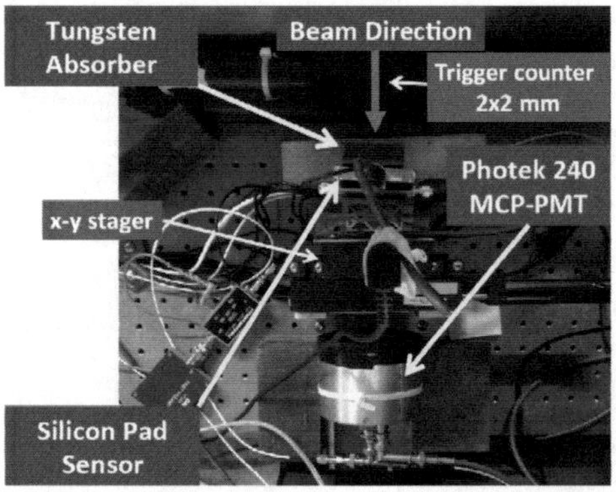

Figure 1. Experimental setup for the precision timing measurement of the Si pad sensor. The Si pad sensor is encapsulated in a metal box and placed as close as possible to the absorber. The beam impact point is defined with a size of $(2\text{mm})^2$. The beam is aligned transversely such that the response of the the Si pad sensor is maximized. The Photek reference counter is placed behind the Si pad sensor with absorber

information from the Si sensor is extracted from the sampled pulses by fitting a straight line to the rising edge of the pulse and determining the time at half the maximum amplitude. The timing information of the reference detector is extracted from a Gaussian fit to the peak of the pulses. A non-linearity from the amplifiers used is corrected in the data analysis. The quoted time resolutions are one sigma width of the differential timing between the reference sensor and the silicon pad sensor.

3. Experimental Results

Data were recorded at beam energies of 4, 8, 16 and 32 GeV with an electron beam, as well as with 120 GeV protons for calibration and alignment purposes. Showers are induced with a tungsten absorber of varying thickness to study the impact on the timing measurement. We have demonstrated in [2] that the temporal evolution of showers is very coherent. This is of crucial importance to exploit calorimeters as precision timing detectors. In Fig.: 2 we show the response of the Si pad sensor for the four beam energies we used. The measurement was performed after 6 radiation lengths of tungsten, approximately at the shower max. The mean value of the response is well correlated with the beam energy, demonstrating that even a single sensor features a calorimeter like response despite sampling only a very small part of the shower.

The spread of the distributions in Fig.: 2 illustrates the RMS spread of the response. Due to

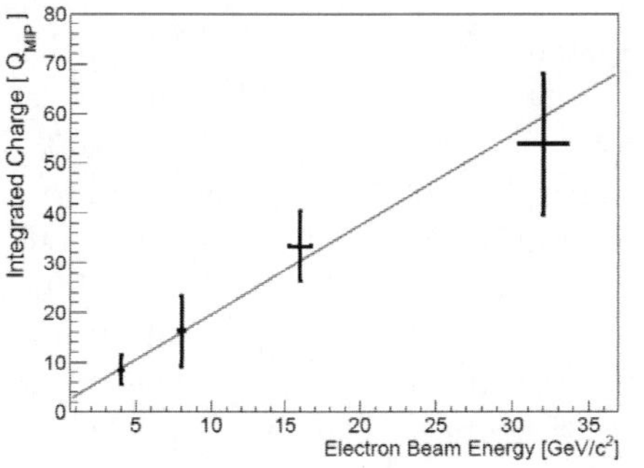

Figure 2. Mean signal amplitude and spread of the signal amplitude as measured with the Si pad sensor. The measurement was done for electron beams of 4, 8, 16 and 32 GeV. We find a good correlation between the beam energy and the mean signal amplitude. This demonstrates that despite the very limited containment of the Si sensor we have a clear pattern of a calorimetric measurement.

the limited containment of the single sensor the spread is considerable.

In Fig.: 3 we show the time resolution of the Si pad sensor measured beam energies of 4, 8, 16 and 32 GeV. The measurement was performed with an absorber thickness of 6 radiation lengths, approximately at shower max. The timing resolution improves with increased signal amplitude and reaches about 22 ps for beam energies of 32 GeV. The contribution of the reference timing sensor is not unfolded from this. It was determined to be less than 10 ps for an equivalent setup [2]. As we have shown in [2] one can improve the timing precision for electromagnetic showers

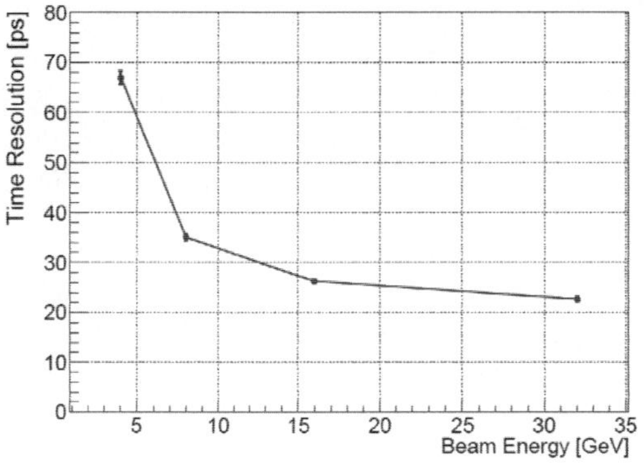

Figure 3. Time resolution of the Si pad sensor as a function of the incident beam energy. As shown in Fig.: 2 the signal amplitude measured in the Si pad sensor correlates well with the beam energy. We see an improved timing resolution as a function of the signal amplitude, reaching about 22 ps for an electron beam with an energy of 32 GeV

by performing multiple measurements on a single shower. Si pads such as the one we use can easily be arranged dense enough to perform 10 or more such measurements on a single shower. This is expected to further improve the performance. Such studies are being carried out and will be reported in the future. In Fig.: 4 we show the same data as used in Fig.: 3. However here the data is more finely binned in signal amplitude. The data from the four different beam energies is shown in different colors. We find that events with the same signal amplitude but different beam energy show the same time resolution. As we see from Fig.: 2 the signal amplitude in the Si pad fluctuates widely. Due to the small size of the pad the local shower fluctuations are substantial. This demonstrates that the time resolution is dominated by the signal amplitude in the sensor. Local shower fluctuations do not impact the timing measurement. This is further evidence to the very coherent temporal evolution of electromagnetic showers as we discussed in [1, 2]

4. Summary

We have measured the timing precision of a single silicon pad sensor in an electromagnetic shower. We find that the timing precision improves with the signal amplitude, reaching 22 ps on average measured with 32 GeV electrons after 6 radiation lengths of tungsten absorber. For events in the tail of the signal amplitude distribution we measure down to 16 ps. This suggests

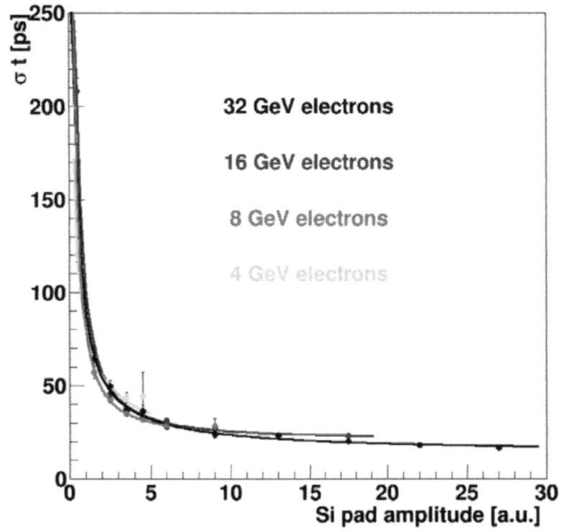

Figure 4. Time resolution as a function of the amplitude in the Si pad sensor. The data from the four different beam energies are separated by different colors. Due to the large spread of the signal amplitude the data from different beam energies overlaps. We find that the timing resolution scales only with the signal amplitude. Local shower fluctuations do not affect the timing measurements at the precision of our measurement.

that for higher energies the timing precision would improve further. We conclude that a silicon sampling calorimeter would allow to provide very precise timing information for electromagnetic showers.

4.1. Acknowledgments

Operated by Fermi Research Alliance, LLC under Contract no. DE-AC02-07CH11359 with the United States Department of Energy. Supported by funding from California Institute of Technology High Energy Physics under Contract DE-SC0011925 with the United States Department of Energy. We thank the FTBF personnel for very good beam conditions during our test beam time. We also appreciate the technical support of the Fermilab SiDet department for the production of high quality silicon samples.

References

[1] LYSO based precision timing calorimeters, A. Bornheim, CALOR2016, these proceedings
[2] Precision Timing with shower maximum detectors based on pixelated micro-channel plates, A. Bornheim, CALOR2016, these proceedings
[3] HGCAL: A High-Granularity Calorimeter for the Endcaps of CMS at HL-LHC, C. Ochando, CMS Collaboration, CALOR 2016, these proceedings
[4] www.hamamatsu.com
[5] https://www.psi.ch/drs/
[6] Test Beam Studies Of Silicon Timing for Use in Calorimetry, A. Apresyan, G. Bolla, A. Bornheim, H. Kim, S. Los, C. Pena, E. Ramberg, A. Ronzhin, M. Spiropulu, S. Xie,NIM-A, doi:10.1016/j.nima.2016.04.031
[7] A. A. Derevshchikov, V. Yu. Khodyrev, V.I. Kryshkin, V.E. Rakhmatov, A. I. Ronzhin, On possibility to make a new type of calorimeter: radiation resistant and fast. Preprint IFVE 90-99, Protvino, Russia, 1990.
[8] A 5 ps TOF-counter with an MCPPMT, K. Inami, , N. Kishimoto, Y. Enari, M. Nagamine, T. Ohshima, NIM-A, doi:10.1016/j.nima.2015.05.029
[9] Development of a new fast shower maximum detector based on microchannel plates photomultipliers (MCP-PMT) as an active element, A. Ronzhin, S. Los, E. Ramberg, M. Spiropulu, A. Apresyan, S. Xie, H. Kim, A. Zatserklyaniy, NIM-A, doi.org/10.1016/j.nima.2014.05.039
[10] Study of the timing performance of micro-channel plate photomultiplier for use as an active layer in a shower maximum detector, A. Ronzhin, S. Los, E. Ramberg, A. Apresyan, S. Xie, M. Spiropulu, H. Kim, NIM-A, dx.doi.org/10.1016/j.nima.2015.06.006
[11] Direct tests of micro channel plates as the active element of a new shower maximum detector, A. Ronzhin , S. Los, E. Ramberg, A. Apresyan, S. Xie, M. Spiropulu, H. Kim, NIM-A, dx.doi.org/10.1016/j.nima.2015.05.029
[12] Calorimeters for Precision Timing Measurements in High Energy Physics, A. Bornheim, CALOR2014, doi:10.1088/1742-6596/587/1/012057
[13] On timing properties of LYSO-based calorimeters, D. Anderson, A. Apresyan, A. Bornheim, J. Duarte, C. Pena, A. Ronzhin, M. Spiropulu, J. Trevor, S. Xie, NIM-A, doi:10.1016/j.nima.2015.04.013, 2015

Electromagnetic calorimeter of the Belle II detector

B.Shwartz, on behalf of BELLE II calorimeter group
Budker Institute of Nuclear Physics Novosibirsk 630090, Russia
Novosibirsk State University, Novosibirsk 630090, Russia

E-mail: shwartz@inp.nsk.su

Abstract.
At present new SuperKEKB collider is under commissioning at KEK (Japan) while the Belle II detector for experiments at this collider is at the final stage of the construction. The luminosity of this collider will exceed the previous one by about 40 times, amounting to $8 \times 10^{35}\mathrm{cm}^{-2}\mathrm{s}^{-1}$. The electromagnetic calorimeter is a very important component of the BELLE II detector. This calorimeter is described in this report.

1. Introduction

An energy-asymmetric e^+e^- collider, SuperKEKB, is under commissioning now at KEK (Japan) [1]. The luminosity of this collider will exceed the previously achieved by KEKB [2] by about 40 times, amounting to $8 \times 10^{35}\mathrm{cm}^{-2}\mathrm{s}^{-1}$. New experiments with the Belle II detector in the center-of-mass (CMS) energy range within Υ-meson family (9 – 11.5 GeV) will continue and widen the studies that began by the previous experiments. However, high luminosity which requires high circulation beam currents is unavoidably accompanied by the high event rate and background. Then the detector should be drastically upgraded. Schematic view of the Belle II detector is presented in Fig. 1.

The Belle II detector contains new vertex detector, central drift chamber and particle identification system. The K_L-meson and muon identification subsystem (KLM) is partially upgraded. The ECL scintillation crystals and mechanical structure of the electromagnetic crystal calorimeter (ECL) is kept unchanged from the previous experiment. The calorimeter is based on CsI(Tl) crystals which provide highest energy resolution for medium energy photons and widely used in the experiments on particle physics [3, 4, 5].

The first priority of the ECL upgrade is the calorimeter electronics modification following the general strategy of the Belle upgrade. The main idea is to shorten the shaping time from 1 μsec to 0.5 μsec and to use a pipe-line readout with waveform analysis. An option for later stage of the replacement of the part or all CsI(Tl) crystals in the end caps to pure CsI crystals with much shorter scintillation decay time is under development.

2. Design and properties of the Belle II electromagnetic calorimeter (ECL)

The electromagnetic calorimeter includes of a 3 m long barrel section with an inner radius of 1.25 m and the annular endcaps at $z = 2.0$ m (Forward part) and $z = -1.0$ m (Backward part) from the interaction point. The calorimeter covers the polar angle region of $12.4^\circ < \theta < 155.1^\circ$ except two gaps $\sim 1^\circ$ wide between the barrel and endcaps. The barrel part has a tower structure projected to the vicinity of the interaction point. It contains 6624 CsI(Tl) elements of 29 types.

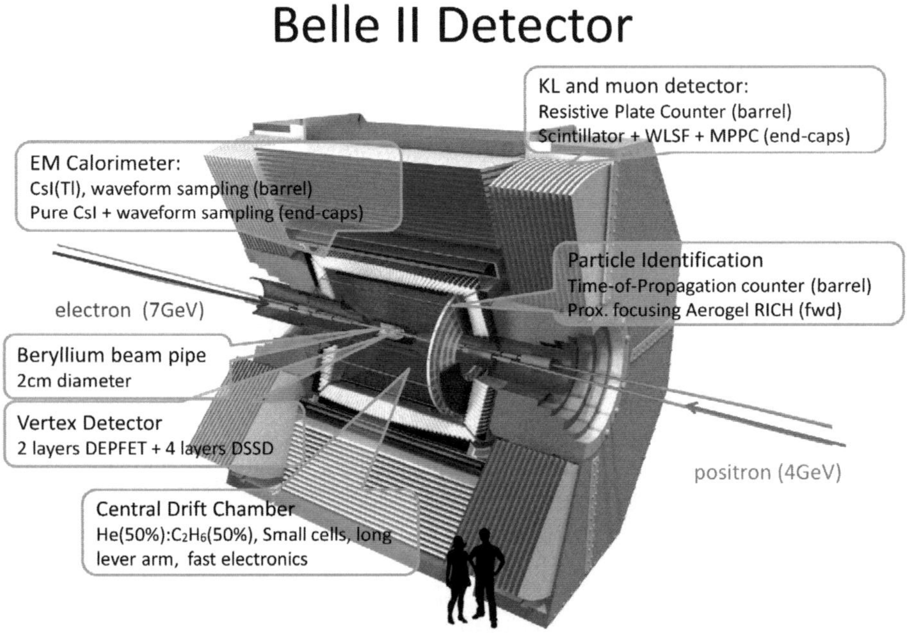

Figure 1. Schematic view of the Belle II detector.

Each crystal is a truncated pyramid of the average size about 6×6 cm^2 in cross section and 30 cm ($16.2X_0$) in length. The endcaps contain altogether 2112 CsI crystals of 69 types. The total number of the crystals is 8736 with a total mass of about 43 tons.

Each crystal is wrapped with a layer of 200 μm thick Gore-Tex porous teflon and covered by the 50 μm thick aluminized polyethylene. For light readout two $10 \times 20\,\text{mm}^2$ Hamamatsu S2744-08 photodiodes are glued to the rear surface of the crystal and connected to the attached preamplifiers. The average output signal of the crystal measured in a calibration run with cosmic rays is about 5000 photoelectrons per 1 MeV while the noise level is equal to about 200 keV. The energy resolution for photons, σ_E/E, slightly changed from about 2.5% at 100 MeV to 1.7% at 5 GeV. Detail description of the Belle ECL and its performance in the previous experiments can be found in [6].

3. ECL performance in a high background environment

Electron and positron beam currents in the SuperKEKB collider will have to be up to 3–4 A to achieve the project luminosity. This inevitably results in much higher background in comparison with previous experiment.

An obvious concern is about a radiation induced degradation of the parameters of scintillation crystals. The main effect of the degradation is caused by low energy gamma flux. The measured integrated dose absorbed by Belle ECL crystals at integrated luminosity 1000 fb^{-1} is about 100 rad for the barrel part and about 500 rad in the most loaded part of the endcaps. The light output of the crystals is around 7% in the barrel and up to 13% in the endcap parts close to the beam pipe. These results are in a good agreement with previous measurements of the crystal radiation hardness [7]. Since these studies showed the loss of the light output to be less or about 30% at 3.6 krad, an increase of the absorbed dose by one order of magnitude will not provide serious problems. According to simulation the absorbed dose at CsI(Tl) crystals

IOP Conf. Series: Journal of Physics: Conf. Series **928** (2017) 012021 doi:10.1088/1742-6596/928/1/012021

at Belle II will be below 500 rad per year or less than 5 krad for five years. A simulation of the radiation background at Belle II detector shows the neutron component with the flux up to $2 \times 10^{10}/(\text{year}\cdot\text{cm}^2)$ in the forward endcap. Such a background does not induce considerable effect for crystals but causes an increase of the dark current of photodiodes. However, the dark current induced by the expected neutron flux will be still below 1 μA and corresponding noise contribution should be below 1 MeV, still much smaller than pile-up noise.

High-energy photon part of the background appears as random clusters in the calorimeter which can overlap with the triggered events. At the Belle experiment each event contained in average 6 background clusters (3 in barrel and 3 in the both endcaps) of the energy exceeding 20 MeV. We expect that the background cluster rate at Belle II will not be higher due to much better time resolution of the upgraded ECL electronics.

Soft background photons with average energy of about 1 MeV and high rate overlap with signal pulses producing so called pile-up noise. The fluctuation of the number of these photons coming during the integration time contributes to the total noise level. According to the background simulation the pile-up noise level at the Belle II ECL will be from 3 to 8 MeV even with upgraded electronics which deteriorates considerably the energy resolution for the photons of the energy below 300 MeV.

4. Readout electronics of the Belle II calorimeter

The ECL readout electronics channel includes Shaper-Digital Signal Processor (Shaper-DSP) and Collector modules. Shaper-DSP boards contain 16 channels per module. Each channel consists of a shaping amplifier CR-(RC)^4 ($\tau_s = 0.5\mu$s) and 18-bit flash ADC (Analog Devices AD7641) which digitizes the signal with 2 MHz clock frequency. The ADC data are read out by the digital processor realized in the XILINX FPGA. Initiated by the trigger pulse, 16 points within the signal are fit to the signal shape function $F(t)$: $F(t) = A_0 \times f(t - t_0)$ where A_0 is a pulse height and t_0 is an event time.

The data from the Shaper-DSP, pulse height and pulse time, are sent to the Collector module which packs the data of 8-12 Shaper-DSP modules and sends this via ROCKET-I/O line to COPPER FINNESSE module of Belle II DAQ. The collector module contains also a test pulse generator for the calibration of the response of each channel. Each Shaper-DSP module generates also a fast analog pulse which is a sum of gain-corrected fast shaped ($\tau_d = 0.2\,\mu$s) signals from all 16 channels corresponding to 4x4 crystals Trigger Cell (TC). Detail description of the new ECL electronics can be found in [8, 1].

By now all 576 Shaper-DSP modules as well as all Collectors have been produced and tested. All tested modules showed the proper characteristics. All barrel counters are connected to DAQ and tested with cosmic rays.

5. Development of the pure CsI option

As mentioned earlier even after ECL electronics upgrade the pile-up noise expectation is up to 8 MeV in the backward endcap. The drastic way to improve ECL characteristics is a replacement of slow CsI(Tl) crystals with faster ones, at least in the end cap parts of the calorimeter. Then the natural option is to use pure CsI crystals of the same shape and size as the presently used CsI(Tl) counters. The advantages of this crystal are a short scintillation decay time, 30ns, and moderate cost in comparison to other fast scintillation crystals. Since physical properties of the pure CsI are the same as CsI(Tl) the present sizes of the calorimeter element as well as the mechanical structure can be kept. However, the light output of pure CsI crystals is approximately ten times lower than that of doped crystals. Then one needs to use photo detectors with internal gain. Proposed solution is to use vacuum photopenthodes (PP), i.e. PM tubes with three dynodes. Since pure CsI emission wavelength is about 300 nm the photosensor should be UV sensitive. Such a device of 2-inch diameter with a low output capacity C≈10 pF has been recently developed by the Hamamatsu Photonics. The low capacity is important since the noise level depends on this value. The new developed PP have a quantum efficiency of about 20-25% and internal gain factor of 120-200 at zero magnetic field which reduces by factor 3.5 times for the B=1.5 T

In the tests with the counter we obtained the signal of about $2 \cdot 10^4$ electrons per 1 MeV (without magnetic field) and the noise of about 900 electrons. These results can be extrapolated to the 1.5 T magnetic field as 6000 e/MeV and 200 keV of the noise energy equivalent. The beam test with 20 pure CsI crystals was carried out at BINP photon beam [9]. Measured energy resolution ($\sigma_E/E \approx$2.5% in the energy range 70-150 MeV) is consistent with the CsI(Tl) ECL energy resolution [10] and MC predictions. The time resolution better than 1 ns for energy more than 20 MeV has been obtained. The alternative method of the scintillation light readout using silicon avalanche photodiodes (APD) is under study as well. Recently very promising results were obtained [11]: the electronics noise level was measured better than 0.5 MeV when a readout with wavelength shifter bar and four APDs were used.

We performed a study of radiation hardness of the full size pure CsI crystals produced in Kharkov (Ukraina) [12, 13]. It was found that all the crystals which satisfy the requirements on the light output and fast/total ratio lose less than 20% of the light output after irradiation by the 14.3 krad of absorbed dose. This applied dose is well above the expectation for the Belle II after ten years of running.

6. Conclusion and Acknowledgments

- New electronics modules of the Belle II electromagnetic calorimeter are produced and tested. The barrel calorimeter is completely connected to the electronics and tested with cosmic rays. All channels shows proper parameters.
- R&D works on the endcap calorimeter based on pure CsI counters with modified electronics is going on. This modification should provide drastic suppression both pile-up noise as fake clusters rate in the endcaps.

I would like to thank organizers for inviting me to this very interesting and well organized Conference.

We acknowledge support from the Ministry of Education and Science of the Russian Federation. Work on the measurements of the crystal characteristics was supported by the Russian Science Foundation (project N 14-50-00080).

References

[1] T. Abe et al., Belle II technical design report, KEK Report 2010-1 (2010).
[2] S. Kurokawa et al., *Nucl. Instr. and Meth.* A**499** (2003) 1.
[3] B.G. Cheon et al., *Nucl. Instr. and Meth.* A**494** (2002) 548.
[4] BaBar Collaboration (B. Aubert et al.) *Nucl. Instr. and Meth.* A**479** (2002) 1.
[5] V.M. Aulchenko et al., *JINST* **10** (2015) P10006.
[6] A. Abashian et al., 2002 *Nucl. Instr. and Meth.* A**479** (2002) 117.
[7] K. Kazui et al., *Nucl. Inst. and Meth.* A**394** (1997) 46
D.M Beylin et al., *Nucl. Inst. and Meth.* A**541** (2005) 501.
[8] A. Kuzmin, Proceedings of the "Calorimetry for High Energy Frontiers - CHEF 2013", (2013) 83.
[9] B. Shwartz *Nucl. Inst. and Meth.* A**598** (2009) 220.
[10] H. Ikeda et al., *Nucl. Instrum. and Meth.* A**441** (2000) 401.
[11] Y. Jin et al., *Nucl. Instrum. and Meth.* A**441** (2016) 691.
[12] I.V. Bedny et al., *Nucl. Inst. and Meth.* A**598** (2009) 273.
[13] A. Boyarintsev et al., *JINST* **11** (2016) P03013.

Status of the electromagnetic calorimeter trigger system at Belle II.

SungHyun Kim[1]*, InSoo Lee[1], Yuuji Unno[1], ByungGu Cheon[1]

[1] Department of Physics, Hanyang University, Seoul, 133-791, Korea

E-mail: shkim@hep.hanyang.ac.kr

Abstract. The Belle II experiment at SuperKEKB collider in Japan has been under the construction toward physics run in 2017 with 40 times higher instantaneous luminosity than the KEKB collider. The main physics goal of the Belle II is to search for the New Physics in rare B decays and τ lepton decays. We have developed an Electromagnetic Calorimeter hardware trigger system that generates a trigger signal by two main triggers, total energy and cluster counting. Analog signals from 16 CsI(Tl) crystals are merged at ShaperDSP, and which are sent to FAM consisting of Flash-ADC and FPGA. ETM receives 52 FAM data from TMM and makes a ECL trigger signal which is sent to GRL and GDL, then the final Belle II trigger signal is generated by GDL which analyzes all trigger signals from each Belle II detector. In this article, the overall design scheme and current status of the ECL trigger system is reported.

1. Introduction

The Belle experiment [1] at KEK in Japan took 772 million pairs of B- and $\overline{B}$-mesons produced by the KEKB accelerator [2] to study CP violations in the neutral B-meson system. Most of the results are in good agreement with the Standard Model (SM) predictions of the Cabibbo-Kobayashi-Maskawa (CKM) structure of quark flavor mixing and CP violation in B decay [3], D-$\overline{D}$ mixing [4] and so on. The SuperKEKB accelerator [5] and the Belle II experiment are expected to find new physics beyond SM in flavor physics. Figure 1 shows the schematic of the Belle II detector.

The total cross sections and trigger rate of physical processes at the target luminosity of $8 \times 10^{35} cm^{-2}s^{-1}$, which is 40 times higher than the peak value of the KEKB collider (Table 1), are listed in Table 2. Since the high trigger rate for Bhabha and $\gamma\gamma$ at this luminosity, a 1/100 pre-scale factor is applied to these triggers. The level 1 trigger requirements of the Belle II experiment are shown in Table 3.

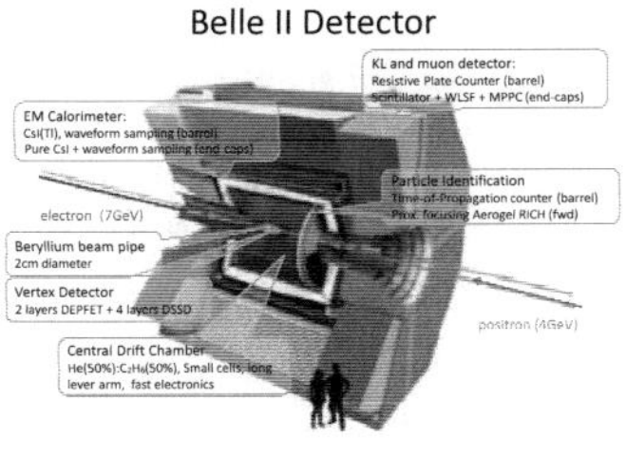

Figure 1: The Belle II detector

Parameters (e^+ / e^-)	KEKB (Achieved)	SuperKEKB (Expected)
Energy (GeV)	3.5 / 8.0	4.0 / 7.0
β_y^*	5.5 / 5.9	0.27 / 0.30
β_x^*	120 / 120	3.2 / 2.5
I(A)	1.64 / 1.19	3.60 / 2.62
Luminosity ($10^{34} cm^{-2}s^{-1}$)	2.11	80

Table 1: Fundamental parameters of the KEKB and SuperKEKB accelerators.

Physics process	Cross section (nb)	Rate (Hz)
$e^+e^- \to \Upsilon(4S) \to B\overline{B}$	1.1	880
$e^+e^- \to$ continuum	3.4	2700
$e^+e^- \to \mu^+\mu^-$	1.1	880
$e^+e^- \to \tau^+\tau^-$	0.9	720
$e^+e^- \to$ Bhabha [a]	44	350 [c]
$e^+e^- \to \gamma\gamma$ [a]	2.4	19 [c]
$e^+e^- \to e^+e^- + 2\gamma$ [b]	13	10000 [d]
Total	66	~15000

[a] $\theta_{lab} \geq 17°$
[b] $\theta_{lab} \geq 17°$, $p_t \geq 0.1\mathrm{GeV}/c$
[c] The rate is pre-scaled by a factor of 1/100
[d] The rate is estimated by the luminosity component in Belle L1 trigger rate

Table 2: Total cross section and trigger rates for $L = 8 \times 10^{35}$ cm^{-2}s^{-1} from possible physics processes at the $\Upsilon(4S)$ energy region.

Trigger efficiency ($\Upsilon(4S)$)	100%
Max. Trigger Rate	30kHz
Trigger Latency	$< 5\ \mu$s
Timing precision	< 10 ns
Min. event separation	200 ns

Table 3: Level 1 trigger requirements of Belle II.

2. Belle II Trigger System

The schematic overview of the Belle II trigger system is shown in Figure 2. It consists of four sub-trigger systems : (1) Charged track information such as momentum, position, and charge from CDC trigger. (2) Energy deposit information, energy cluster information, and Bhabha identification from ECL trigger. (3) Precise timing and hit topology information from Barrel PID (BPID) trigger. (4) μ track information from KLM trigger. The global decision logic (GDL) receives all sub-trigger information and makes the final trigger.

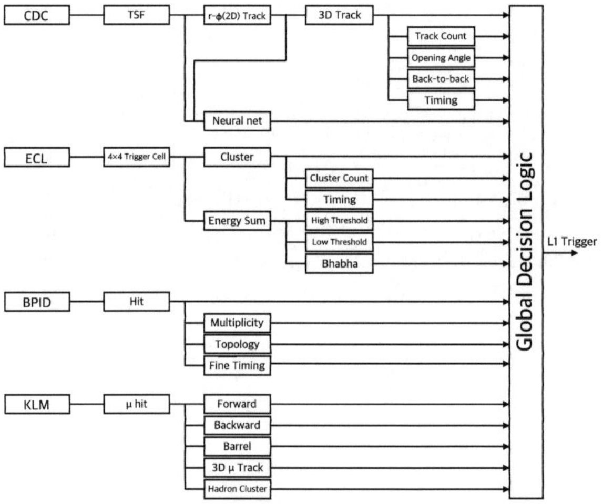

Figure 2: The Belle II trigger system schematic. GDL, where the final online trigger decision is performed, receives information from four sub-trigger.

3. ECL Trigger System

Since π^0 or other particles produce photons in a wide energy range, a high resolution electromagnetic calorimeter (ECL) is a very important part of the Belle II detector for a precise

IOP Conf. Series: Journal of Physics: Conf. Series **928** (2017) 012022 doi:10.1088/1742-6596/928/1/012022

energy measurement of photon and trigger signal generation. ECL, which contains 8763 CsI(Tl) crystals in total, consists of barrel, forward and backward endcap sections.

The basic framework and idea are same as Belle trigger system [1][6]. The different point between Belle and Belle II is that Belle II must deal with a much higher trigger rate due to high luminosity and beam background. To make the ECL trigger performance more robust and flexible, we adopted a new trigger scheme that utilizes a readout electronics architecture with Flash-ADC (FADC) and FPGA components and serial data transmission.

The block diagram of ECL trigger readout electronics system is shown in Figure 3. We develop four components of the trigger electronics : a fast shaper circuit built in the digital signal processing shaper module (ShaperDSP), a FADC analysis module (FAM), a trigger merger module (TMM), and an ECL trigger master (ETM).

The data of electromagnetic energy deposition in CsI(Tl) crystals counter is sent to ShaperDSP through photo-diodes (PD) attached to one CsI(Tl) crystal and pre-amplifier. 16 fast-shaping signals from neighboring 4×4 crystals are merged in the fast shaper circuit. This is implemented on the main shaper board to form an analog trigger sum called the trigger cell (TC). TC is a basic unit in ECL trigger system and 576 TCs used in total. The analog TC informations are transmitted to FAM.

On FAM, as shown in Figure 4, the analog TC data is digitized by FADC and the digitized data are passed to FPGA(XC7K160T). In order to measure energy and decide timing, Belle experiment used the *simple method.* While for Belle II, we prepare a wave form analysis in FPGA, so called *fit method.* The *simple method* also used for debugging. The *fit method* is applied to 12 data points at every 8 MHz clock and performs the following chi-square fit :

$$\chi^2 = \sum_{i,j} \{y_i - Af(t_i - \delta t - t_0) - P\} S_{ij}^{-1} \{y_j - Af(t_j - \delta t - t_0) - P\}$$

where $y_{i(j)}$ is data point and S_{ij} is noise covariance matrix for $i(j)$th sampled data. f denote fit function which consists of pedestal with first 4 points and signal with last 8 points, and P is pedestal. By performing chi-square fit, we measure A and t_0 which represent reconstructed signal amplitude and signal start timing on minimum δt, respectively. In the firmware logic, A, t_0 and P are computed on FPGA.

Table 4 shows TC energy and timing resolutions for two methods estimated by a ECL trigger simulator (TSim-ECL) which include the expected beam background in Belle II. According to this result, timing resolution improvement is performed by *fit method.*

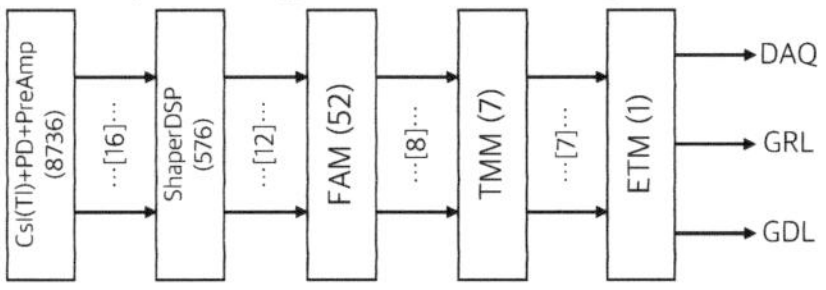

Figure 3: Hardware configuration of the ECL trigger system. The number in a parentheses(bracket) shows the number of set of "CsI(Tl) + PD + PreAmp" and ECL trigger electronics boards (the number of inputs to the one board).

Method	Input E (MeV)	Resolution (RMS)	
		E (MeV)	T (ns)
simple	100 - 200	2.7	35.52
	1000 - 1100	2.8	10.15
	8000 - 8100	2.8	4.83
fit	100 - 200	2.2	5.52
	1000 - 1100	2.2	0.61
	8000 - 8100	3.7	0.25

Table 4: TC energy and timing resolution comparison between *simple* and *fit method* for different input energies.

All FAM send the hitted TC energy and timing informations to TMM by GTX serial link. TMM, as shown in Figure 5, merge all TC informations by FPGA(XC7K325T) and send it to ETM which is shown in Figure 6. All FAM and TMM board mass production were done.

By receiving all TC energy and timing information from TMM, ETM generate two trigger signals on FPGA (XC6VHX380T) : physics trigger and Bhabha trigger. Basic strategy for physics trigger is same as Belle. Physics trigger condition is as following : (1) total energy trigger (E_{tot}), which is a total energy sum in barrel and forward-endcap excluding most inner layer, greater than 1 GeV. (2) the number of isolated clusters (ICN), which is to count the number of tracks that deposit the energy in ECL, is bigger than 3. If one event satisfies the condition (1) or (2) without Bhabha trigger condition, ETM generate physics trigger signal. In order to suppress Bhabha process in Belle, 2-Dim(r-θ) back-to-back topology is required for the two clusters together with an energy threshold for the sum of the two clusters. However, Belle II trigger system require an improved Bhabha veto logic because of 40 times higher luminosity, so 3-Dim(r-θ-ϕ) back-to-back topology will be used for Bhabha veto trigger. And we can separate charged and neutral tracks by performing ECL and CDC trigger matching at GDL state.

Figure 4: Photograph of the FAM board.

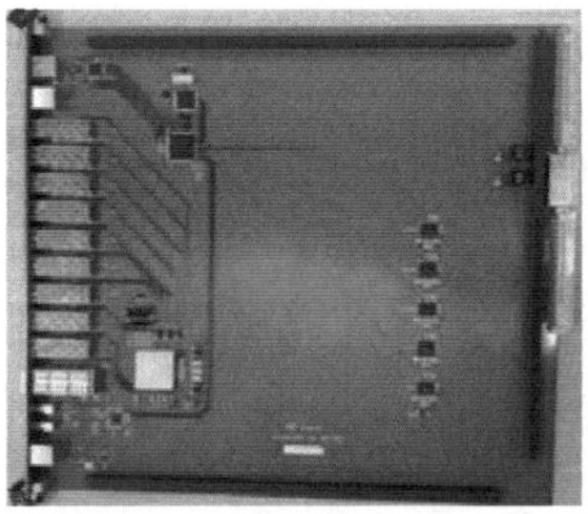

Figure 5: Photograph of the TMM board.

Figure 6: Photograph of the ETM board.

4. Conclusion

The SuperKEKB collider and the Belle II detector will probe for new physics phenomena in rare B meson, D meson and τ lepton decays. The online trigger system is required for reducing high trigger rates from beam backgrounds during data taking from the super KEKB collider to the Belle II detector. The ECL trigger should be robust and flexible to satisfy all trigger system requirements in high trigger rate environment. All ECL trigger electronics, FAM, TMM, and ETM, mass production and test were done. In this article, we have described the status of ECL trigger system.

References

[1] A. Abashian *et al.*(Belle Collaboration), 2002 Nucl. Instr. Meth. A, **479** 117
[2] S. Kurokawa, E. Kiktani, 2003 Nucl. Instr. Meth. A, **499** 1
[3] K. Abe *et al.*(Belle Collaboration), 2005 Phys. Rev. D, **71** 072003
[4] M. Starič *et al.*(Belle Collaboration), 2007 Phys. Rev. Lett., **98** 211803
[5] Y. Funakoshi, 2006 Beam Dynam. Newslett, **40** 27
[6] B.G. Cheon *et al.*, 2002 Nucl. Instr. Meth. A, **494** 548

IOP Conf. Series: Journal of Physics: Conf. Series **928** (2017) 012023 doi:10.1088/1742-6596/928/1/012023

LYSO based precision timing calorimeters

A Bornheim[1], A Apresyan[1], A Ronzhin[2], S Xie[1], J Duarte[1], M Spiropulu[1], J Trevor[1], D Anderson[1], C Pena[1] and M H Hassanshahi[3]

[1]Caltech, 1200 E. California Blvd., 91125 Pasadena, USA

[2]FNAL, Kirk Road and Pine Street, 60510 Batavia, IL, USA

[3]Sharif University, Azadi Avenue, Tehran, Iran

E-mail: bornheim@hep.caltech.edu

Abstract. In this report we outline the study of the development of calorimeter detectors using bright scintillating crystals. We discuss how timing information with a precision of a few tens of pico seconds and below can significantly improve the reconstruction of the physics events under challenging high pileup conditions to be faced at the High-Luminosity LHC or a future hadron collider. The particular challenge in measuring the time of arrival of a high energy photon lies in the stochastic component of the distance of initial conversion and the size of the electromagnetic shower. We present studies and measurements from test beams for calorimeter based timing measurements to explore the ultimate timing precision achievable for high energy photons of 10 GeV and above. We focus on techniques to measure the timing with a high precision in association with the energy of the photon. We present test-beam studies and results on the timing performance and characterization of the time resolution of LYSO-based calorimeters. We demonstrate time resolution of 30 ps is achievable for a particular design.

1. Introduction

High energy particle collider experiments are facing ever more challenging conditions, operating at todays accelerators, such as the Large Hadron Collider (LHC), capable of providing instantaneous luminosities of $10^{34}\,\mathrm{cm}^{-2}\mathrm{s}^{-1}$ and above. The high center of mass energy, the large number of simultaneous collisions of beam particles in the experiments and the very high repetition rates of the collision events pose huge challenges. Current upgrade plans for the LHC, the High Luminosity LHC (HL-LHC), future energy upgrades of the LHC or next generation hadron colliders such as the Future Circular Collider (FCC) will push the boundaries even further. These operational conditions result in extremely high particle fluxes, causing very high occupancies in the particle physics detectors operating at these machines. A precise timing information with a precision of around 10 ps and below is seen as a major aid in the reconstruction of the physics events under such challenging conditions.

The rate of simultaneous interactions per bunch crossing (pileup) at the HL-LHC is projected to reach an average of 140 to 200. The large amount of pileup increases the likelihood of confusion in the reconstruction of events of interest, due to the contamination from particles produced in different pileup interactions. The ability to discriminate between jets produced in the events of interest, especially those associated with the vector boson fusion processes, and jets produced by pileup interactions, will be degraded, the missing transverse energy resolution will deteriorate, and several other physics objects performance metrics will suffer.

One way to mitigate pileup confusion effects, complementary to precision tracking methods, is to perform a time of arrival measurement associated with a particular layer of the calorimeter, allowing for a time assignment for both charged particles and photons. Such a measurement with a precision of about 20 to 30 ps, when unambiguously associated to the corresponding energy measurement, will significantly reduce the inclusion of pileup particles in the reconstruction of the event of interest given that the spread in collision time of pileup interactions is about 200 ps. The association of the time measurement to the energy measurement is crucial, leading to a prototype design that calls for the time and energy measurements to be performed in the same active detector element. We have demonstrated in the past that full absorption crystal calorimeters as well as crystal sampling calorimeters can achieve such a precise timing measurement [1, 2].

2. Experimental Setup

Our previous results on precision timing with LYSO based calorimeters [2] were obtained with multi-channel plate photo multiplier (MCP-PMT) as photo detectors. MCP-PMTs feature a signal rise time on the order of 100 ps. For laser pulses with a FWHM of 50 ps we achieved a differential timing resolution of 7 ps [3], corresponding to a single device timing resolution of 5 ps. This performance is limited by the DRS based digitizer we use as a DAQ system, operating at 5 GS/s. In the studies described in this paper we use silicone multi pixel photon counters (SiPM). While they feature a slower rise time they allow a very good timing performance for large, coherent signals [4]. For our setup we use four different types of SiPMs with 10, 15 and 25 μm pixel size and 1×1 mm and 3×3 mm sensor size [5]. They are all read out with a DRS based digitizer through a clipping circuit as described in [4]. We do not amplify the output signal of the SiPMs, only exploiting the very large light yield of the LYSO scintillator and the intrinsic amplification of the SiPMs. Using the same fast laser pulses and DAQ system we obtain a differential timing resolution between two SiPMs of better than 10 ps, very similar to the MCP-PMT performance.

The experimental setup we use for the calorimetric timing measurements consists of a single cell of a sampling calorimeter with 29 layers of LYSO crystal and tungsten absorber. The lateral dimensions are $14 \times 14\,\mathrm{mm}^2$. The total depth of the cell is about 11.5 cm with the LYSO plates having a thickness of 1.5 mm. The scintillation light from the LYSO plates is extracted with four wave length shifting (WLS) fibers. The same cell has been used to measure the timing performance in comparison to the timing performance of a single cube of LYSO [2]. More details of the calorimetric performance of the Shashlik configuration are discussed in references [6] and [7].

In addition to plastic WLS fibers we also tested quartz capillaries filled with liquid wave length shifter using DSB as a wave length shifting agent [8]. To optically couple the quartz capillaries to the SiPMs we use a clear plastic fiber light guide which is connected to the end of the quartz capillary with a metal sleave tube. The same clear fiber coupler is used for the plastic WLS fibers to maintain equivalent light collection efficiency. The ratio of the light collection efficiency between the plastic fibers and the quartz capillaries approximately scales with the ratio of the diameter of the plastic fiber and the liquid core of the quartz capillary. This ratio is about 3 for the fibers and capillaries we used.

As a timing reference we use a Photek 240 MCP-PMT. It is placed behind the calorimeter cell and detects secondary shower particles escaping from the Shashlik calorimeter cell as we did in our previous studies [2]. To extract the time of flight we measure the time difference between the reference counter and the calorimeter cell. The time stamp is extracted from a Gaussian fit to the peak of the reference counter pulse which resembles a Gaussian shape. For the time stamp from the calorimeter cell we perform a straight line fit to the rising edge of the pulse and extract the time at half of the maximal signal amplitude.

We measure the timing performance of the calorimeter cell with high energy electrons in a range between 20 GeV and 200 GeV in the CERN North Area test beam. The impact point of the electrons onto the calorimeter cell is measured with a fiber hodoscope with a precision of better than 1 mm. The timing results are extracted from events impacting in the center of the calorimeter cell in an area of 2x2 mm. As we are using a single calorimeter cell the containment outside this region is limited which affects the timing measurement.

3. Experimental Results

The timing precision extracted from a pulse improves as the rise time of the pulse gets shorter and as the signal over background improves. The rise time of the pulses from the shashlik calorimeter cell are driven by the time constants of the wave length shifter as we demonstrated previously [2]. The rise time of the LYSO scintillation signal and the rise time of the previously used MCP-PMT is much faster than the WLS time constants. As SiPMs feature a slower rise time than MCP-PMTs we studied the signal rise time with our new setup. In figure 1 we show the rise time of the pulses as a function of the signal amplitude. We observe that the the rise time

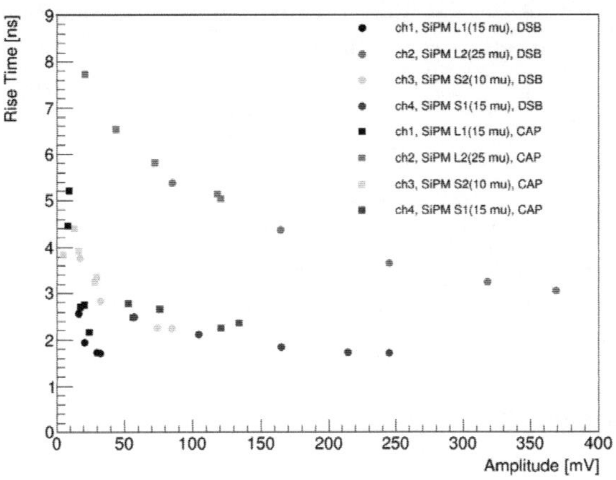

Figure 1. Rise time of the pulses as a function of the amplitude for the four different SiPMs. The data recorded with the plastic fibers and the quartz capillaries are distinguished as dots and squares. For each of the two data sets with plastic fibers and capillaries there are 5 measurements corresponding to beam energies of 20, 50, 100, 150 and 200 GeV.

of the pulses becomes shorter for larger pulses, reaching around 2 ns for the largest amplitudes in each channel. The amplitude dependency of the rise time is similar between different types of SiPMs but reaches is shortest value at different signal amplitudes. The bias voltage of the SiPMs is set to the same voltage, no individual optimization has been performed. The absolute signal amplitude in the SiPMs is different for the same beam energy which is expected as pixel count, sensor size and the light guide coupling efficiency varies. The clipping circuit used for all SiPMs is the same while the capacitance of the SiPMs is not. In figure 2 we show the time resolution of the four individual fibers as a function of their respective signal amplitude. Lines indicate the trend of the time resolution. We see a similar trend as for the rise times. However we also see that the improvement of the timing resolution is not just due to the faster rise times. The increased signal amplitude and the resulting improvement in the signal over background results in additional improvement.

For the same amplitude we find the same timing resolution with the capillaries and the plastic WLS fibers, both using DSB WLS agent. This further supports our previous findings that the the light propagation in the optical elements of a scintillation based calorimeter dos not significantly affect the performance at the level of 50 ps.

The light extraction efficiency of capillaries with liquid WLS remains sufficiently high for dose rates of 100 Mrad and beyond and for fluences of 10^{14}protons/cm^2 and beyond [7]. The energy

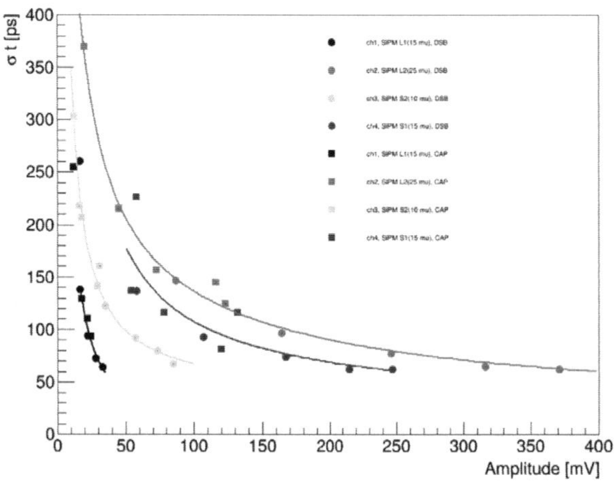

Figure 2. Time resolution achieved with the calorimeter cell using the signal of each of the SiPMs individually. The data for each SiPM consists of two sets, one with the DSB WLS plastic fiber and one with the capillaries with a liquid DSB based WLS. The time resolution improves with increasing signal amplitude. The best time resolution per fibre is around 60 ps for all of the channels. The amplitude at which this performance is achieved varies.

resolution performance of the LYSO/tungsten cell is not limited by photo statistics but the sampling fraction. The timing performance at lower energies could be improved by increasing the raw signal. This could be achieved by using larger diameter capillaries.

4. Summary

We present timing performance studies on a LYSO/tungsten Shashlik calorimeter cell with wave length shifting fiber and capillary readout with SiPMs as photo detectors and a DRS based fast digitizer data acquisition. We find a timing performance which reaches around 60 ps per individual readout fiber and better than 50 ps when combining the information from four fibers. Our results demonstrate that timing resolution of a few 10 ps can be achieved with scintillator based calorimeters even when complex optical readout schemes are employed. SiPMs are suitable as photo detectors for precision timing calorimeters.

4.1. Acknowledgments

Supported by funding from California Institute of Technology High Energy Physics under Contract DE-SC0011925 with the United States Department of Energy. We thank Randy Ruchti from the University of Notre Dame for providing the wave length shifting capillaries. We thank CERN for providing access to the test beam facilities and our colleagues from Rome and ETH for fruitful collaboration and help with the DAQ system.

References

[1] Calorimeters for Precision Timing Measurements in High Energy Physics, A. Bornheim, CALOR2014, doi:10.1088/1742-6596/587/1/012057
[2] On timing properties of LYSO-based calorimeters, D. Anderson, A. Apresyan, A. Bornheim, J. Duarte, C. Pena, A. Ronzhin, M. Spiropulu, J. Trevor, S. Xie, NIM-A, doi:10.1016/j.nima.2015.04.013, 2015
[3] Precision Timing Calorimetry for High Energy Physics, A. Bornheim et al, 13th Pisa meeting on Advanced Detectors, Elba 2015, NIM-A doi:10.1016/j.nima.2015.11.129
[4] Timing Performance of new Hamamatsu Silicon Photomultipliers, A. Mangu et al, IEEE2014
[5] www.hamamatsu.com
[6] Longevity of the CMS ECAL and Scintillator-Based Options for Electromagnetic Calorimetry at HL-LHC, H. Li, CMS Collaboration, CMS-CR-2015-127
[7] A Crystal Shashlik Electromagnetic Calorimeter for Future HEP Experiments, R. Zhu, CALOR2016, these proceedings
[8] Studies of Wavelength-Shifting Liquid Filled Quartz Capillaries for Use in a Proposed CMS Calorimeter, B. W. Baumbaugh, R. Ruchti, et al, IEEE/NSS2015, Conference Record.

Upgrade of Tile Calorimeter of the ATLAS Detector for the High Luminosity LHC.

Eduardo Valdes Santurio (On behalf of the ATLAS Tile Calorimeter System.)

Stockholm University Department of Physics, Stockholm, Sweden

E-mail: eduardo.valdes@cern.ch eduardo.valdes@fysik.su.se

Abstract.
The Tile Calorimeter (TileCal) is the hadronic calorimeter of ATLAS covering the central region of the ATLAS experiment. TileCal is a sampling calorimeter with steel as absorber and scintillators as active medium. The scintillators are read out by wavelength shifting fibers coupled to photomultiplier tubes (PMT). The analogue signals from the PMTs are amplified, shaped and digitized by sampling the signal every 25 ns. The High Luminosity Large Hadron Collider (HL-LHC) will have a peak luminosity of $5 \times 10^{34} cm^{-2}s^{-1}$, five times higher than the design luminosity of the LHC. TileCal will undergo a major replacement of its on- and off-detector electronics for the high luminosity programme of the LHC in 2026. The calorimeter signals will be digitized and sent directly to the off-detector electronics, where the signals are reconstructed and shipped to the first level of trigger at a rate of 40 MHz. This will provide a better precision of the calorimeter signals used by the trigger system and will allow the development of more complex trigger algorithms. Three different options are presently being investigated for the front-end electronic upgrade. Extensive test beam studies will determine which option will be selected. Field Programmable Gate Arrays (FPGAs) are extensively used for the logic functions of the off- and on-detector electronics. One hybrid demonstrator prototype module with the new calorimeter module electronics, but still compatible with the present system, may be inserted in ATLAS at the end of 2016.

1. Introduction

ATLAS (A Toroidal LHC ApparatuS) [1] is a general purpose detector designed to study proton-proton and heavy ions collisions at the Large Hadron Collider (LHC) at CERN (Geneva, Switzerland). The ATLAS hadronic Tile Calorimeter (TileCal) [2] is a sampling calorimeter composed of steel plates and approximately 460000 plastic scintillator tiles. It is divided longitudinally in four cylindrical sections; the two central "long barrel" segments (LBA, LBC) and shorter "extended barrel" segments (EBA, EBC) at each end (figure 1). Each cylindrical section is composed of 64 wedge-shaped modules, and the scintillators in each module are divided into individual cells. Light from the side of each cell is collected by wavelength shifting fibers, and read out by two photomultiplier tubes (PMTs) located in insertable "drawers" within the back girders of each module, along with the high-voltage and readout electronics. A total of 9852 PMTs are needed for the complete readout of TileCal (figure 1).

In order to maximize the physics potential of the LHC, the accelerator and detectors are being upgraded [4] for an instantaneous luminosity of around $5 \times 10^{34}\ cm^{-2}s^{-1}$ and a total integrated luminosity of 3000 fb^{-1} over ten years of running. This will present significant challenges for

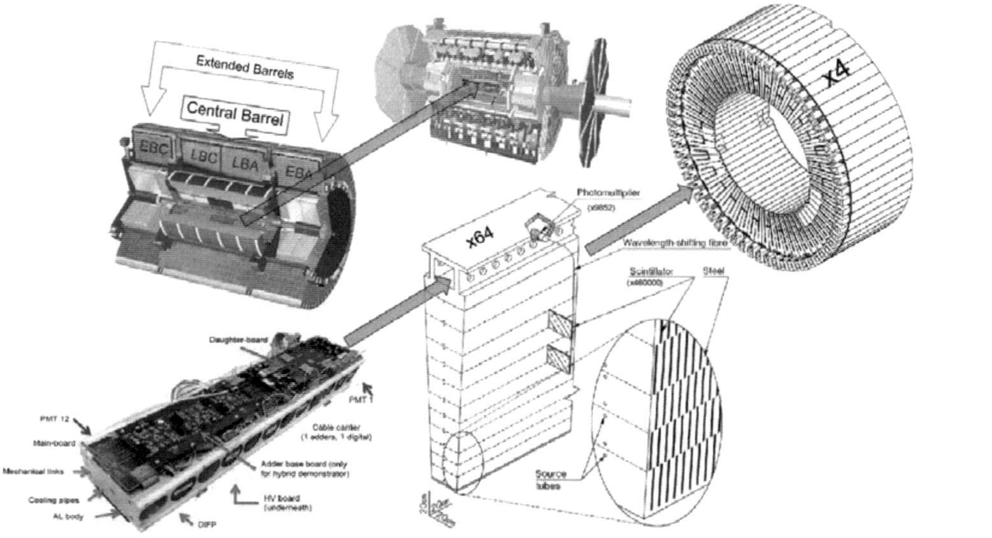

Figure 1. ATLAS detector, Tile Calorimeter subsystem mechanics, Demonstrator minidrawer mechanics.

ATLAS data taking, with increased pileup from close to 200 simultaneous events per bunch crossing. An upgrade program to replace the current TileCal on- and off-detector electronics is underway in order to meet the new requirements for high-luminosity running. The upgrade aims include improving the latency, achieving high readout rates, helping to improve event selection and coping with the higher radiation levels.

2. The demonstrator

The on-detector TileCal electronics [3] are installed along with the PMTs in one so-called “super drawer” per module. In the current system, each super drawer is mechanically divided into two “drawers” that function together as a single unit. Analog PMT signals are amplified and shaped in front-end boards and then sent to boards where the pulses are digitized and stored in pipeline buffers to await readout upon a level-1 trigger accept. In a separate signal path, projective “towers” of calorimeter cells are summed into single analog pulses that are sent to the Level-1 Calorimeter Trigger (L1Calo) for event selection. Data from events selected by the Level-1 trigger are read out over optical links by interface boards to the off-detector Read Out Drivers (ROD). Optical fibers from the Timing Trigger and Control (TTC) system provide each module with clocks, trigger signals, and control and configuration capabilities. The High Voltage (HV) system and Low Voltage Power Supply (LVPS) are controlled and monitored by the Detector Control System (DCS).

For Phase II the aim is to replace the current system with a more modularized one, in which each superdrawer (figure 2) comprises four independent “mini-drawers” for simplified access and service (figure 1). The signals from the PMTs will be shaped and amplified in the front-end boards (FEB) behind each PMT and digitized in the “main board” present in each mini-drawer. A “daughter board” will read out the sampled data over high-speed optical links to off-detector “Tile Preprocessor” boards. All cell data from every bunch crossing will be read out continuously, which will mean transistioning from the current 256 optical links and total data rate of 165 *Gbps* to 8192 optical links and approximately 80 *Tbps*.

To evaluate the upgrade concept in realistic conditions, a demonstrator [11] is being designed for an upgraded system that can be run in selected ATLAS modules while maintaining compatibility with the current system (figure 2) to avoid compromising data taking.

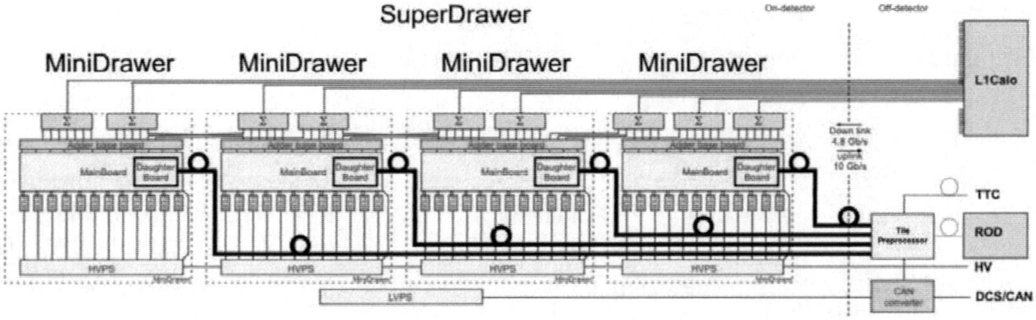

Figure 2. Superdrawer and demonstrator concept. In yellow new electronics, in green: adaptor boards for compatibility with the legacy system and in blue: legacy system electronics.

3. Low Voltage Power Supply and High Voltage Power Supply solutions

The Low Voltage Power Supply (LVPS) [5] system is located at the end of the modules in the outer side of the calorimeter and supplies a +10V power (figure 3) to the on-detector electronics. Each on-detector board then uses Point-of-Load regulators (POLs) to supply the locally needed voltages. For increased system reliability, each Mainboard (MB) and Daughterboard (DB) is functionally separated into independent halves. Each half processes the PMTs from one side of the module and additionally has a separate, redundant low voltage feed that is capable of powering both halves of the mini-drawer if one feed fails. In the case of one side failing completely, each of its cells is still read out by the other side.

Two different approaches are under study for the High Voltage Power Supply (HVPS). An on-detector solution [6] uses one HV board per minidrawer, that provides up to 12 channels of high-voltage control. This card is radiation hard and communicates with the off-detector side by slow control signals via the daughterboard. An alternative, remote solution [7] distributes the HV for each PMT from off-detector by routing 12 HV cables to each mini-drawer. Since the HV source is placed off detector, the radiation hardness of the electronic components for this solution is not a concern.

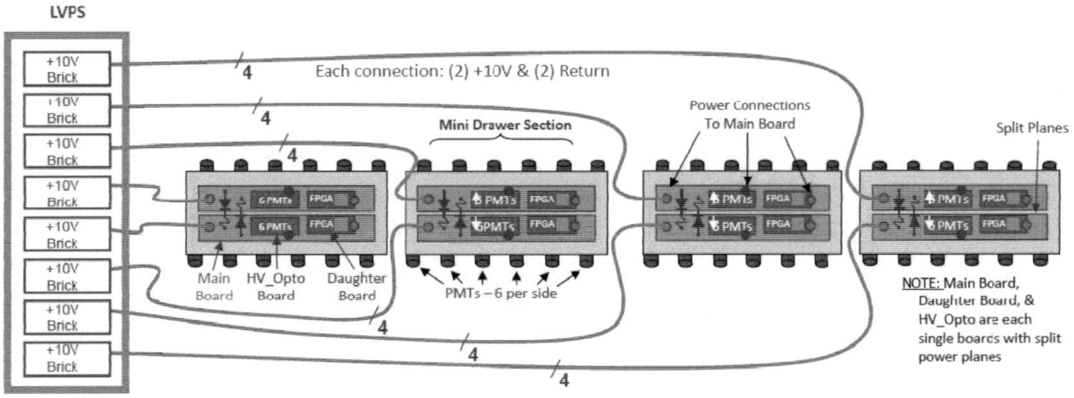

Figure 3. Power connectivity block diagram using a LVPS and the on-detector HV solution.

4. Front-end solutions

Three different solutions for the front-end boards (FEBs) are being evalulated, each with their own corresponding motherboard:

(i) the modified 3-in-1 card[8] which is a redesign of the legacy front-end boards, but with better linearity and lower noise

(ii) the FATALIC-ASIC solution [9] which includes ADCs in a front-end ASIC

(iii) the charge-integrating QIE ASIC solution [10]

The three alternatives have similar approaches based on "all in one cards" that hold different electronic designs for each solution. The modified 3-in-1 card and the FATALIC alternatives are pulse shaper solutions, based on simple electronic components and a complex ASIC design respectively; while the QIE design is based on a gated integrator ASIC. The present demonstrator system is based on the modified 3-in-1 solution, for its ability to provide compatibility with the legacy system.

The mainboard solutions provide the control, monitoring and readout of the FEBs, and deliver digitized data to the daughterboard. All three mainboards are designed with double redundancy in mind. The 3-in-1 mainboard prototype [12] is divided into halves, each half hosting six FEBs from one side of the module. Each half mainboard reads out and supplies low voltage to the FEBs, sends digitized signal and distributes power to its corresponding daughterboard, and sets gains and controls the charge injectors on each 3-in-1 channel.

All three solutions are being extensively studied in test beams and the results will help determinine which optimum solution will be selected for the insertion in ATLAS.

5. The Daughterboard

Control and communication between the front-end and off-detector electronics are performed by the daughterboard [13], which also has a double redundant design. The daughterboard (figure 4) is compatible with all three mainboard designs, the on-detector HV board and the Cs calibration system. It is also in charge of recovering the system clock and distributing it to the front-end system via the GBTx ASICs [15]. The main functionalities are programmed in dual Kintex 7 FPGAs. Current and temperature monitoring of both FPGAs and the mini-drawer environment are also performed. Communication is provided by a redundant pair of QSFP optical transceivers, each alllowing high speed communication at 4×9.6 *Gbps* uplink and 4×4.8 *Gbps* downlink.

6. The TilePreprocessor

The Tile Preprocessor (TilePPr) [16] is the main component of the back-end electronics. The current prototype for the demonstrator project provides compatibility between the current and the upgraded system. The TilePPr prototype (figure 4) can control and read out one complete TileCal super drawer (four mini-drawers). Its functionality includes receiving and processing digital data from the daughterboards, decoding and distributing Trigger Timing and Control (TTC) signals to the front-end electronics for configuration and synchronization with the LHC clock. It also provides control and monitoring of the high voltage power supplies through a Detector Control System (DCS) interface. Through the demonstrator phase and up to the Phase-II upgrade the TilePPr prototype will also serve as the interface between the legacy Read Out Driver (ROD) and the demonstrator, as well as a test bench for the development of new real time data processing algorithms for reconstruction of energy, timing and quality factors for the L0/L1 trigger in Phase II and as a test bench for the proposed Felix readout concept [14].

The TilePPr is a complex board with three different FPGAs for implementing all the needed functionalities. SFP modules are used for receiving TTC signals and communicating with the legacy system, while a minipod tx module provides ribbon fiber output links to the L0/L1 trigger.

IOP Conf. Series: Journal of Physics: Conf. Series **928** (2017) 012024 doi:10.1088/1742-6596/928/1/012024

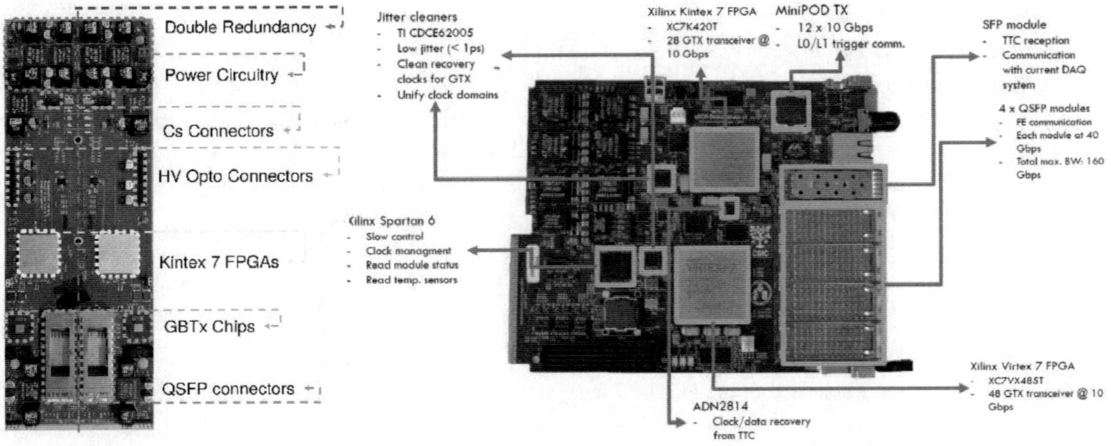

Figure 4. Left: Daughterboard revision 4. Right Tile Preprocessor Prototype.

Four QSFP modules provide communication links to the daughterboards. Another important design consideration is the low-jitter clocking circuitry for assuring clean clock domains for the FPGAs and high-speed optical links.

The first TilePPr prototypes have been designed, produced and tested as part of the TileCal Demonstrator project.

7. Radiation Tolerance

All parts of the front-end system need to be qualified for running in a radiation environment. Since the demonstrator is a project under constant development, a several parts are being or will soon be tested for verifying the radiation requirements of TileCal (figure 4).

The system has been designed to be radiation tolerant by using multiple redundancy on the electronic design [3], radiation tolerant components, Triple Mode Reduncancy (TMR) in the FPGAs, redundant links, cyclic redundancy check (CRC) data protection, and the use of the radiation-hard GBT ASIC [15] with forward error correction (FEC) for sending the downstream data off-detector.

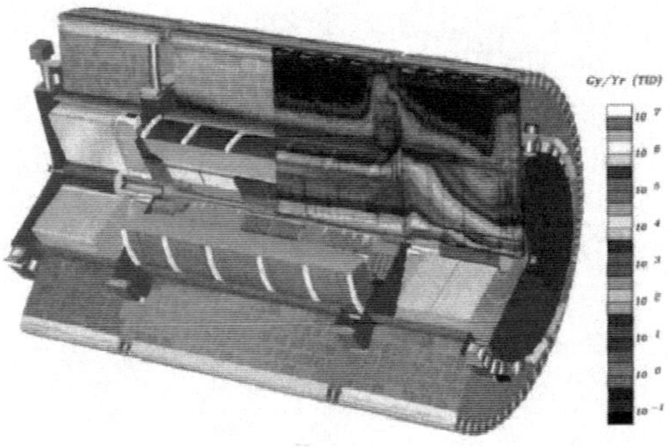

	TID	NIEL	SEE	
COTS regulators	Done - OK	Not done	Not done	Need different -5V
3-in-1	Preliminary	Pending	Preliminary	More testing needed
Main board	Not done	Pending	Not Done	Size limitations
Daughter board	Not done	Pending	Done – OK	More testing needed
Modulator	Done – OK	Pending	Done – OK	Good to go
HV_Opto	Done – OK	Done – OK	Done – OK	Good to go
LVPS	Done (v7.5)	Done (v7.5)	Done (v7.5)	Needs full testing
Adders	Not needed	Not needed	Not needed	Testing not needed
Active bases	Done	Done	Not needed	Good to go
FATALIC	Not done	Not done	Not done	Not started
QIE	Not done	Not done	Not done	Not started
Scintillators	Preliminary			More tests needed

Figure 5. Left: Radiation levels in ATLAS. Right: Ongoing work on the different radiation tests for the demonstrator. (TID: Total Ionizing Dose, NIEL: Non-Ionizing Energy Loss and SEE: Single Event Upsets)

8. Conclusions

The demonstrator project is still under development. Some testing remains to be done of both the electronics functionality and the radiation radiation hardness of the system. The strategies followed for the design of the system are meant to cover the possible difficulties that could arise given the high requirements to be met for Phase-II. The results of the 2016 testbeams will guide the final design of the hybrid demonstrator system to be installed in ATLAS.

References

[1] ATLAS Collaboration, *The ATLAS Experiment at the CERN Large Hadron Collider*, 2008 JINST 3 S08003.
[2] ATLAS Collaboration, Readiness of the ATLAS Tile Calorimeter for LHC collisions, Eur. Phys. J. C 70 (2010) 1193, [arXiv:1007.5423].
[3] ATLAS Tile Calorimeter Technical Design Report, arXiv:1007.5423CERN/LHCC 96-42
[4] ATLAS Collaboration, Letter of Intent for the Phase-II Upgrade of the ATLAS Experiment, CERN-LHCC-2012-022 (2012).
[5] G. Drake. Design of a new switching power supply for the ATLAS TileCAL front-end electronics. 2013 JINST 8 C02032.
[6] Atlas-tile-demonstrator, ATLAS TileCal demonstrator project, CERN HV OPTO Demonstrator webpage.
[7] F. Vazeille on behalf of the ATLAS Tile Calorimeter System, Performance of a remote High Voltage power supply for the Phase II upgrade of the ATLAS Tile Calorimeter. 2016 JINST 11 C02050
[8] F. Tang, K. Anderson, G. Drake, J.-F. Genat, M. Oreglia, J. Pilcher and L. Price. Design of the Front-End Readout Electronics for ATLAS Tile Calorimeter at the sLHC, IEEE TRANSACTIONS ON NUCLEAR SCIENCE, VOL. 60, NO. 2, APRIL 2013
[9] L. Royer. FATALIC: a dedicated front-end ASIC for the ATLAS TileCal upgrade, in Topical Workshop on Electronics for Particle Physics 2015, Lisbon Portugal, 28 Sep2 Oct 2015. ATL-TILECAL-SLIDE-2015-764
[10] G. Drake, A. Paramonov, J. Proudfoot, R. Stanek and S. Chekanov. QIE12: A New High-Performance ASIC for the ATLAS TileCal Upgrade, in Topical Workshop on Electronics for Particle Physics 2015, Lisbon Portugal, 28 Sep2 Oct 2015. ATL-TILECAL-SLIDE-2015-771
[11] Atlas-tile-demonstrator, ATLAS TileCal demonstrator project, CERN Demonstrator webpage.
[12] F. Tang, H. Åkerstedt, K. Anderson, C. Bohm, K. Hildebrand, S. Muschter and M. Oreglia. Design of Main Board for ATLAS TileCal Demonstrator, MB_ALT-COM-TILECAL-2014-023.
[13] S. Muschter. Readout link and control board for the ATLAS Tile Calorimeter upgrade. diva2:805818
[14] J. Anderson, A. Borga, H. Boterenbrood, H. Chen, K. Chen, G. Drake, D. Francis, B. Gorini, F. Lanni, G. Lehmann Miotto, L. Levinson, J. Narevicius, C. Plessl, A. Roich, S. Ryu, F. Schreuder, J. Schumacher, W. Vandelli, J. Vermeulen and J. Zhang. FELIX: a High-Throughput Network Approach for Interfacing to Front End Electronics for ATLAS Upgrades. doi:10.1088/1742-6596/664/8/082050
[15] S. Baron, J. P. Cachemiche, F. Marin, P. Moreira and C. Soos. Implementing the GBT data transmission protocol in FPGAs. CERN-2009-006
[16] F. Carrio, P. Moreno and A. Valero on behalf of the ATLAS Tile Calorimeter System, Performance of the Tile PreProcessor Demonstrator for the ATLAS Tile Calorimeter Phase II Upgrade 2016 JINST 11 C03047.
[17] H. Åkerstedt, S. Muschter, G. Drake, K. Anderson, C. Bohm, M. Oreglia and F. Tang. Reliable and Redundant FPGA Based Read-Out Design in the ATLAS TileCal Demonstrator. diva2:874222.

IOP Conf. Series: Journal of Physics: Conf. Series **928** (2017) 012025 doi:10.1088/1742-6596/928/1/012025

HGCAL: A High-Granularity Calorimeter for the Endcaps of CMS at HL-LHC

Christophe Ochando, on behalf of the CMS Collaboration

Laboratoire Leprince-Ringuet, UMR7638, Ecole Polytechnique 91128 Palaiseau Cedex, France

E-mail: ochando@cern.ch

Abstract. Calorimetry at the High Luminosity LHC (HL-LHC) faces two enormous challenges, particularly in the forward direction: radiation tolerance and unprecedented in-time event pileup. To meet these challenges, the CMS experiment has decided to construct a High Granularity Calorimeter (HGCAL), featuring a previously unrealized transverse and longitudinal segmentation, for both electromagnetic and hadronic compartments. This will facilitate particle-flow-type calorimetry, where the fine structure of showers can be measured and used to enhance particle identification, energy resolution and pileup rejection. The majority of the HGCAL will be based on robust and cost-effective hexagonal silicon sensors with about 1cm^2 or 0.5cm^2 hexagonal cell size, with the final 5 interaction lengths of the hadronic compartment being based on highly segmented plastic scintillator with on-scintillator SiPM readout. We present an overview of the HGCAL project, including the motivation, engineering design, readout concept and simulated performance.

1. Introduction

The discovery in 2012 of the Higgs boson by the ATLAS and CMS experiments at the LHC [1, 2] is a major breakthrough in the understanding of the fundamental interactions. The measurements of its properties are compatible, within large uncertainties, with the ones expected by the standard model (SM) predictions. However unrevealing the true nature of the electroweak symmetry breaking will require additional measurements: percent level precision on the couplings to fundamental fields, observation of rare decays ($H \to \mu\mu$, $H \to Z\gamma$, ...), di-Higgs production and constraints on the self-coupling, forbidden decays (e.g. $H \to e\mu$), unitarization of the scattering amplitudes of longitudinally polarized vector bosons, ... In addition, if new particles are found during the Phase-I data taking period of the LHC, aiming at accumulating 300 fb^{-1} by 2023, their properties will have to be determined precisely to disentangle between various models beyond SM. This physics program, in addition to precision studies in top or heavy flavour physics, provides a powerful demand on high luminosity.

The goal of the the so-called Phase-II of the LHC machine is thus to accumulate 3000 fb^{-1} after ten years of data taking, starting after the Long Shut-down 3 (LS3) following the end of Phase-I. The proposed operating scenario is to level the instantaneous luminosity at 5×10^{34}cm^{-2}s^{-1}, with potential peak value at 2×10^{35}cm^{-2}s^{-1}. Under these conditions, the additional interactions per bunch crossing, referred to as pile-up (PU) become a tremendous challenge for the experiments with a foreseen average of 140, possibly 200, per beam crossing. Moreover, the particles emerging from the collisions and the radioactivity they induce in the material of the detectors and the on-board electronics will cause significant damage. For instance,

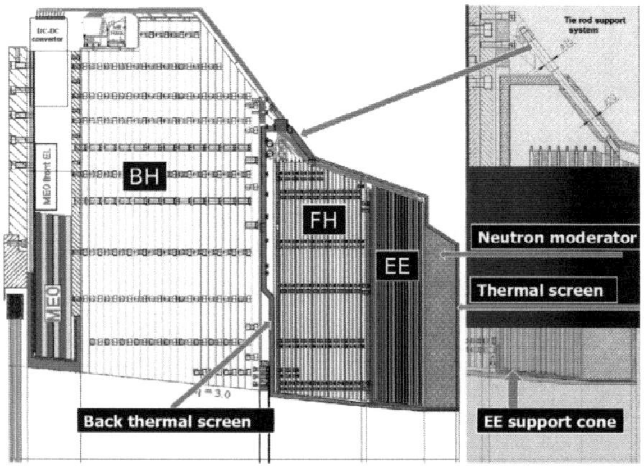

Figure 1. Technical drawing of the endcap calorimetry overall structure.

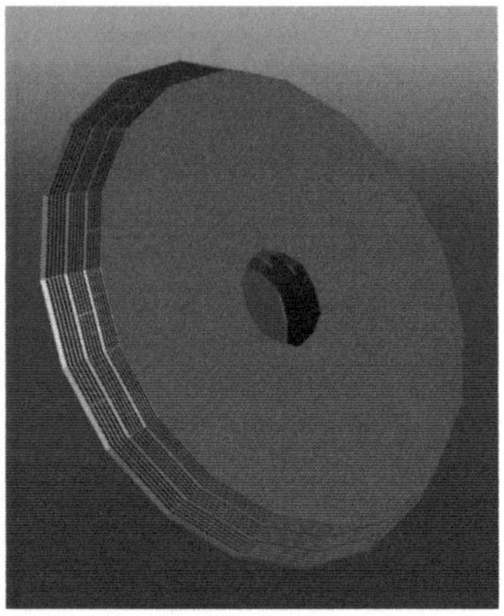

Figure 2. EE carbon-fibre structure integrating tungsten absorber plates alternating with empty slots.

it has been shown with FLUKA simulations that the endcap electromagnetic (resp. hadron) calorimeter will have to sustain up to 1.5 MGy (resp. 30 kGy) of integrated dose and a fluence of 10^{16}n/cm^2. The ageing and radiations studies [3] have demonstrated the need to replace the entire calorimetric system in the endcaps during LS3.

2. HGCAL Design Overview

The new system has to be radiation-tolerant and has capability to cope with the very high PU, especially at the trigger level, where one has to preserve low thresholds for the Higgs physics while exploring the TeV scale. To address these challenges, CMS proposes the replacement of the endcap calorimeters with a new high-granularity sampling calorimeter, covering the range $1.5 < \eta < 3.0$, referred as HGCAL. His concept is inherited from ILC/CALICE concepts for 3D measurement of shower topologies, adding as well very fast timing capabilities. Its electromagnetic section (EE) consists of 28 layers of tungsten and copper plates interleaved with silicon sensors as active material, for a total of 26 radiation lengths (X_0) and 1.5 interaction length (λ). The hadronic part (FH) has 12 layers of brass and copper plates interleaved with silicon sensors for a total of about 3.5 λ. HGCAL will be cooled down to -30 $^\circ$ C via evaporating CO2 system to mitigate leakage current in silicon sensors due to radiation damage. HGCAL will be complemented by a brass-plastic scintillator backing-hadron calorimeter (BH) adding 5 λ to ensure the full containment of showers. The overall structure can be seen in Fig. 1. Only a few key aspects are described in the following.

2.1. Sensors, Modules, Cassettes and Mechanics

Active elements in HGCAL are 8 inches (or pairs of 6 inches) hexagonal 320 μm-thick silicon sensors. Size of cells varies from 1.05cm^2 to 0.53cm^2 and Si active depletion thickness from 300 to 100 μm as the expected neutron fluence increases in order to maintain the ability to see minimum ionizing particle signal (MIP) with a decent signal-to-noise ratio. It is foreseen to read about six millions channels.

Sensors are mounted on printed circuit boards (PCB), with front-end chip bonded to it, and glued on the other face to a copper-tungsten baseplate to form a module (Fig. 3). The connection to the sensors cells will be made with wire-bonding through suitable holes in the PCB.

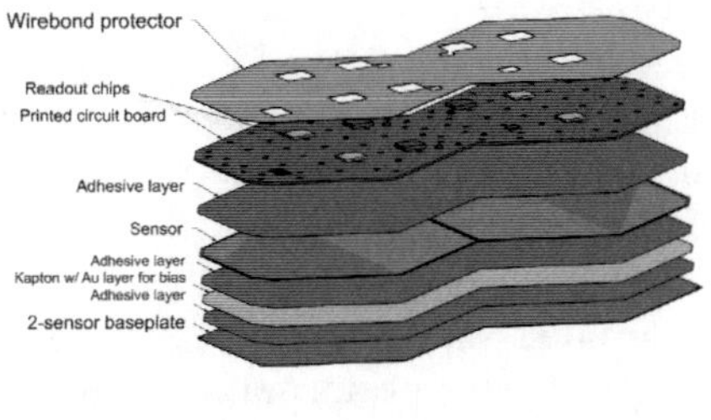

Figure 3. Module, consisting of printed circuit board, silicon sensors and baseplate.

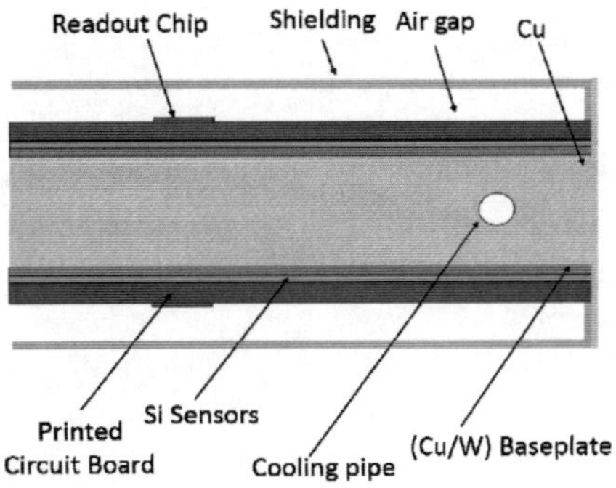

Figure 4. Sketch of a cassette with modules mounted on either side of the copper cooling plate.

Modules will be mounted on a 6mm-thick copper plate with embedded stainless steel pipes for cooling, thus making a 30 ° wedge "cassette" (Fig. 4). Services and power will be routed in a 2mm air-gap. Cassettes in EE (resp. FH) are double-sided (resp. single-sided).

Cassettes are finally inserted into the mechanical structures. The overall mechanical design of the EE is derived from the prototypes for the ILD [4], although adapted to the CMS needs. It is a carbon-fibre alveolar structure with embedded tungsten plates made from twelve 30 ° sectors. The geometry is designed so that the radial sector boundaries can be arranged to occur in three different orientations, each rotated by 10 ° with respect to the others, to mitigate the degradation of performance due to dead areas (Fig. 2). The overall mechanical structure of the FH follows the one of the current CMS endcap hadron calorimeter, with bolted plates of brass, machined to host the cassettes.

2.2. Front-End Electronics

One of the most challenging aspect of this project is the front-end chip as large dynamic range (up to 10 pC, corresponding to about 3000 MIP in 300 μm sensors) is needed together with low noise and low power (about 10 mW per channel).

The baseline architecture consists of a pre-amplifier and shaper DC-coupled to the sensors and a time-over-threshold (ToT) measurement with a TDC for digitization of large signals, while a standard charge readout with a 10-bits ADC for pulses below 100 fC. The timing precision per cell is expected to be 50 ps.

Developments on various aspects of this chip are on-going. The production of a test-chip, SKIROC2_CMS, derived from SKIROC2 [5], was launched in January 2016 and is currently under tests. It includes some of the features specifics to HGCAL: 20ns shaping time, sampling at 40 MHz, p-on-n and n-on-p readout options, ADC and ToT, ... These functionalities will be tested in test beams foreseen in Fall 2016 at CERN. In the mean time, test vehicles on specific blocks are launched (various flavour of pre-amplifiers, shapes, discriminators...). A first iteration of the full chip is expected by Spring 2017.

2.3. Expected Performance

The expected performance of HGCAL were first evaluated with a standalone GEANT4 simulation, benchmarked against CALICE published test-beam results. Figure 5 shows the electron energy resolution as a function of the electron energy for various active thickness of the silicon sensors. The stochastic term ranges from 20 to 24 % but the constant term is targeted to be low (1 %).

IOP Conf. Series: Journal of Physics: Conf. Series **928** (2017) 012025 doi:10.1088/1742-6596/928/1/012025

The Figure 6 shows the energy containment of electromagnetic showers as a function of the EE layer number. Although it is shown the Molire radius (R_M) is 28mm, one can notice that the shower radius is much smaller than R_M in the first layers. Thus, longitudinal segmentation together with high granularity give additional power handles for particle identification or energy reconstruction. It is possible to exploit the difference in the start of shower between electrons and hadrons, restore projectivity, make 3-dimensional fit of shower profile, subtract PU layer-by-layer...

The HGCAL geometry was implemented into a fully detailed simulation of CMS, including the new Phase-II tracker. It was demonstrated that similar or better performance than Phase-I configuration with 50 PU could be achieved with Phase-II simulations with 140 PU. Note that not all the capabilities of HGCAL were exploited, e.g. timing, thus leaving room for future improvements.

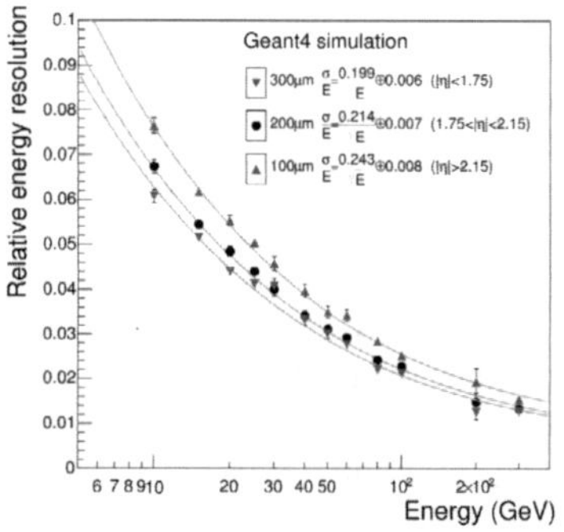

Figure 5. Energy resolution as a function of energy from a standalone simulation of incident electrons on HGCAL, for different thickness of silicon sensors.

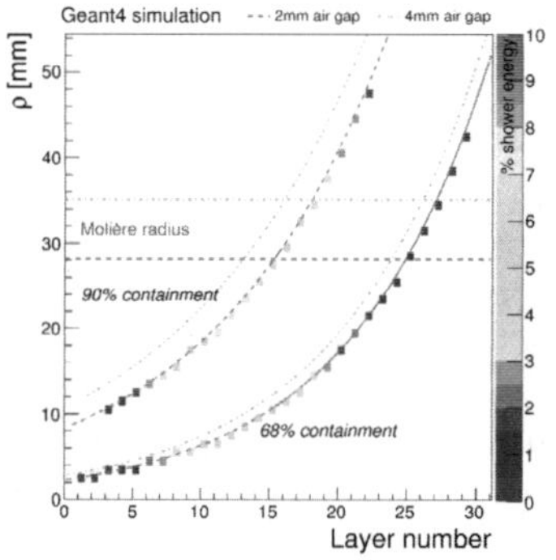

Figure 6. Radii ρ of the energy deposited in an individual silicon layer by a shower, layer by layer.

3. Conclusion and Perspective

The High-Granularity calorimeter is on the critical path towards physics discoveries and measurement in the High Luminosity Phase of the LHC. It has all ingredients to sustain the high radiation field, deal with the high rates of particles and mitigate pile-up interactions. However, major and exciting challenges must be faced for the next decade in all areas: engineering, including the routing of thousands of power and signal cables in the very constraint space of the existing detector, front-end electronics and trigger capabilities, software and computing.

HGCAL is now in R&D phase. Much progress has been done since the release of the Technical Proposal. First prototypes are being tested with electron and proton beams. A Technical Design Report is foreseen in late 2017. It will include key technical choices and improved design. The construction should start in 2020 in order to be ready for an installation during LS3.

References

[1] G. Aad *et al.* [ATLAS Collaboration], *Phys. Lett. B* **716** (2012) 1
[2] S. Chatrchyan *et al.* [CMS Collaboration], *Phys. Lett. B* **716** (2012) 30
[3] CMS Collaboration, CERN-LHCC-2015-010, CMS-TDR-15-02 (2015)
[4] T. Behnke *et al.* [ILD Collaboration], "ILC Reference Design Report Volume 4 - Detectors,"
[5] S. Callier *et al.*, *JINST* **6** (2011) C12040.

IOP Conf. Series: Journal of Physics: Conf. Series **928** (2017) 012026 doi:10.1088/1742-6596/928/1/012026

Concepts and design of the CMS high granularity calorimeter Level-1 trigger

Jean-Baptiste Sauvan, on behalf of the CMS Collaboration
CERN, CH-1211 Geneva 23, Switzerland
E-mail: jean-baptiste.sauvan@cern.ch

Abstract. The CMS experiment has chosen a novel high granularity calorimeter for the forward region as part of its planned upgrade for the high luminosity LHC. The calorimeter will have a fine segmentation in both the transverse and longitudinal directions and will be the first such calorimeter specifically optimised for particle flow reconstruction to operate at a colliding beam experiment. The high granularity results in around six million readout channels in total and so presents a significant challenge in terms of data manipulation and processing for the trigger; the trigger data volumes will be an order of magnitude above those currently handled at CMS. In addition, the high luminosity will result in an average of 140 to 200 interactions per bunch crossing, giving a huge background rate in the forward region that needs to be efficiently reduced by the trigger algorithms. Efficient data reduction and reconstruction algorithms making use of the fine segmentation of the detector have been simulated and evaluated. They provide an increase of the trigger rates with the luminosity significantly smaller than would be expected with the current trigger system.

1. Introduction

The Compact Muon Solenoid (CMS) detector [1] has been designed to study proton–proton and heavy ion collisions produced by the LHC. Only a few hundreds of events per second can currently be recorded offline and a trigger system is used to select events of interest [2]. The trigger is organised in two stages: the Level-1 (L1) hardware-based trigger reduces the rate to 100 kHz, and the High-Level Trigger (HLT), based on a farm of computers, reduces the rate to 300 to 600 Hz. This system has been performing extremely well during the Phase 0 (from 2010 to 2012) and the beginning of the Phase 1 (since 2015) of the LHC, even under the highest luminosity conditions achieved. During the high-luminosity phase of the LHC (HL-LHC, or Phase 2) that will start around 2025, the instantaneous luminosity will be 3 to 4 times higher than during Phase 1, with an average number of simultaneous interactions per bunch crossing (pile-up) up to 200. For the desired energy thresholds the current trigger system would give a L1 rate of at least 1500 kHz, while only 100 kHz are available. To avoid a significant increase in the energy thresholds, which would be detrimental for physics, an upgrade of the L1 trigger system is required [3]. The future high granularity calorimeter (HGCal) of CMS [3] will provide new inputs to this upgraded trigger system with detailed information for the reconstruction and selection of objects of interest.

In the following, the conceptual design of the HGCal L1 trigger architecture and the associated data flow are discussed. Concepts of algorithms for the reconstruction and identification of electromagnetic objects and jets are described and their expected performance are shown.

2. The HGCal trigger architecture and data processing

Though the precise architecture of the system is not yet known, the first part of the data processing will be performed inside on-detector ASICs in order to reduce the amount of data to be sent off-detector. The second part of the data processing related to the reconstruction algorithms, such as energy clustering and pile-up estimation, will be done in off-detector FPGAs. This off-detector processing will be shared between several layers connected together via high-speed optical links. The result will be the production of HGCal trigger primitives, which will be used, together with primitives from other sub-detectors, to reconstruct higher-level objects such as electrons, photons, tau leptons and jets.

In order to fit within the available bandwidth for transferring trigger data out of the detector, a reduction factor of about 20 in data size is needed. This reduction factor is obtained in several ways:

- The dynamic range and resolution of the measured energies are reduced and the timing information is discarded.
- The HGCal sensor cells are grouped into larger trigger cells.
- Only the most energetic trigger cells are selected and sent off-detector.

The last two of these steps are illustrated in Figure 1. The reduced data will be sent off-detector via a mixture of optical and electrical links. Optical links at 10 Gb/s will be used in the low pseudorapidity region where radiation hardness requirements are less stringent. Electrical links at 5 Gb/s will be used in the high pseudorapidity region, with electrical to optical conversion located in the back of the calorimeters.

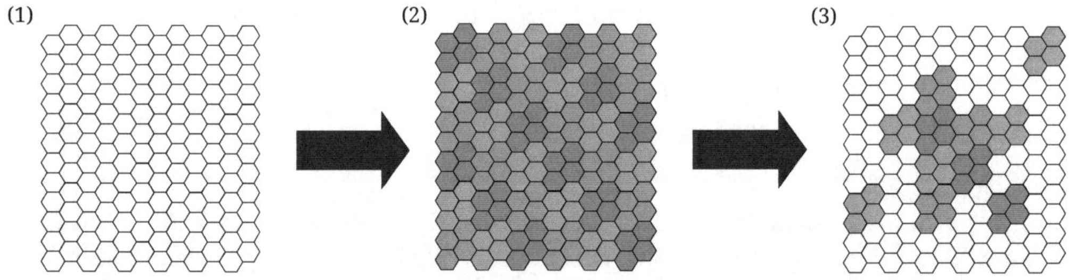

Figure 1. Illustration of trigger data reduction performed in the front-end ASICs. HGCal sensor cells (1) are grouped into trigger cells (2). Only the most energetic ones (3) are selected and transfered.

The back-end electronics will decode the incoming trigger data and perform a subsequent and more sophisticated reduction of the information with a three-dimensional clustering of the energies. This process can be done in different ways and the final algorithm has not been chosen yet. The choice of the type of algorithm will depend on both the triggering performance and the hardware feasibility. The formation of clusters can be done in two steps, starting with two-dimensional clusters built in each layer independently, followed by a linking procedure of these two-dimensional clusters. The clusters can also be built directly using the full three-dimensional information, which would benefit from more information but would need more hardware resources. The latter scenario has been explored so far.

In the process of energy clustering the estimation of the level of pile-up is of paramount importance in order to reduce the sensitivity of the trigger to pile-up. This pile-up estimation will be done event-by-event and regionally, in order to better catch local fluctuations (in time and space). The number of trigger cells above a given energy threshold is highly correlated with the number of additional interactions and is very simple and fast to compute. It is therefore a good estimator of the level of pile-up. In order to improve further this correlation and reduce the bias coming from the presence of energetic objects, this cell counting is limited to the first layers of the HGCal, which are dominated by the energy deposits of low-energy particles coming from minimum-bias interactions. This number of cells is then used to compute several quantities, such as clustering thresholds and energy corrections in order to reduce their sensitivity to additional energy not coming from the hard interaction. It can be mentioned that, though not studied so far, timing information could eventually provide an additional handle for the mitigation of pile-up, if propagated through the trigger processing chain.

Clusters will suffer from energy non-containment and inefficiencies and will not be able to catch the total energy in the events. Therefore, in addition to three-dimensional clusters, projective trigger towers with a coarser transverse granularity will be produced. This will provide a complete map of the energy and complement the localized aggregated energy.

3. From trigger primitives to object reconstruction

The HGCal trigger system will produce three-dimensional clusters and trigger towers and send them to the global CMS trigger system. Together with the energy of the clusters, information on their shape and quality will be sent. Given the fine segmentation of the calorimeter, both transverse and longitudinal, the shape information provides an efficient discrimination between electromagnetic and hadronic showers. In particular, information on the start, on the maximum and on the tail of the shower can be condensed into simple variables easily used at the L1 trigger for cluster identification. The physics object reconstruction (such as electrons, photons and jets) will be performed outside of the HGCal trigger system and combine information from several sub-detectors. But algorithms using solely the HGCal trigger primitives described above have been developed and studied in order to evaluate the triggering performance of this detector.

Electrons and photons are reconstructed from clusters identified as electromagnetic energy deposits. Electrons and photons can produce several nearby reconstructed clusters, spread in the azimuthal direction due to the magnetic field, in case of bremsstrahlung and photon conversion. The granularity of the HGCal L1 trigger will be sufficiently fine to separate these nearby clusters; it is therefore necessary to sum their energy together in subsequent trigger stages in order to recover this energy spread. Electromagnetic clusters will also be matched to tracks reconstructed by the future track trigger in order to separate electrons and photons. The results presented in the next section do not include this matching though; the HGCal is used in a standalone way, without combination with other subdetectors.

Jets are objects very sensitive to pile-up due to their large transverse size. In order to reduce the number of reconstructed jets, mostly dominated by pile-up energy, the jet reconstruction is seeded by high-density clusters and built from trigger towers around these seeds. The size of the jets is also optimized such that the energy fluctuations coming from pile-up energy and non-containment are as small as possible. A small jet radius of the order of $\Delta R = 0.2$ is obtained with this procedure.

4. Performance

The efficiencies and background rates obtained with the electron and photon and jet reconstruction algorithms using only the HGCal information have been compared with those obtained with the algorithms currently used during the Phase 1 of the LHC. The Phase 2 conditions have been used for the HGCal case, with an average of 140 interactions per bunch crossing, while the Phase 1 conditions have been used for the latter case, with an average of 40 simultaneous interactions. In both cases a full simulation of the detectors has been performed for the estimation of the performance of the trigger algorithms. In the Phase 1 case the rate computation has been restricted to the endcaps for a direct comparison with the HGCal results.

In all the considered scenarios the algorithms are giving selection efficiencies of signal objects close to 100%. For these high efficiencies, the background rates are given in Figure 2. For an increase of the instantaneous luminosity by a factor of the order of 3.5 the expected rates are increased by a factor of 2.5 to 3 at most. In some regions this increase of rates is even limited to a factor less than 1.5. This good control of the rate despite the large increase in luminosity is obtained in particular with the usage of the longitudinal segmentation of the detector, giving an efficient estimation and rejection of pile-up energy contributions.

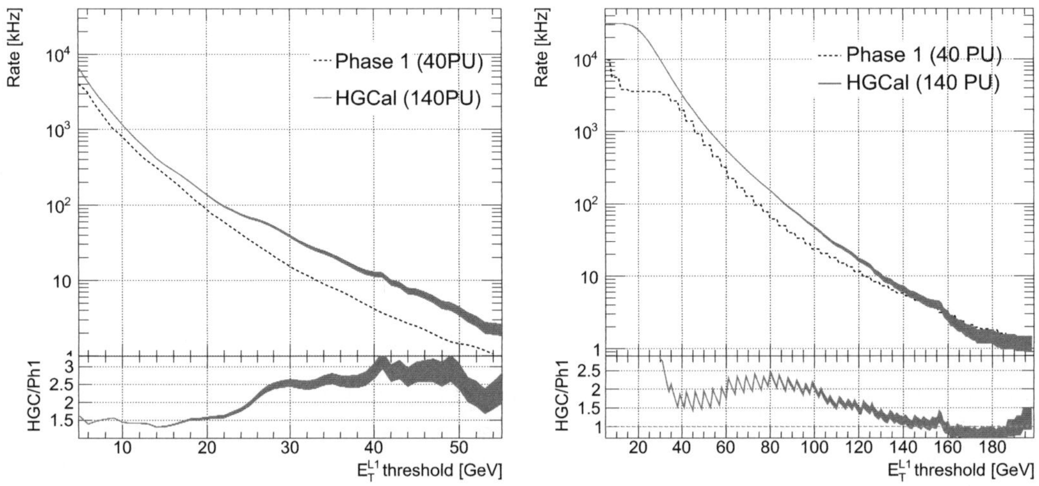

Figure 2. Background rates obtained with electron and photon (left) and jet (right) triggers. Rates obtained with the HGCal with 140 simultaneous interactions are compared with those obtained with the Phase 1 trigger, restricted to the endcaps, with 40 simultaneous interactions. The stepwise decrease of the Phase 1 jet rate is due to the discontinuous digital values of energy.

5. Conclusion

Due to the intense instantaneous luminosity and the harsh pile-up conditions expected during the high luminosity phase of the LHC, the L1 trigger system has to be upgraded. In particular the new HGCal subsystem must be designed to provide inputs to the upgraded L1 trigger. Preliminary design concepts for the HGCal L1 trigger, particularly in terms of data handling and processing, have been developed and evaluated. It has been shown that simple techniques can provide effective data reduction in the front-end ASICs, and efficient pile-up mitigation can be obtained using the three-dimensional information of the particle showers. For similar efficiencies on signal objects (electrons, photons and jets) as those obtained with the Phase 1 system, the background rate is increased by a factor of between 2.5 to 3 at most, and down to less than 1.5, for an increase of the instantaneous luminosity by a factor of around 3.5.

References

[1] CMS Collaboration, *The CMS experiment at the CERN LHC*, JINST 3 (2008) S08004
[2] CMS Collaboration, *CMS TriDAS project: technical design report, vol. 1: the trigger systems*, CERN-LHCC-2000-038, CMS-TDR-6-1 (2000)
[3] CMS Collaboration, *Technical Proposal for the Phase-II Upgrade of the CMS Detector*, CERN-LHCC-2015-010, LHCC-P-008, CMS-TDR-15-02 (2015)

IOP Conf. Series: Journal of Physics: Conf. Series **928** (2017) 012027 doi:10.1088/1742-6596/928/1/012027

Challenges of Particle Flow reconstruction in the CMS High-Granularity Calorimeter at the High-Luminosity LHC

Frank Chlebana, on behalf of the CMS Collaboration.

Fermi National Accelerator Laboratory, P.O. Box 500, MS 205, Batavia, IL 60510-5011

E-mail: chlebana@fnal.gov

Abstract. The challenges of the High-Luminosity LHC (HL-LHC) are driven by the large number of overlapping proton-proton collisions (pileup) in each bunch-crossing and the extreme radiation dose to detectors at high pseudorapidity. To overcome this challenge CMS is developing an endcap electromagnetic+hadronic sampling calorimeter employing silicon sensors in the electromagnetic and front hadronic sections, comprising over 6 million channels, and highly-segmented plastic scintillators in the rear part of the hadronic section. This High-Granularity Calorimeter (HGCAL) will be the first of its kind used in a colliding beam experiment. Clustering deposits of energy over many cells and layers is a complex and challenging computational task, particularly in the high-pileup environment of HL-LHC. Baseline detector performance results are presented for electromagnetic and hadronic objects, and studies demonstrating the advantages of fine longitudinal and transverse segmentation are explored.

1. Introduction

The Large Hadron Collider (LHC) is expected to be upgraded to the High Luminosity LHC (HL-LHC) starting in 2024, resulting in an average of 140-200 interactions per crossing (pileup), and is projected to deliver an integrated luminosity of 3000 fb^{-1} over a ten year running period. Not all subdetectors of the Compact Muon Solinoid (CMS) will be able to survive in the high radiation environment and will have to be replaced. In order to achieve the full benefit of the HL-LHC, the physics objects should be reconstructed with high efficiency, low fake rate, and high resolution in the higher luminosity environment of the HL-LHC. CMS has selected a highly granular sampling calorimeter (HGCal) to replace the current endcap calorimeter. The HGCal consists of a electromagnetic section (EE) of 28 sampling layers of silicon, followed by the front hadronic section (FH) consisting of 12 sampling layers of silicon, and finally the back hadronic section (BH) consisting of 12 sampling layers of scintillator. The design uses silicon wafers of three different types. Wafers with an active thickness of 300 μm and 200 μm have 128 channels with a cell size of about 1 cm^2. Wafers with a 100 μm active thickness have 256 channels with a cell size of about 0.5 cm^2. Wafers with smaller cells are used at higher pseudorapidity providing finer segmentation and better radiation tolerance. More details can be found in Reference [1].

Particle flow (PFlow) reconstruction [2] was used in CMS with great success for the analyses in Run 1. The reconstruction attempts to measure and identify each particle originating from the collision and makes optimum use of the tracker, calorimeter, and muon detectors. Detector features that help improve the performance of PFlow reconstruction include: efficient track reconstruction with high purity, a strong magnetic field to separate charged and neutral particles, and high calorimeter transverse granularity to separate overlapping showers. Significant improvements to the tracking system are planned for the phase 2 upgrades which will reduce the overall tracker mass, improve the track reconstruction efficiency, and extend the tracking coverage. Further details on the phase 2 upgrades can be found in the CMS technical proposal [3].

2. Simulation and Clustering

Performance studies for the technical proposal [3] were done using a GEANT4 [4] based simulation of the absorber structure as well as a detailed full detector simulation of the HGCal using square cells and the CMS PFlow clustering algorithm. Recently the simulation has been updated to more closely reflect the geometry described in the technical proposal. Changes include using hexagonal cells for EE and FH sections, updating the absorber structure, and improvements to the electronics simulation including the digitization and readout of time information. The active layers are simulated as a plane of hexagonal sensors and uninstrumented areas necessary for mechanical assembly are not currently simulated. A finer segmentation for the BH section was implemented with the ability to gang readout into different transverse sizes allowing studies on the effects of segmentation. The reconstruction of clusters and the energy sharing between adjacent clusters is a crucial input to the PFlow algorithm and the performance of several clustering algorithms have been investigated. The CMS PFlow algorithms that have been used for the much coarser granularity Run 1 detector, and were adapted for the technical proposal studies, use layer by layer clustering [2] that is seeded by the local maximum energy. The Pandora software development kit [5] provides a flexible framework to test different clustering algorithms. A forward projective cone algorithm from Pandora was studied as well as the ARBOR clustering approach which models showers as following the topology of a tree [6]. The clustering algorithms were originally developed for low occupancy environments and when used in the high pileup environment of the HL-LHC the execution times were found to be slow. Clustering algorithms employ nearest neighbor searches and require efficient identification of the outermost points of the clusters. Modern computing techniques were used to improve both the CPU and memory usage. Utilization of a k-d tree provided several orders of magnitude in speedup. Techniques adapted from graph theory such as the use of quick unions allow efficient association of hits that form clusters. Data structures were optimized for more efficient memory management and the code was rewritten to be thread safe. These software improvements reduced execution times from several hours to several minutes per event. Performance of the clustering algorithms was good, however, when examining the hit patterns using an event display it could be seen that the topological based clusters did not efficiently resolve closely adjacent energy deposits. Experience gained from these studies is being used to modify the clustering algorithms to better separate multiple narrow shower cores.

3. Reconstruction and Performance

The electron resolution is shown in Figure 1 for silicon sensors having different active layer thicknesses and for different regions of pseudorapidity as obtained from the GEANT4 simulation with no pileup. The e/γ resolution is $20-24\%/\sqrt{E}$. In the presence of pileup, electron clusters will contain subclusters originating from pileup which can be identified and removed. Low p_T neutral photons can be resolved using the transverse granularity, late interacting charged hadrons can be rejected using the longitudinal segmentation, measurement of the shower profile helps reduce pileup contamination, and the lower density design of the tracker results in more efficient tracking. The reconstructed energy of the electron supercluster is shown as a function of the number of vertices in Figure 2 before and after pileup subtraction was applied. The dependence of the energy measurement on the pileup is clearly removed.

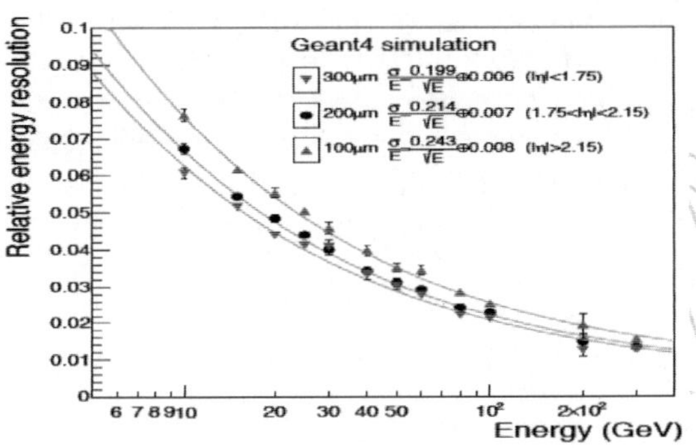

Figure 1. The electron resolution for sensors with different active thickness and for different pseudorapidity regions.

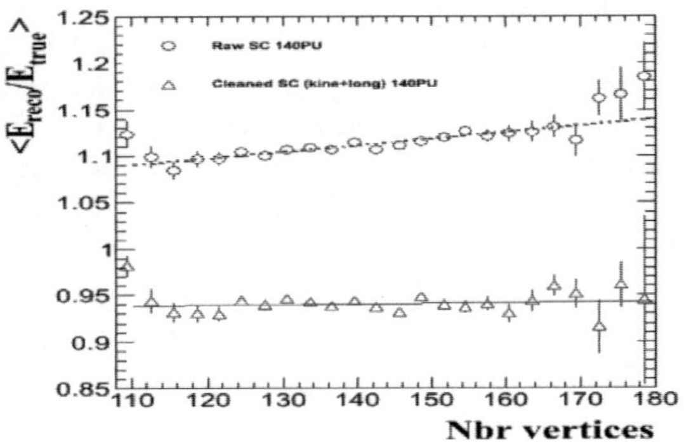

Figure 2. Energy of the electron super cluster before and after applying pileup subtraction.

The electron reconstruction efficiency for the phase 1 detector is degraded as the electromagnetic crystals are damaged from radiation and with increasing pileup as is shown in Figure 3. The performance of the HGCal for 140 pileup events is comparable with the aged phase 1 endcap performance for 50 pileup and the increased coverage of the phase 2 tracker extends the acceptance of electrons. Pure electromagnetic showers have a higher energy density than hadronic showers. The high transverse and depth granularity allows us to measure the cluster density and a correction can be applied to the energy of the hadronic cluster to help compensate for electromagnetic fluctuations and improve the jet energy resolution. The results shown in Figure 4 were obtained by applying the method developed by the CALICE collaboration [7] and improves the jet resolution by 10%. Development is ongoing to further exploit the benefits that the high granularity provide and to improve the reconstruction performance.

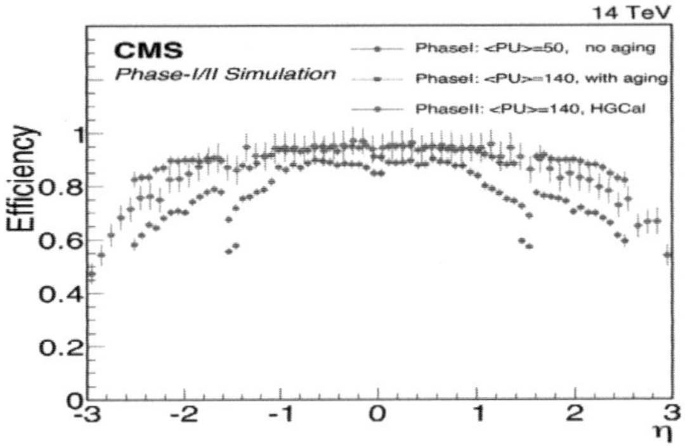

Figure 3. The electron reconstruction efficiency for the current endcap calorimeter and the HGCal with different pileup events.

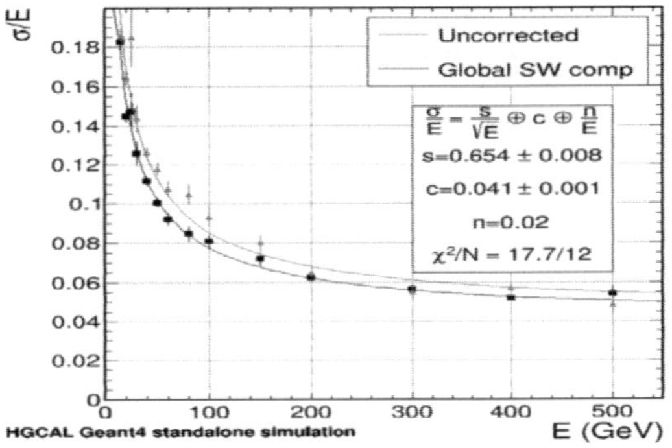

Figure 4. The jet energy resolution, before and after software compensation is applied.

4. Timing

A collision can occur anywhere in the overlapping region as the colliding bunches pass through each other and can occur at the same z location but with different times. The actual beam spot distribution in time and space depends on the LHC operating conditions. Precision timing can be used to mitigate pileup by helping to associate energy deposits with the originating vertex. This is of particular importance for clusters originating from neutral particles such and photons. The HGCal provides precise timing from the silicon. A study was done to estimate the time resolution of electromagnetic clusters for different assumptions of the time resolution for the individual cells. The results are shown in Figure 5 where a cell time resolution of 50 ps for deposits with 60 fC results in a cluster resolution of 20 ps for 10 GeV photons. A time resolution of 20 ps in silicon has been achieved in test beams [8].

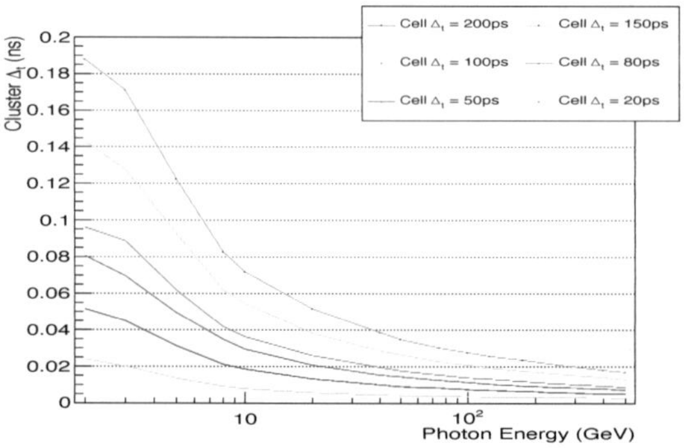

Figure 5. The photon cluster time resolution obtained for different assumptions of the cell time resolution and as a function of the photon energy.

5. Summary

The preliminary studies for the CMS phase 2 technical proposal indicate that the HGCal will be able to maintain the necessary physics performance in the high pileup environment of the HL-LHC. The energy clustering in the calorimeter is an essential input to particle flow reconstruction and we investigated several existing algorithms that were developed for imaging calorimeters of the ILC. Modern computational techniques were used to dramatically improve clustering algorithm execution times. The clustering performance was good, however we did see that nearby clusters were not fully resolved in very dense environments and are exploring methods of energy weighting the hits to help with the separation of clusters. The simulation of the geometry has been updated to better reflect the description in the technical proposal, and we have changed from using square cells to hexagonal cells which required a major update of the geometry description and reworking the clustering algorithms. The HGCal design is still evolving and we will need to incorporate the details of uninstrumented areas necessary for the mechanical support. We have only started to exploit the full potential that the high granularity provides for pileup mitigation and improved event interpretation.

References

[1] C. Ochando, "HGCAL: A High-Granularity Calorimeter for the Endcaps of CMS at HL-LHC", these proceedings.
[2] F. Beaudette, "The CMS Particle Flow Algorithm", arXiv:1401.8155.
[3] D. Contardo et al., "Technical Proposal for the Phase-II Upgrade of the CMS Detector", CERN-LHCC-2015-010.
[4] S. Agostinelli et al., "GEANT4 - a simulation toolkit", Nuclear Instruments and Methods in Physics Research A 506 (2003) 250-303.
[5] M. A. Thomson, "Particle Flow Calorimetry and the PandoraPFA Algorithm", Nucl. Instrum. Meth. A611: 25-40, 2009.
[6] M. Ruan, "Arbor, a new approach of the Particle Flow Algorithm", arXiv:1403.4784.
[7] C. Adloff et al., "Hadronic energy resolution of a highly granular scintillator-steel hadron calorimeter using software compensation techniques", JINST 7 P09017 (2012), arXiv:1207.4210.
[8] N. Akchurin, "Fast-timing capabilities of silicon sensors for the CMS High-Granularity Calorimeter at the High-Luminosity LHC", these proceedings.

IOP Conf. Series: Journal of Physics: Conf. Series **928** (2017) 012028 doi:10.1088/1742-6596/928/1/012028

Fast-timing Capabilities of Silicon Sensors for the CMS High-Granularity Calorimeter at the High-Luminosity LHC

Nural Akchurin
(on behalf of the CMS Collaboration)
Texas Tech University, Department of Physics, Lubbock, TX, 79416 USA
E-mail: nural.akchurin@ttu.edu

Abstract. We report on the signal timing capabilities of thin silicon sensors when traversed by multiple simultaneous minimum ionizing particles (MIP). Three different planar sensors, 133, 211, and 285 μm thick in depletion thickness, have been exposed to high energy muons and electrons at CERN. We describe signal shape and timing resolution measurements as well as the response of these devices as a function of the multiplicity of MIPs. We compare these measurements to simulations where possible. We achieve better than 20 ps timing resolution for signals larger than a few tens of MIPs.

1. Introduction

Event reconstruction at future hadron colliders will be challenged by the dramatic increase in the number of concurrent interactions (pileup) per beam crossing in the experiments. At the High-Luminosity Large Hadron Collider (HL-LHC) –with up to 200 pileup events per beam crossing–, current detectors will show limitations in the reliance on purely spatial information to resolve interactions and associate particles to vertices. Moreover, the random overlap of energy deposits from neutral particles, which cannot be associated via a track to any vertex, will deteriorate the calorimeter performance in terms of energy measurement and particle identification as particles appear to be less isolated. A compact and highly granular sampling calorimeter based on silicon sensor technology has been chosen to replace the end-cap calorimeter in the CMS experiment at the HL-LHC [1]. Precision time measurement of the energy deposits could provide an additional means to resolve interactions [1, 2] by exploiting the spread in the time domain of the collision vertices predicted to be about 150 ps RMS, within the 25 ns bunch crossing structure of the colliding beams at the HL-LHC [3].

The silicon trackers have played crucial roles in many experiments because they afford excellent spatial resolutions at high rates and are robust. There are also a handful of examples of sampling calorimeters based on active silicon sensors. The SICAPO collaboration studied salient features of electromagnetic calorimeters with layers of high-Z absorber and silicon sensors for what were then future colliders [4]. The ALEPH and OPAL collaborations employed luminosity calorimeters consisting of tungsten and silicon sensors [5, 6]. The CALICE collaboration has been investigating high-granularity calorimeters with different types of active elements as well as silicon for precision physics at future colliders [7]. In recent times, silicon sensors have also been used for timing applications in high intensity environments. The NA62 Gigatracker is one example [8]. At a rate of 1 GHz, in addition to excellent single track resolution, the achieved timing resolution is better than 120 ps in particle fluxes of over 10^{13} particles/cm^2.

In this paper, we report on the results of our first study with small planar sensors with focus on their precision timing capability, in particular for the case of multiple particles impinging simultaneously on the sensors. We briefly describe the experimental setup in Section 2 and discuss the pulse shape reconstruction and single minimum ionizing particle (MIP) calibration in Section 3. We present sensors' multi-MIP response in Section 4 and evaluate the timing resolution in Section 5 in some detail.

2. Experimental Setup

The test setup is shown in Figure 1. Two silicon sensors were installed in the beam line at all times. A micro-channel plate (MCP), viewing a Cherenkov radiator, provided a precise timing reference for each event within $\sim$ 20 ps, with full efficiency for single MIPs at -2750 V [9]. A scintillator counter (2×2 cm^2) upstream of the detectors defined the area of triggered events. The impact position of particles on the sensors was estimated by projecting the hit positions, with better than a millimeter resolution, obtained from a set of delay wire chambers (Figure 1). Lead sheets with different thickness were positioned in front of the sensors to generate electromagnetic showers to be able to investigate the multi-MIP response of these sensors.

The measurements described in this paper were performed at the H2 beam line of the super proton synchrotron (SPS) at CERN in July 2015. 150 GeV muons and 50 GeV electrons were used for the majority of the measurements. The beam particle rates were typically several thousand per spill. Each spill lasted 4.9 s and was repeated twice a minute for most of the data taking period. Although the beam spot size on the sensors varied slightly with the beam tune and settings, the FWHM was about 1 cm in horizontal and vertical directions – larger than the sensors (Figure 1).

All tested sensors were p-type (n-on-p), 5×5 mm^2 in the effective area and physically 320 μm thick. They were biased at 600 V and fully depleted. They were produced by deep-diffused float-zone (dd-FZ) technique by Hamamatsu within the framework of the CMS tracker upgrade project in three different depletion thickness: 133, 211, and 285 μm with capacitances of 22.5, 13.6, and 9.9 pF, respectively [10, 11]. The radiation effects on these sensors are currently under study; however, none of the sensors used here was previously irradiated.

The electrical signals from the sensors were amplified by a broadband (2 GHz/40 dB) amplifier[1], and the waveforms were digitized at 5 GHz by a Domino Ring Sampler (DRS) unit from CAEN (V1742). The signals from MCP and the event trigger were also fed to the same digitizer unit in order to remove trigger jitter off-line. The intrinsic timing resolution of the digitizer was $\sim$ 5 ps and did not contribute to the systematic uncertainty of our measurements in any significant way [12].

3. Pulse Shape and MIP Calibration

The rise-times from 10% to 90% of the pulse amplitude for the three sensors were measured to be 1.1 ns, as shown in Figure 2. The pulse time was defined as the time when the pulse reaches its 50% amplitude. The offline algorithm searched for a pulse in the data stream, fitted its peak with a Gaussian function requiring at least 5 samples, and calculated the time at which the pulse reached its half amplitude.

We measured the responses of the three different type of sensors to single MIPs. A 50 GeV electron beam was used to perform this measurement. Without the lead plates in front of the sensors, 50 GeV electrons were effectively equivalent to MIPs. The results of the calibration

[1] CIVIDEC Instrumentation, Vienna, Austria

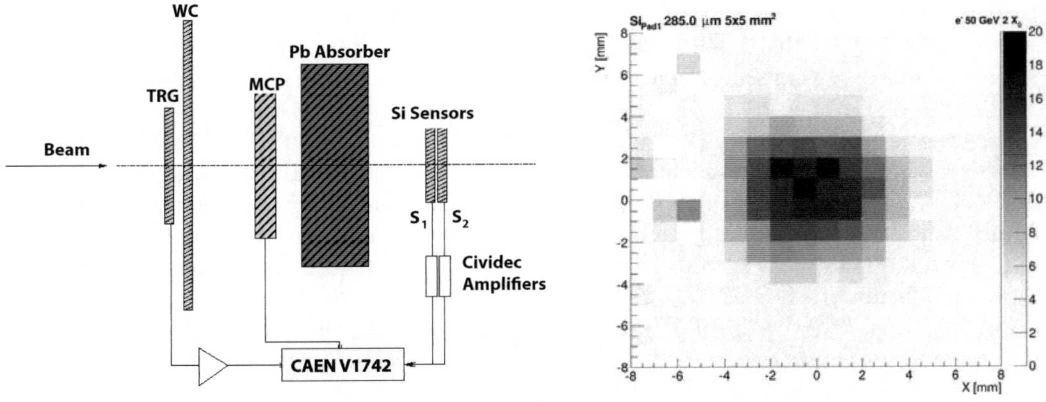

Figure 1. The schematic of the layout displays the main components and the readout scheme on the left. Downstream of the trigger counter (TRG) and wire chambers (WC), a micro-channel plate (MCP) photomultiplier tube was positioned to provide a timing reference in front of the silicon sensors. Various lead plates were placed in between the MCP and the sensors to evaluate their response to multi-MIPs. A typical response pattern of a 285-μm thick silicon sensor (5×5 mm^2) to 50 GeV electrons when normalized to the MIP signal is displayed on the right. Note that the sensors were placed behind $2X_o$ of lead absorber in this case.

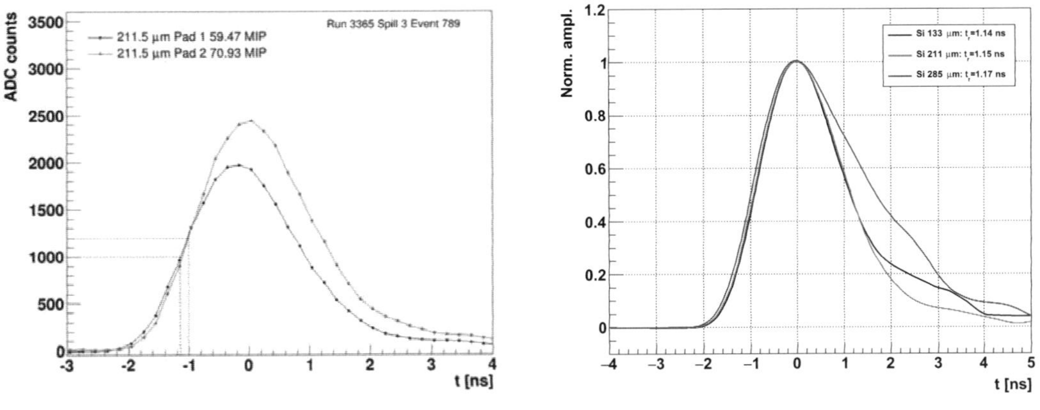

Figure 2. Examples of single pulse shapes from two 211-μm thick sensors for the same event are shown on the left. 50% of the peak amplitude is indicated by dotted lines and is used in all the timing measurements discussed in Section 5. The average pulse shapes from different thickness sensors are shown on the right for comparison. The peak of the pulses is arbitrarily set at $t = 0$.

with electrons have been found to be in agreement with the ones obtained using a 150 GeV muon beam. We selected events with a signal from the MCP and from the second sensor. The presence of a signal in the second sensor ensured that most tracks did pass through the first sensor whose response was being measured. The signal from the first sensor included a correction for a 20% noise correlation between the two sensors. The signal distribution from the first sensor included contributions from zero (*e.g.* no signal) and single-track (1 MIP) events, as well as from a small number of events with 3 charged tracks (3 MIPs) where a photon emitted by the interacting initial electron produced an electron-positron pair. We modeled this spectrum with a linear sum of a Gaussian and two Landau distributions convolved with a Gaussian distribution, representing contributions from 1 MIP and 3 MIPs. Figure 3 shows the signal distributions. The most probable value (MPV) of the Landau distribution establishes the MIP measurement. The precision of the measurement was limited by the amount of data and by the systematic uncertainties related to the noise correlation between the two sensors.

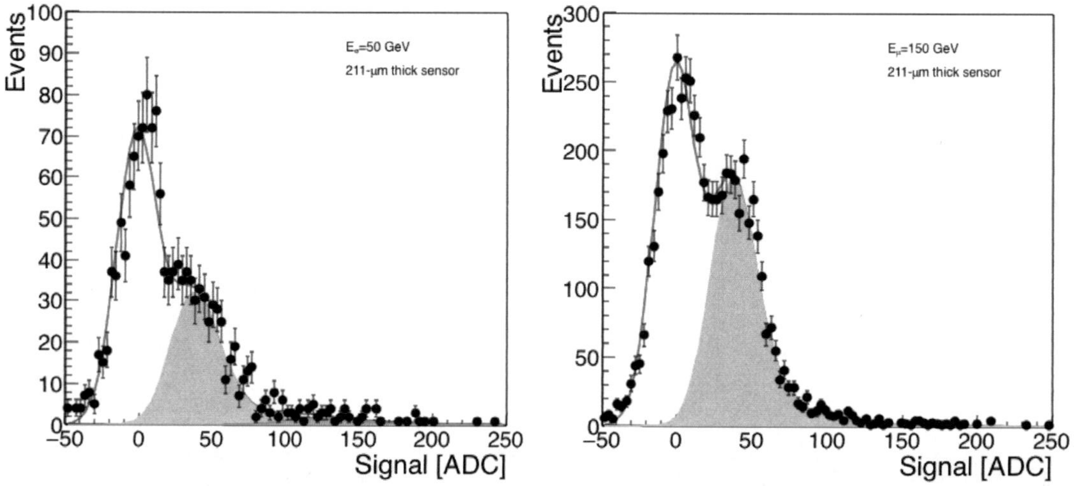

Figure 3. The response of a 211-μm thick sensor to 50 GeV electrons (left): the red solid line depicts the fitted curve, which includes a Gaussian (pedestal centered around zero) and two Landau distributions convoluted with Gaussian distributions accounting for one and three charged particles traversing the sensors. The distribution on the right is the same but for 150 GeV muons. Note the absence of events due to more than one charged track in this case.

The MIP response of the three sensor thicknesses is compared in Figure 4. The signal strength scales nearly linearly with the depletion zone thickness, as expected. We compare the MPVs (solid black squares) for each sensor for a MIP as well as for the signal-to-noise ratios (open red circles).

4. Multiple MIP Signal

While clearly separated single-MIP response from the background is desirable and maybe used to set the energy scale in calorimetry, measurement of the sensor response to multiple MIPs gives us a sense of response linearity. We consider a signal from $1, \ldots, N$ tracks simultaneously passing through the silicon sensor, where N is small enough so the signals from different track multiplicities are distinguishable. In the absence of independent information providing the number of charged particles passing through the sensor, this study was particularly well-suited for our setup. Thus, we analyzed the spectrum containing events with different track multiplicities. We modeled the observed spectrum with an analytic function and then evaluated

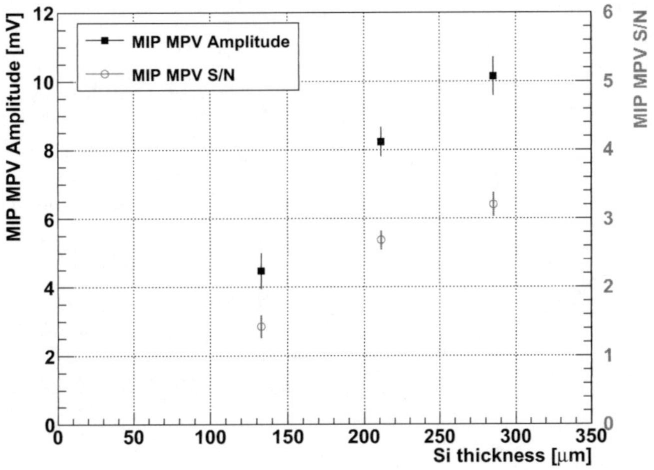

Figure 4. Comparison of the sensor responses to MIPs as a function of their depletion-zone thickness. In open red circles, the signal-to-noise ratio for single MIPs are given (right scale).

the linearity of the response from the parameters of this function. This procedure was also validated with a GEANT4 [13] simulation.

Two radiation lengths of lead were placed in front of the sensors with a 50 GeV electron beam initiating electromagnetic showers. We also required a signal from the MCP to ensure proper timing/jitter correction in signal reconstruction, as noted earlier.

The signal distribution was fitted with an analytic function (Figure 5) with multiple components. Each component comprised a Landau distribution to describe ionization loss by a charged particle in a thin silicon layer convolved with a Gaussian distribution to model electronics noise. $N+1$ components were added linearly to describe a spectrum with track multiplicities ranging from 1 to N. The pedestal component of the distribution (no charged particle going through the sensor) was modeled by a Gaussian distribution, and its width was used to constrain the Gaussian component in the MIP signals. The only constraint in the fitting procedure was that the widths of the individual distributions were limited to a range. All other parameters were left free in the minimization process.

In order to evaluate the linearity of response, shown in Figure 6, we formed a ratio of the MPVs of the Landau distributions from the measurements and simulations and found that for $N < 4$, the signal linearity was within 3–4% for the 285-μm-thick sensor.

5. Timing Resolution

For three different sensors (133, 211, and 285 μm thick) and three ranges of signal ($3 < S_1$ and $S_2 < 8$ MIPs; $8 < S_1$ and $S_2 < 20$ MIPs; and $S_1 > 20$ and $S_2 > 20$ MIPs), Figure 7 depicts distributions of time differences between the first (S_1) and the second (S_2) sensors. The timing resolutions are determined from the standard deviations resulting from the Gaussian fits to these distributions. This difference is divided by $\sqrt{2}$ because the two measurements are assumed to be completely independent. For large signals ($S_1 > 20$ and $S_2 > 20$ MIPs), the timing resolution is better than 20 ps for all sensors. In the intermediate range ($8 < S_1$ and $S_2 < 20$ MIPs), the timing resolution varies from 30 to 60 ps, and when we require only a few MIPs ($3 < S_1$ and $S_2 < 8$ MIPs), the performance degrades significantly from 70 to 150 ps.

We evaluated the timing resolution performance of silicon sensors in three different

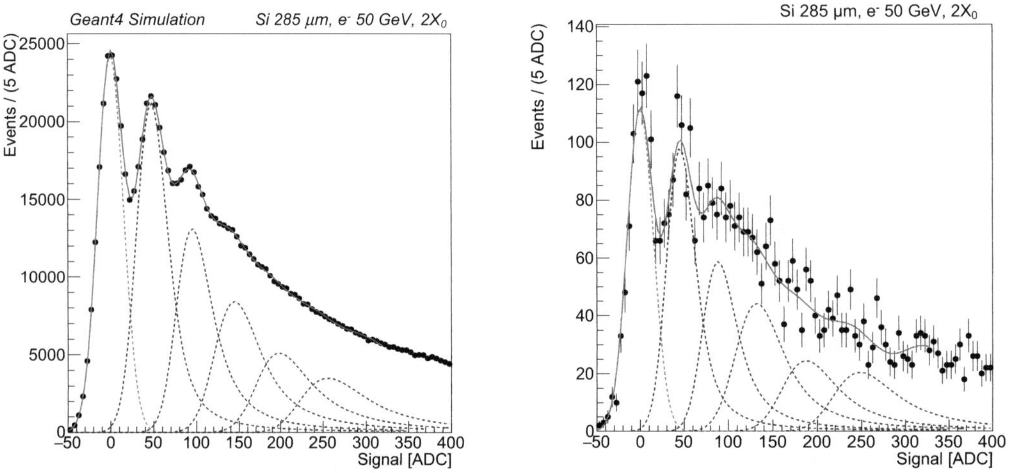

Figure 5. Reconstructed multi-MIP signal from the GEANT4 simulation (left) and the beam data (right). The pedestal distributions are indicated by red dashed lines and centered at zero. The error bars are statistical only.

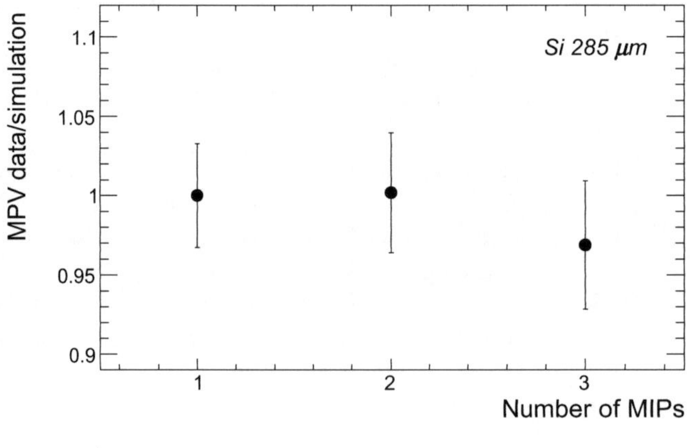

Figure 6. The linearity of response to multiple charged tracks as a function of track multiplicity is within 3 to 4 % in this narrow range. The vertical axis is the ratio of MPVs from the experimental data and simulations for the 285-μm thick sensor. The error bars are largely statistical because we measure a differential non-linearity with respect to one MIP where most of the systematic uncertainties cancel.

combinations. The first measurement (I) was carried out using a pair of same-thickness silicon sensors as described above. For the second measurement (II), one silicon sensor was used in conjunction with an MCP. In the third measurement (III), we used the average of signals from two same-thickness silicon sensors and the signal from the same MCP as in the previous case. The results from each measurement are summarized in Table 1.

Measurement I: Figure 8 presents the timing resolution as a function of the effective signal strength in units of MIPs and the effective signal-to-noise ratio. We defined the effective signal strength as $S_{\text{eff}} = S_1 S_2 / \sqrt{S_1^2 + S_2^2}$. It can be seen that the timing performance improves with

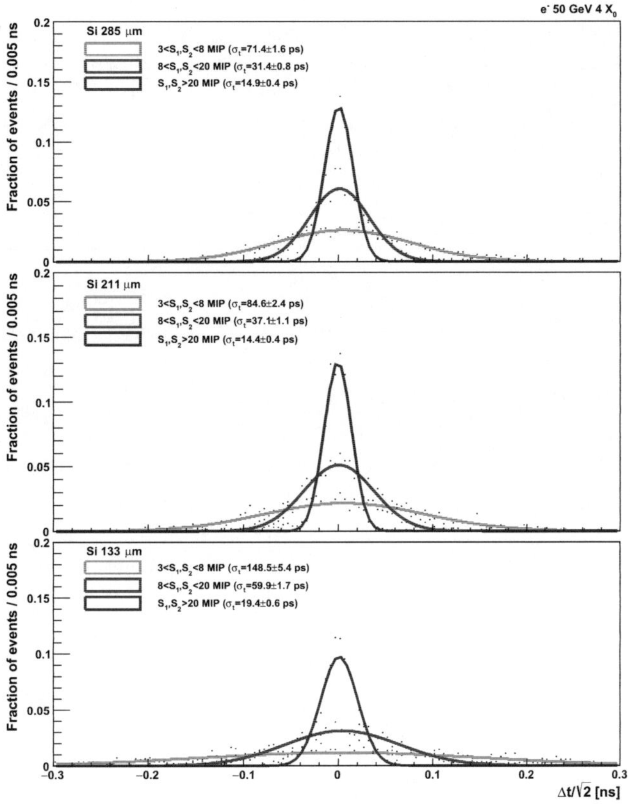

Figure 7. The distribution of the time difference between the signals from a pair of silicon sensors ($\Delta t = t_1 - t_2$), 133-μm (bottom), 211-μm (middle), and 285-μm (top) thick, as a function of three different signal ranges as indicated on the upper left corner of each plot. The solid lines represent Gaussian fits to the data points. The sensors were placed behind a $4X_o$ lead absorber and the electron beam energy was 50 GeV.

increasing signal strength (Figure 8-left), but that for equal $S_{\rm eff}/N$ the timing performance of the three sensor types is similar (Figure 8-right). The solid lines in Figure 8 represent the fits to a form

$$\frac{\sigma(t_1 - t_2)}{\sqrt{2}} = \frac{A}{\sqrt{2}S_{\rm eff}} \oplus C \tag{1}$$

to quantify the timing resolution of the two-sensor system where A represents the 'noise' and C represents the 'constant' terms. The two terms on the right side of Eq.1 are summed in quadrature as symbolized by $\oplus$, and the $\sqrt{2}$ factor accounts for the two independent sensors. When the signal strength is large, $S_{\rm eff} \geq 80$ MIPs, the constant term dominates, and the timing resolution is about 10 ps. The toy simulation results are also plotted in Figure 8 as dotted lines. This simulation was performed by using the average pulse shapes for each sensor and adding a Gaussian noise corresponding to the measured noise level. No noise correlation was introduced among the samples. The overall 'noise' terms are well reproduced in this way whereas the 'constant' term is underestimated, mainly because only 5 ps timing uncertainty contribution from the V1742 digitizer is accounted for in this simulation.

Measurement II: When the timing resolution was evaluated with one silicon sensor and an

Table 1. The summary of timing resolutions achieved by combinations of detectors are shown for three different cases. The first and second column indicate which two detectors are used in the measurement. In *Measurement III*, the average of the timing information from the two indicated silicon sensors is treated as the effective first detector, and the MCP as the second. The mean time is defined as $t_{\rm mean} = \frac{1}{2}(t_{S1} + t_{S2})$. The error on A and C are statistical and derived from fits.

Det 1	Det 2	Fit Function	A [ns × ADC]	C [ns]
Measurement I				
S_1(133-μm)	S_2(133-μm)	$\frac{\sigma(t_1-t_2)}{\sqrt{2}} = \frac{A}{\sqrt{2}S_{\rm eff}} \oplus C$	0.69 ± 0.01	0.010 ± 0.001
S_1(211-μm)	S_2(211-μm)		0.38 ± 0.01	0.009 ± 0.001
S_1(285-μm)	S_2(285-μm)		0.34 ± 0.01	0.010 ± 0.001
Measurement II				
S_1(133-μm)	MCP	$\sigma(t_1 - t_{\rm MCP}) = \frac{A}{S_1} \oplus C$	0.62 ± 0.01	0.018 ± 0.001
S_1(211-μm)	MCP		0.40 ± 0.01	0.022 ± 0.001
S_1(285-μm)	MCP		0.35 ± 0.01	0.024 ± 0.001
Measurement III				
S_1(133-μm), S_2(133-μm)	MCP	$\sigma(t_{\rm mean} - t_{\rm MCP}) = \frac{A}{2S_{\rm eff}} \oplus C$	0.75 ± 0.01	0.020 ± 0.001
S_1(211-μm), S_2(211-μm)	MCP		0.38 ± 0.01	0.019 ± 0.001
S_1(285-μm), S_2(285-μm)	MCP		0.33 ± 0.01	0.021 ± 0.001

MCP, $\sigma(t_1 - t_{\rm MCP})$, the constant term varied from 18 to 24 ps reflecting the limiting performance of the MCP at large signals as shown in Figure 9.

Measurement III: We also studied the average signal from two same-thickness silicon sensors versus the signal from the MCP, $\sigma(t_{\rm mean} - t_{\rm MCP})$. In this case (Figure 10), the timing resolution of the system was dominated by that of the MCP as in measurement II. However, the comparison of the noise terms reveals the potential power and the higher level of performance of the independent timing measurements from multiple silicon sensors. The noise term is a factor of $\sqrt{2}$ better in *Measurement I* compared to *Measurement III* for this reason.

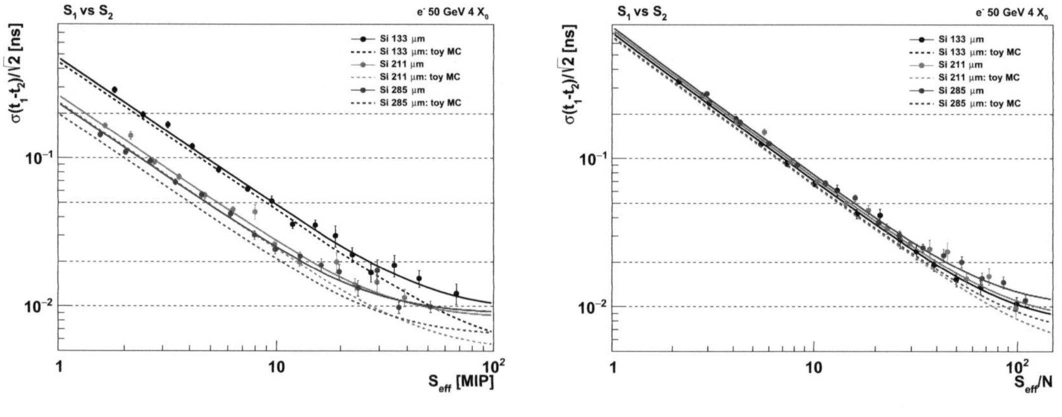

Figure 8. The timing resolution based on two silicon sensors as a function of the effective signal strength in units of MIPs (left) and as a function of the signal-to-noise ratio (right). The fitted resolution functions with a noise A and a constant term C are also shown as solid lines. The dashed lines represent toy simulation results (see text for details).

6. Conclusions

In this study, we have shown that it is possible to perform precise timing measurements ($\mathcal{O}(20)$ ps) using silicon sensors for large pulse heights, $S/N \gtrsim 20$. We have also shown that it is possible

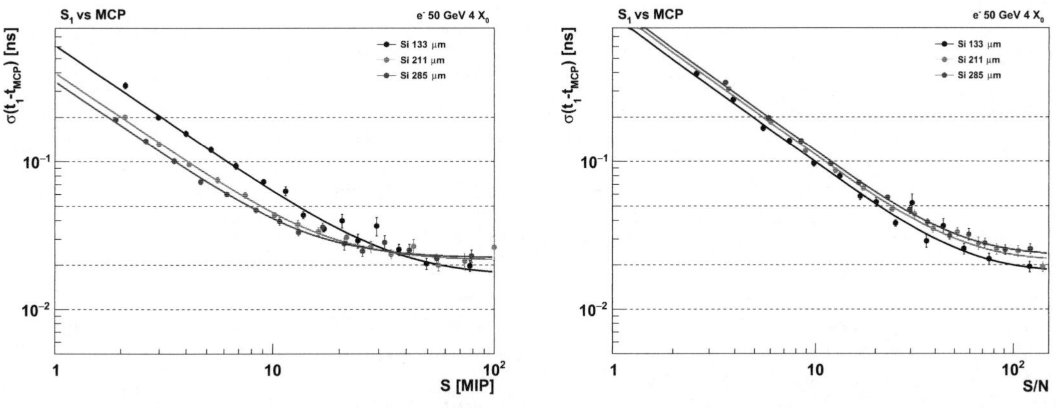

Figure 9. The timing resolution between the first silicon sensor and the MCP as a function of the signal (left) and as a function of the signal-to-noise ratio (right). All three different thickness sensors are compared. The constant term, C, is dominated by the MCP resolution, estimated to be 21 ps.

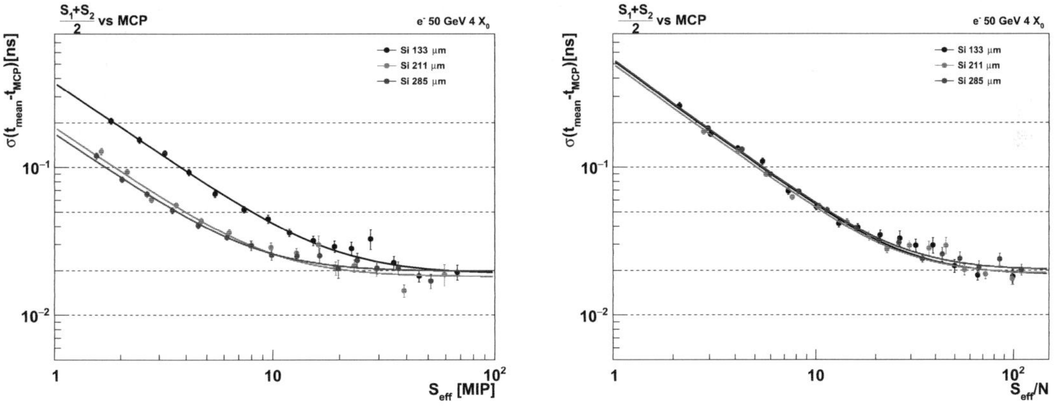

Figure 10. The timing resolution between the mean signal of the two silicon sensors and the MCP as a function of the effective signal (left) and as a function of the signal-to-noise ratio (right).

to further improve the timing measurement when information from several sensors are combined. For example, the possibility of measuring with high precision the shower time evolution offers novel opportunities, in particular it could allow the separation of nearby showers beyond what is achievable from the spatial segmentation and precise time information associated with each reconstructed shower in calorimetry. These pieces of information have the potential to improve the calorimeter performance in very dense environments, such as the forward region of CMS during HL-LHC operation. In the wake of these measurements, similar timing studies exposing silicon sensors to multiple MIPs have recently also been performed at Fermilab [14].

References

[1] CMS Collaboration, "Technical Proposal for the Phase-II Upgrade of the CMS Detector," Tech. Rep. CERN-LHCC-2015-010. LHCC-P-008, CERN, Geneva, Jun 2015.

[2] S. N. White, "R&D for a Dedicated Fast Timing Layer in the CMS Endcap Upgrade," *Acta Phys. Pol. B Proc. Suppl.*, vol. 7, p. 743, 2014.

[3] S. Fartoukh, "Pile up management at the high-luminosity LHC and introduction to the crab-kissing concept," *Physical Review Special Topics-Accelerators and Beams*, vol. 17, no. 11, p. 111001, 2014.

[4] E. Borchi *et al.*, "Silicon Sampling Hadronic Calorimetry: A Tool for Experiments at the Next Generation of Colliders," *Nucl. Instrum. Meth.*, vol. A279, pp. 57–65, 1989.

[5] D. Bederede *et al.*, "SICAL: A High precision silicon - tungsten luminosity calorimeter for ALEPH," *Nucl. Instrum. Meth.*, vol. A365, pp. 117–134, 1995.

[6] G. Abbiendi *et al.*, "Precision luminosity for Z0 line shape measurements with a silicon tungsten calorimeter," *Eur. Phys. J.*, vol. C14, pp. 373–425, 2000.

[7] F. Sefkow, A. White, K. Kawagoe, R. Pöschl, and J. Repond, "Experimental tests of particle flow calorimetry," *Rev. Mod. Phys.*, vol. 88, p. 015003, Feb 2016.

[8] M. Perrin-Terrin, "The NA62 GigaTracker," *PoS*, vol. VERTEX2015, p. 016, 2015.

[9] L. Brianza *et al.*, "Response of microchannel plates to single particles and to electromagnetic showers," *Nucl. Instrum. Meth.*, vol. A797, pp. 216–221, 2015.

[10] A. Dierlamm, "Characterisation of silicon sensor materials and designs for the CMS Tracker Upgrade," Tech. Rep. AIDA-CONF-2014-006, CERN, Geneva, Sep 2012.

[11] E. Curras, M. Fernandez, C. Gallrapp, L. Gray, M. Mannelli, P. Meridiani, M. Moll, S. Nourbakhsh, C. Scharf, P. Silva, G. Steinbrueck, T. T. de Fatis, and I. Vila, "Radiation Hardness and Precision Timing Study of Silicon Detectors for the CMS High Granularity Calorimeter (HGC)," *Nucl. Instrum. Meth.*, 2016.

[12] CAEN, "V1742: 32+2 Channel 12bit 5 GS/s Switched Capacitor Digitizer," 2015.

[13] S. Agostinelli *et al.*, "GEANT4: A Simulation toolkit," *Nucl. Instrum. Meth.*, vol. A506, pp. 250–303, 2003.

[14] A. Apresyan, G. Bolla, A. Bornheim, H. Kim, S. Los, C. Pena, E. Ramberg, A. Ronzhin, M. Spiropulu, and S. Xie, "Test beam studies of silicon timing for use in calorimetry," *Nucl. Instrum. Meth.*, vol. 825, pp. 62 – 68, 2016.

IOP Conf. Series: Journal of Physics: Conf. Series **928** (2017) 012029 doi:10.1088/1742-6596/928/1/012029

Performance of Crystal Scintillators in a Severe Radiation Environment Caused by Protons

Fan Yang, Liyuan Zhang and Ren-Yuan Zhu
California Institute of Technology, Pasadena, CA 91125, USA

Jon Kapustinsky, Ron Nelson and Zhehui Wang
Los Alamos National Laboratory, Los Alamos, NM 87545, USA

E-mail: zhu@hep.caltech.edu

Abstract. Future high energy physics experiments at the energy and intensity frontiers will face challenges from a severe radiation environment caused by both ionization dose and charged and neutral hadrons. This paper reports an investigation on proton induced radiation damage in various crystal scintillators. Large size BGO, LFS and PWO crystals of 18 to 22 cm long were irradiated by 800 MeV protons up to 3×10^{15} p/cm^2 at the WNR facility of LANSCE with longitudinal transmittance measured *in situ*. LYSO plates of $14 \times 14 \times 1.5$ mm^3 were irradiated by 24 GeV protons up to 8.2×10^{15} p/cm^2 at the IRRAD facility of CERN. Degradations in both transmittance and light output are reported. The results show an excellent radiation hardness of LYSO/LFS crystals against charged hadrons.

1. Introduction

Because of their superb energy resolution and detection efficiency, crystal scintillators are widely used in HEP experiments. The CMS lead tungstate ($PbWO_4$ or PWO) crystal calorimeter, for example, has played an important role in the discovery of the Higgs boson [1]. One crucial issue, however, is its radiation damage in a severe radiation environment, which requires precision monitoring to correct variations of crystal's transparency [2]. The proposed HL-LHC with 5×10^{34} $cm^{-2}s^{-1}$ luminosity and up to 3,000 fb^{-1} integrated luminosity will present an extreme severe radiation environment, where up to 3×10^{14} charged hadrons/cm^2 and 5×10^{15} neutrons/cm^2 in addition to an ionization dose of up to 130 Mrad are expected for forward calorimeters (FCAL) at $\eta = 3$ [3]. Cerium doped lutetium yttrium oxyorthosilicate ($Lu_{2(1-x)}Y_{2x}SiO_5$:Ce or LYSO) [4] and cerium fluoride (CeF_3)[5] crystals were proposed to construct a sampling Shashlik calorimeter for the CMS FCAL upgrade for the HL-LHC. The survivability of these crystals is an crucial issue. In this paper, we report proton induced radiation damage in large size BGO, LYSO and PWO crystals measured *in situ* before and immediately after 800 MeV proton irradiation at the Weapons Neutron Research facility of Los Alamos Neutron Science Center (WNR of LANSCE). Results of thin LYSO plates of $14 \times 14 \times 1.5$ mm^3 irradiated by 24 GeV protons at CERN are also reported.

IOP Conf. Series: Journal of Physics: Conf. Series **928** (2017) 012029 doi:10.1088/1742-6596/928/1/012029

Table 1. Samples Irradiated at the WNR Facility of LANSCE by 800 MeV Protons

Position	Samples	Dimensions (mm^3)	*in-situ* Measurement	Fluence (p/cm^2)
1	Shashlik cell	34×34×215	420 nm LED, Al	1.24×10^{15}
2	BaF_2	30×30×20	Al foil activation	2.94×10^{14}
2	LuAG ceramic	25×25×0.4	Al foil activation	2.94×10^{14}
2	10 PWOs	25×25×10	Al foil activation	2.94×10^{14}
2	BGO	17×17×17	Al foil activation	2.94×10^{14}
2	20 LFS plates	14×14×1.5	Al foil activation	2.94×10^{14}
3	2 Capillaries, 2 Y11s	$\phi1\times200$	Al foil activation	3.05×10^{15}
4	PWO	$28.5^2\times30^2\times220$	LT (350-700 nm)	1.80×10^{14}
5	LFS	25×25×180	LT (350-700 nm)	2.87×10^{15}
6	BGO	25×25×200	LT (350-700 nm)	1.77×10^{14}

2. Samples and experiment

Table 1 lists the samples loaded at six positions on a remotely controlled linear stage with a travel distance of 1 m at the WNR facility of LANSCE. The 800 MeV proton beam has a Gaussian shape with a FWHM of about 2.5 cm. Because of the 20 cm spacing between the samples multiple scattering effect to neighboring crystal is negligible. Optical and scintillation properties of the crystal samples were measured before irradiation at Caltech and again after samples are returned to Caltech after cooled down. The fluence listed in the table is calculated by using beam current data provided by LANSCE, and is cross-checked with the activation data measured by Al foils.

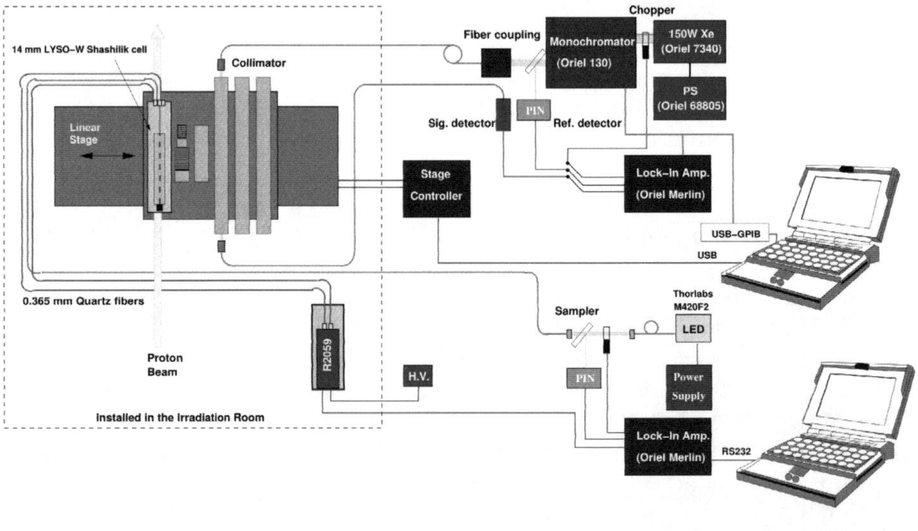

Figure 1. A schematic showing the experimental setup used to measure crystal longitudinal transmittance *in situ* at WNR of LANSCE.

Fig. 1 shows an optical fiber based spectrophotometer used to measured longitudinal transmittance (LT) for long crystal samples *in situ* before, during and after irradiation. It uses a lock-in amplifier to mitigate the noise caused by phosphorescence in crystals after irradiation. To avoid multiple Coulomb scattering and shower leakage in long crystals, LYSO and LFS plates of $14 \times 14 \times 1.5\ mm^3$ were also irradiated by 24 GeV protons at the IRRAD facility of CERN up to $8.2 \times 10^{15}\ p/cm^2$. Transmittance was measured at Caltech by using PerkinElmer Lambda 950 spectrophotometer with 0.15% precision. The corresponding precision for the radiation induced absorption coefficient (RIAC) is about 3.5 m^{-1} for 1.5 mm thick LYSO plates. Light output was measured as the intensity of photoluminescence excited by a 365 nm LED for LYSO and LFS plates before and after irradiation with a systematic uncertainty of about 1.5%, where the sample was surrounded by a Teflon box and read out by four Y-11 wavelength shifting (WLS) fibers. Similarly, a lock-in amplifier was used to mitigate the residual phosphorescence.

3. Results and discussions

Fig. 2 and 3 show the LT spectra measured *in situ* before and after $1.8{\times}10^{14}$ p/cm^2 irradiation for the BGO and PWO samples respectively. The BGO and PWO samples are completely black below 400 and 450 nm respectively. Recovery of radiation damage was observed in both BGO and PWO with the RIAC values at 480 and 450 nm recovered from 15 and 36 to 10 and 32 m^{-1} in 37 and 38 h respectively. LYSO was found radiation hard against proton irradiation in the 2014 test [6]. Fig. 4 shows the LT spectra measured for the LFS sample up to $2.9{\times}10^{15}$ p/cm^2 in five steps with RIAC values at 430 nm of 3.7 and 14.1 m^{-1} after $3.6{\times}10^{14}$ and $2.9{\times}10^{15}$ p/cm^2 respectively, confirming excellent radiation hardness.

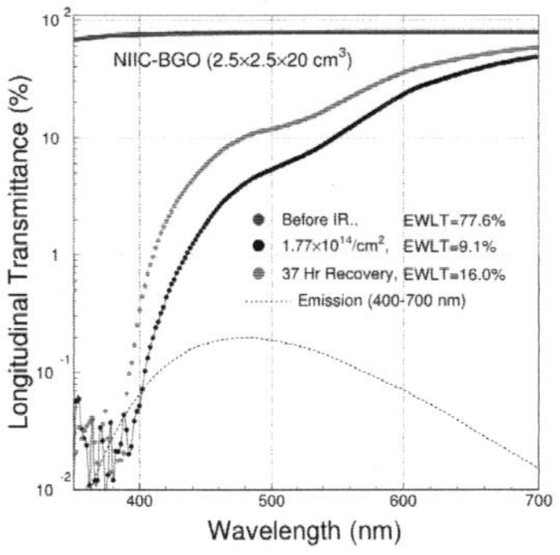

Figure 2. LT of NIIC-BGO before and after $1.8{\times}10^{14}$ p/cm^2, as well as after 37 hours recovery.

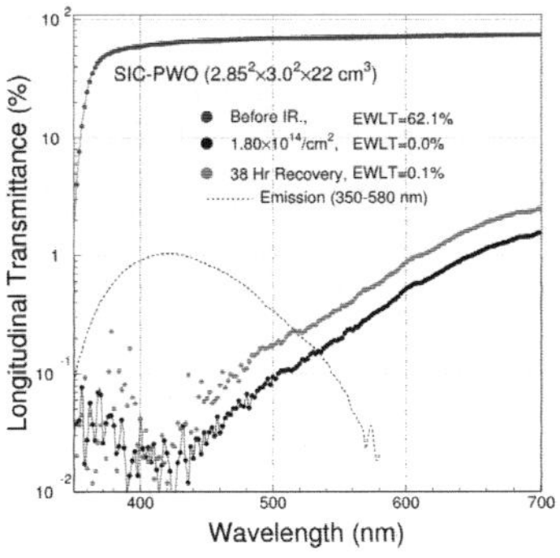

Figure 3. LT of SIC-PWO before and after $1.8{\times}10^{14}$ p/cm^2, as well as after 38 hours recovery.

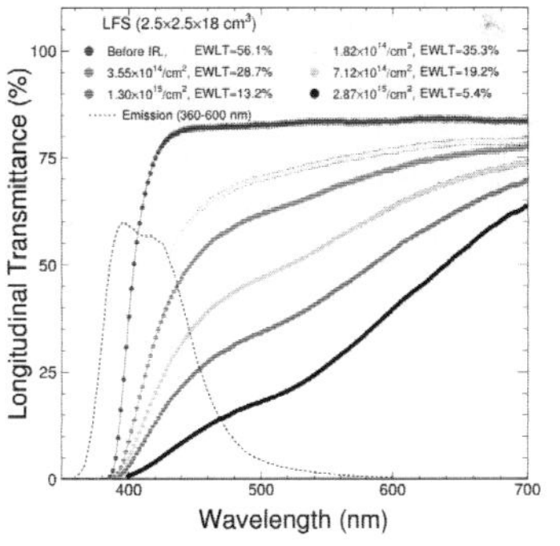

Figure 4. LT of LFS before and after the irradiation in five steps, 0.18, 0.36, 0.71, 1.3 and $2.9{\times}10^{15}$ p/cm^2.

Consistent RIAC values at 430 nm are observed in LYSO and LFS plates irradiated by 24 GeV protons up to $8.2{\times}10^{15}$ p/cm^2 at CERN in 2014 and 2015 as shown in Fig. 5, indicating an excellent radiation hardness of these crystals against charged hadrons. Fig. 6 shows the normalized light output (LO) of the LYSO/LFS plates measured with four Y11 WLS fibers as a function of the RIAC values at 430 nm. The exponential fit for the LO versus the RIAC values at 430 nm indicates an average of 2.4 cm path length, consistent with what observed under γ-ray irradiation [7]. A LO loss of about 6% was observed for LYSO/LFS plates with WLS readout after $3{\times}10^{14}$ p/cm^2 irradiation.

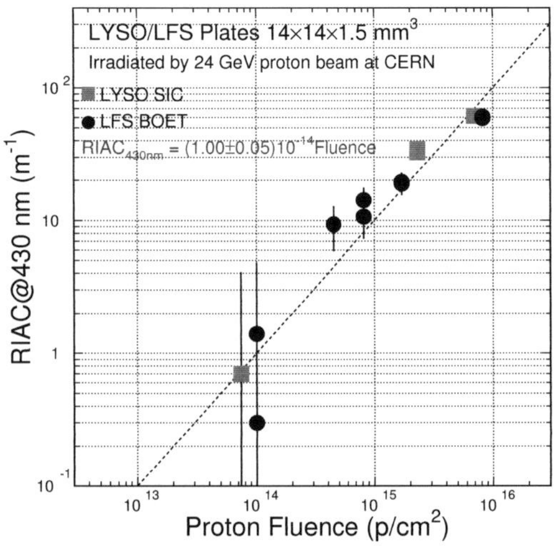

Figure 5. RIAC values at 430 nm are shown as a function of 24 GeV proton fluence for LYSO/LFS plateso f $14 \times 14 \times 1.5$ mm^3.

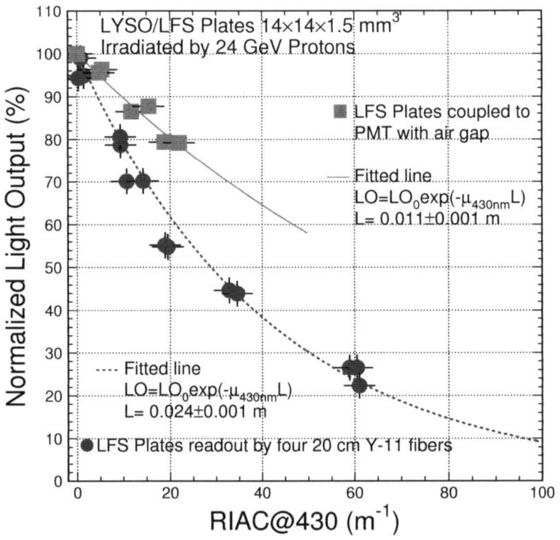

Figure 6. LO of LYSO/LFS plates of $14 \times 14 \times 1.5$ mm^3 is shown as a function of the RIAC values at 430 nm.

4. Summary

A 20 cm long BGO crystal, a 18 cm long LFS crystal and a 22 cm long PWO crystal were irradiated by 800 MeV protons at LANSCE up to 1.8, 1.8 and 29 $\times 10^{14}$ p/cm^2 respectively with crystal's longitudinal transmittance measured *in situ*. After 1.8×10^{14} p/cm^2, PWO became completely dark at its emission peak 420 nm, BGO and LFS showed RIAC values of 14.7 and 2.4 m^{-1} respectively at their emission peak of 480 and 430 nm. After 2.9×10^{15} p/cm^2, the long LFS sample showed a RIAC value of 14 m^{-1} at 430 nm. LYSO and LFS plates of $14 \times 14 \times 1.5$ mm^3 show consistent RIAC values at 430 nm after proton irradiation with a loss of light output about 6% after 3 $\times 10^{14}$ p/cm^2 with Y-11 readout. These results provide important information for understanding charged hadron induced radiation damage in fast crystal scintillators and their use in future HEP experiments at the energy and intensity frontiers.

Acknowledgments

This material is based upon work supported by the U.S. Department of Energy, Office of Science, Office of High Energy Physics program under Award Number DE-SC0011925. This work has benefited from the use of the Los Alamos Neutron Science Center at LANL. This facility is funded by the U.S. Department of Energy under Contract No. DE-AC52-06NA25396. The authors would also like to thank Drs. F. Ravotti and D. Bailleux, who helped us to irradiate the LYSO/LFS plates at CERN.

References

[1] Chatrchyan S et al. 2012 *Phys. Lett B.* **716** 30
[2] Zhu R-Y 2004 *IEEE T Nucl. Sci.* **51** 1560
[3] B. Bilki and the CMS Collaboration 2015 *J. Phys.: Conf. Ser.* **587** 012014
[4] Zhang L, Mao R, Yang F and Zhu R-Y 2014 *IEEE T Nucl. Sci.* **61** 483
[5] Becker R et al. 2015 *J. Phys.: Conf. Ser.* **587** 012039
[6] Yang F, Zhang L, Zhu R-Y, Kapustinsky J, Nelson R and Wang Z 2016 *Nucl Instrum Meth A* **824** 726
[7] Yang F, Zhang L and Zhu R-Y 2016 *IEEE T Nucl. Sci.* **63** 612

IOP Conf. Series: Journal of Physics: Conf. Series **928** (2017) 012030 doi:10.1088/1742-6596/928/1/012030

Beam tests of proton-irradiated $PbWO_4$ crystals and evaluation of double-ended read-out technique for mitigation of radiation damage effects

Marco Lucchini on behalf of the CMS Collaboration

European Organization for Nuclear Research, CERN, CH-1211 Geveva 23, Switzerland

E-mail: marco.toliman.lucchini@cern.ch

Abstract. The harsh radiation environment in which detectors will have to operate during the High Luminosity phase of LHC (HL-LHC) represents a crucial challenge for many calorimeter technologies. In the CMS forward calorimeters, ionizing doses and hadron fluences will reach up to 300 kGy (at a dose rate of 30 Gy/h) and 2×10^{14} cm^{-2}, respectively, at the pseudo-rapidity region of $|\eta| = 2.6$. To evaluate the evolution of the CMS ECAL performance in such conditions, a set of $PbWO_4$ crystals, exposed to 24 GeV protons up to integrated fluences between 2.1×10^{13} cm^{-2} and 1.3×10^{14} cm^{-2}, has been studied in beam tests. A degradation of the energy resolution and a non-linear response to electron showers are observed in damaged crystals. Direct measurements of the light output from the crystals show the amplitude decreasing and pulse becoming faster as the fluence increases. The evolution of the $PbWO_4$ crystals calorimetric performance has been well understood and parameterized in terms of increasing light absorption inside the crystal volume. A double-ended read-out configuration, in which two identical photodetectors are coupled to the opposite ends of each crystal, has also been tested. The separate and simultaneous read out of the light from the two ends of the crystal allows to correct for longitudinal shower fluctuations and to mitigate the degradation of energy resolution in highly damaged crystals. The non-linear response to electromagnetic showers, arising from high non-uniformity of light collection efficiency along the longitudinal axis of irradiated crystals, can also be corrected by means of the double-ended read-out technique.

1. Introduction

To exploit the full potential of the LHC at CERN a major upgrade will be carried out around 2023 to increase the instantaneous luminosity of the collider. During this High Luminosity LHC phase (HL-LHC), all the detectors will have to face a challenging radiation environment and the understanding of the performance of the current detectors under these conditions is of fundamental importance. The radiation levels are expected to be a factor 5 larger than during the standard LHC running, and the integrated luminosities - thus integrated radiation levels - will become a factor 6 greater than the design values [1]. To understand whether the CMS electromagnetic calorimeter (ECAL) will be able to maintain good performance during HL-LHC, several studies on Lead Tungstate ($PbWO_4$) crystals have been performed, establishing that high-energy hadrons cause a cumulative loss of light transmission in $PbWO_4$ [2, 3, 4, 5]. To evaluate the impact of such degradation on the calorimetric performance of the CMS ECAL, a set of $PbWO_4$ crystals, following exposures to hadron fluences similar to those predicted at the end of HL-LHC, has been tested with electron beam in the energy range from 10 to 250 GeV. A double-ended read-out of the crystals has also been tested to better understand the mechanisms of light propagation inside the crystal volume and the degradation of the signal linearity and resolution to electrons. These experimental data have been used to construct and validate a simulation model in order to predict the detector performance in situ.

2. The CMS ECAL during High Luminosity LHC

The CMS ECAL detector [7] was designed to withstand the radiation levels expected by the end of the nominal LHC running [8], before LS3, when 500 fb^{-1} of data are planned to be collected. The expected ionizing radiation levels correspond to 0.2 Gy/h in EB and 6 Gy/h in EE (at $|\eta| = 2.6$). The hadron fluence after 500 fb^{-1} is expected to be 1.2×10^{11} cm^{-2} for EB and 3×10^{13} cm^{-2} for EE (at $|\eta| = 2.6$) [1]. To further exploit the full potential of LHC, a major upgrade of the interaction region will be required around 2023 (LS3 shutdown) in order to achieve a peak luminosity of $\mathcal{L} = 10^{35}$ $cm^{-2}s^{-1}$ and a leveled value of $\mathcal{L} = 5 \times 10^{34}$ $cm^{-2}s^{-1}$ during the High Luminosity LHC Phase (HL-LHC) which is expected to deliver an integrated luminosity of 3000 fb^{-1} by 2035. According to dedicated simulations, during HL-LHC the ionizing dose rate inside CMS are predicted to be ~5 times higher (1 Gy/h in EB and 30 Gy/h in EE) and the integrated particle fluences to be ~6 times higher (7.6×10^{12} cm^{-2} in EB and 2×10^{14} cm^{-2} in EE) [1]. Such hadron fluences will degrade the crystal performance especially in the forward region of the detector where radiation levels are higher.

3. Radiation damage in $PbWO_4$ crystals

The main consequence of a high radiation environment in the $PbWO_4$ crystals will be the loss of transparency. This is due to two types of damage: the first is induced by ionizing radiation and the second by hadrons. The ionizing damage causes the formation of colour centres that reduce the transparency of the lead tungstate, leading to a light transmission loss which depends on dose rate [3]. The crystal transparency recovers in the periods without irradiation through spontaneous annealing at the ECAL operating temperature of 18°C. Hadrons induce a significantly different type of damage which involves the creation of clusters of colour centres characterized by the following features [9, 10, 11, 12]. First, there is almost no recovery at room temperature, hence the damage effect is cumulative. Second, the band edge of the transmission curve is shifted by several tens of nanometers to higher values leading to an overlap with the $PbWO_4$ emission peak.

Several irradiation campaigns have been performed in recent years to study the correlation between the hadron fluence and the loss of transparency which is usually quantified using the induced absorption coefficient:

$$\mu_{ind}(\lambda) = \frac{1}{L} \cdot \ln \frac{T_{before}}{T_{after}}$$

where T_{before} (T_{after}) is the longitudinal light tranmission value at $\lambda = 420$ nm measured before (after) irradiation and L is the length of the crystal. The increase of induced absorption coefficient with radiation is responsible for a crystal light output loss. Results combining irradiations of ECAL Endcap like crystals performed with 24 GeV protons in 2011 and 2012, as well as irradiation inside the CMS detector, are shown in Fig.1 [14].

4. Test beam results using double-ended read-out

The energy resolution of the CMS ECAL can be parameterized as:

$$\frac{\sigma_E}{E} = \frac{A}{\sqrt{E}} \oplus \frac{B}{E} \oplus C$$

where A represents a stochastic term, B the electronic noise and C a constant term [7].

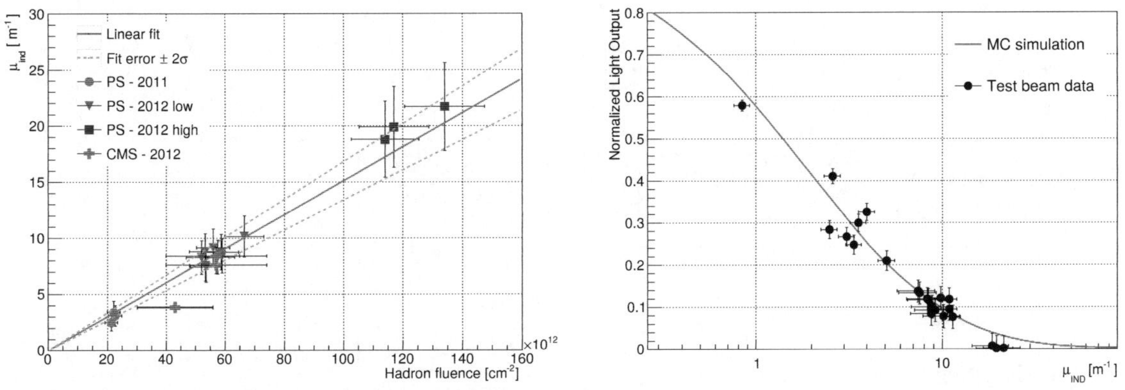

Figure 1. Left: linear fit (red line) of the correlation between the hadron fluence to which the crystals were exposed and the corresponding induced absorption coefficient. Right: loss of light output in proton-irradiated $PbWO_4$ crystals as a function of induced absorption, compared with simulation predictions (red line).

As discussed in [14], damaged crystals show a non linear response to electrons and a degradation of energy resolution. Due to loss of transparency and thus decrease of light output, both the stochastic and the noise terms degrade due to larger photostatistic fluctuations and smaller signal over noise ratio, respectively. These terms dominate at low energies (below 50 GeV) whereas at high energies the energy resolution is dominated by the constant term which also increases in irradiated crystals due to highly non-uniform efficiency of light collection along the longitudinal axis of the crystal. As the longitudinal position of electromagnetic shower maximum fluctuates event-by-event, the total light signal measured by the phototedector attached to the rear end of irradiated crystals will be affected by these fluctuations. This mechanism is also responsible for the non-linearity of response since more energetic showers will develop closer to the rear photodetector and thus the light signal will be less attenuated with respect to low energy electrons.

A double-ended read-out configuration allows to study in detail these effects as it provides the simultaneous read-out of light from both ends. For this reason, following a first set of test beam measurements performed on hadron-damaged $PbWO_4$ crystals [14], a further investigation of the mechanism underlying the performance degradation of irradiated crystals has been carried out by instrumenting a set of lead tungstate crystals with a double-ended read-out configuration. Each crystal was thus read-out from both ends using Vacuum Photo-Triodes (VPTs), identical to those used in the CMS ECAL detector [7], as shown in Fig. 2. The response of each crystal has been measured with electron beam of energy ranging from 10 to 250 GeV. The signal from the two VPTs was combined to estimate the position of shower maximum and thus compensate for its longitudinal fluctuations as illustrated in Fig. 3 according to $S_{corr} = \sqrt{F \cdot R}$, where F and R are respectively the signals measured by the front and rear VPT.

The non-linearity of crystal response due to radiation damage can be defined, by comparison with a non-irradiated crystal, as the change of response between 50 and 200 GeV normalized to beam energy:

$$\Delta NL = \frac{S_i(200)}{S_i(50)} \cdot \frac{S_{ni}(50)}{S_{ni}(200)}$$

where $S_i(200)$ ($S_{ni}(200)$) is the signal measured by the VPT for irradiated (non-irradiated) crystal at a given beam energy. Similarly, the increase of constant term due to radiation damage,

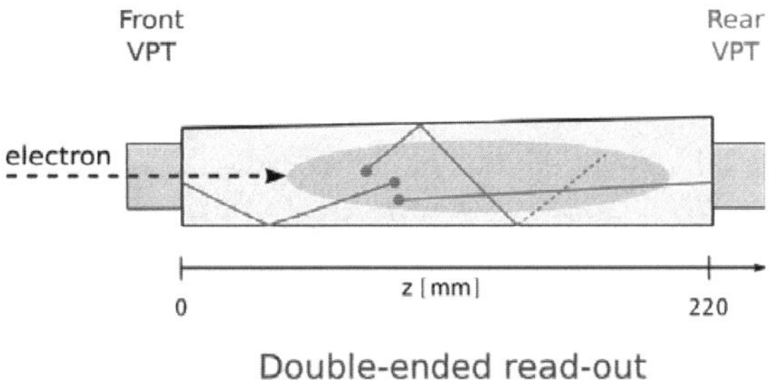

Figure 2. Schematic representation of a double-ended read-out configuration for a CMS ECAL $PbWO_4$ crystal using a pair of Vacuum Photo Triodes (VPTs). Red dots and lines represents optical scintillation photons produced by the electromagnetic shower and their propagation throughout the crystal volume.

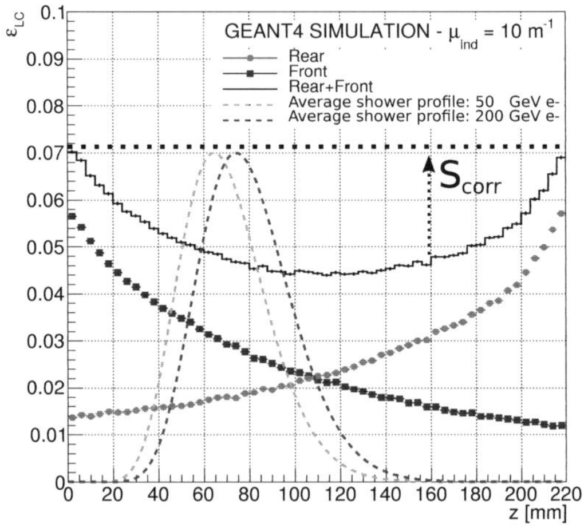

Figure 3. Geant4 simulation of average shower profiles for 50 and 200 GeV electrons in $PbWO_4$ superimposed to the light collection efficiency curves for front (blue) and rear (green) photodetectors corresponding to $\mu_{ind} = 10$ m^{-1}.

can be obtained by substracting the constant term of a non-irradiated crystal

$$\Delta C = \sqrt{C_i^2 - C_{ni}^2}$$

where C_i^2 (C_{ni}^2) is the constant term measured for an irradiated (non-irradiated) crystal. The performance of crystals in terms of non-linearity and degradation of constant term are shown in Fig. 4 for a single read-out configuration (front or rear VPT only) and for the combined signal obtained with a double-ended read-out configuration. Once the double read-out correction is applied, S_{corr}, the signal linearity is restored and the constant term degradation is mitigated with respect to a single read-out configuration from 10% to 3%. A Geant4 [15] simulation tool has been developed to model these effects and its predictions are compared with experimental data in the figures, showing a reasonable agreement with observed results.

5. Conclusions

A campaign of test beam measurement has been carried out to evaluate the performance of hadron-irradiated $PbWO_4$ crystals to electrons in the 10-250 GeV range. The crystals were instrumented with double-ended read-out using VPTs, identical to those used in the CMS ECAL detectors. Test beam results allow to understand the mechanism underlying the degradation of crystal performance due to radiation damage in term of light output loss, response non-linearity and constant term degradation and to build a Geant4 simulation model which describes such effects. Although for practical reasons, a double-ended read-out configuration cannot be implemented in situ in the CMS ECAL detector, the present results can be used to predict the performance of the standard single read-out configuration as used in the CMS ECAL during the harsh radiation environment foreseen for the High Luminosity phase of LHC.

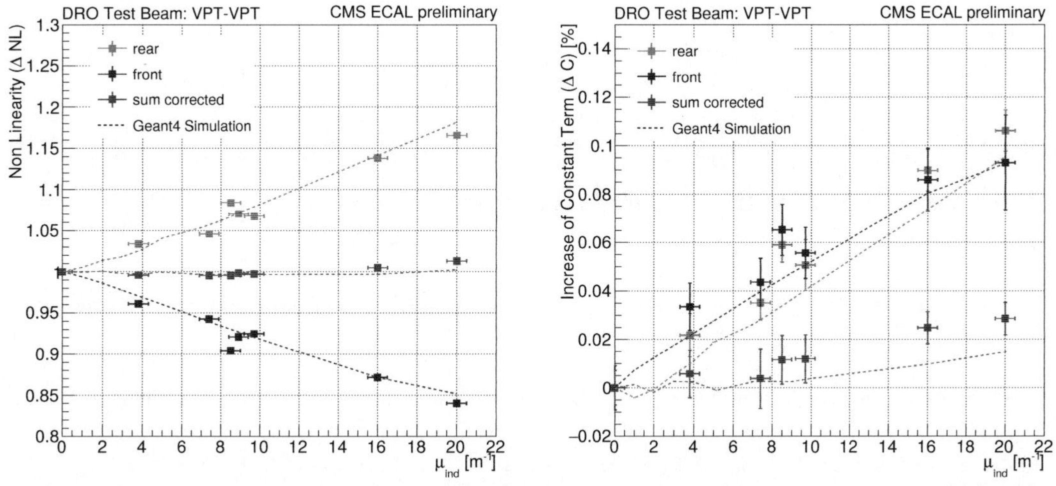

Figure 4. Test beam results showing the performance of hadron irradiated crystals in terms of response linearity (left) and constant term degradation (right). The performance is shown for single-ended (blue and green) and double-ended read-out (red) configurations and compared with prediction from a Geant4 model (dotted lines).

References

[1] The CMS Collaboration, *Technical Proposal for the Phase-II Upgrade of the CMS Detector*, CERN-LHCC-2015-010, CMS-TDR-15-02, LHCC-P-008 (CERN, Geneva, Switzerland, 2015).

[2] E. Auffray and A. Singovski, *Experimental Study of Lead Tungstate Scintillator Proton-Induced Damage and Recovery*, IEEE Transactions on Nuclear Science, Vol. 59, No. 5, 2012.

[3] M. Huhtinen, P. Lecomte, D. Luckey, F. Nessi-Tedaldi, F. Pauss, Nucl. Instr. and Meth. A 545 (2005) 63-87 and CMS Note 2005-010.

[4] P. Lecomte, D. Luckey, F. Nessi-Tedaldi, F. Pauss, D. Renker, Nucl. Instr. and Meth. A 587 (2008) 266 - 271 and CMS Note 2006-131.

[5] P. Lecomte, D. Luckey, F. Nessi-Tedaldi, F. Pauss, Nucl. lnstr. and Meth. A 564 (2006) 164-168.

[6] The CMS Electromagnetic Calorimeter Group, P. Adzic et al., 2010 JINST 5 P0301O.

[7] The CMS Collaboration, *The CMS Electromagnetic Calorimeter Technical Design Report*, CERN/LHCC 97-33, CMS TDR 4 (CERN, Geneva, Switzerland 1997).

[8] The CMS Collaboration, *CMS: The Compact Muon Solenoid – Technical Proposal*, CERN/LHCC/94-38 (CERN, Geneva, Switzerland, 1994).

[9] F. E. Maas, et al., Physical Review Letters 94 (2005) 082001.

[10] A. Annenkov, M. Korzhik, P. Lecoq., Nucl. Instr. and Meth. A 490 (2002) 30.

[11] P. Kozma, R. Bajgar, P. Kozma Jr., Nucl. Instr. and Meth. A 484 (2002) 149.

[12] M. Kobayashi, K. Harada, Y. Hirose, M. Ishii, I. Yamaga., Nucl. Instr. and Meth. A 400 (1997) 392.

[13] G. Dissertori, P. Lecomte, D. Luckey, F. Nessi-Tedaldi, F. Pauss, Th. Otto, S. Roesler, Ch. Urscheler, Nucl. Instr. and Meth. A 622 (2010) 41-48.

[14] T. Adams et al., *Beam test evaluation of electromagnetic calorimeter modules made from proton-damaged PbWO4 crystals*, 2016 JINST 11 P04012.

[15] GEANT4 collaboration, S. Agostinelli et al., *GEANT4: A Simulation toolkit*, NIM A 506 (2003) 250..

Progress in the Development of the Lead Tungstate Crystals for EM-Calorimetry in High-Energy Physics

R. W. Novotny[1], K.-T. Brinkmann[1], A. Borisevich[2], V. Dormenev[1], J. Houzvicka[3], M. Korjik[2], and H.-G. Zaunick[1]

[1]Justus-Liebig-Universität Giessen, II. Physikalisches Institut , Heinrich-Buff-Ring 16 D 35392 Giessen, Germany
[2]Institute for Nuclear Problems, Bobruiskaya11, 220030, Minsk, Belarus
[3]CRYTUR, spol. s.r.o., Palackeho 175, CZ-511 01, Turnov, Czech Republic

Email: r.novotny@exp2.physik.uni-giessen.de

Abstract. Even at present time there is a strong interest and demand for high quality lead tungstate crystals ($PbWO_4$, PWO) for electromagnetic (EM) calorimetry. PWO is implemented into the EM calorimeter of the CMS-ECAL detector at LHC [1] and required for the completion of the PANDA EMC [2] and various ongoing detector projects at Jefferson Lab. The successful mass production of PWO using the Czochralski method was stopped after bankruptcy of the Bogoroditsk Technical Chemical Plant (BTCP) in Russia as major producer so far. The Shanghai Institute of Ceramics, Chinese Academy of Science (China) was considered as an alternative producer using the modified Bridgman method. The company CRYTUR (Turnov, Czech Republic) with good experience in the development and production of different types of inorganic oxide crystals has restarted at the end of 2014 the development of lead tungstate for mass production based on the Czochralski method. An impressive progress was achieved since then. The growing technology was optimized to produce full size samples with the quality meeting the PANDA-EMC specifications for PWO-II. We will present a detailed progress report on the research program in collaboration with groups at Orsay and JLab. The full size crystals will be characterized with respect to optical performance, light yield, kinetics and radiation hardness.

1. Introduction

The lead tungstate ($PbWO_4$, PWO) scintillation crystal is one of the most widely used scintillation materials in electromagnetic calorimetry in modern accelerator experiments. It combines an unique set of physical and scintillation properties, which allow the construction of an affordable compact electromagnetic calorimeter (EMC). The first generation of mass production of PWO-I crystals was used in the electromagnetic calorimeter of the CMS detector at LHC. The technology of mass production was optimized to obtain detector units with high radiation hardness and less stringent requirements to the light yield [1] due to the high photon energies to be expected. The production technology of the next generation of lead tungstate, PWO-II for the PANDA EMC, was optimized to increase the light yield without degradation or even with achieving improvement of the radiation hardness [2]. The operating temperature of the PANDA EMC will be -25° C which enhances the light

yield by a factor of four due to the reduction of the quenching of luminescence but blocks statistical recovery of radiation damage during operation. The crystal material is considered as well as a candidate for some future experiments [3,4].
Crystals for CMS and partly for the PANDA calorimeter were produced by Czochralski method by the Bogoroditsk Technological Chemical Plant (BTCP, Bogoroditsk, Russia) until its shutdown some years ago. As an alternative producer for mass production of PWO crystals, Shanghai Institute of Ceramics, Chinese Academy of Science (SICCAS, Shanghai, China) was considered. It produces lead tungstate crystals by the modified Bridgman method.
The company CRYTUR (Turnov, Czech Republic) has a good experience in the development and production of different types of inorganic oxide crystals and has initiated a research program to develop the technology of mass production of lead tungstate crystals.

Table 1. The relevant parameters defining the PWO quality required for the PANDA-EMC. The parameter dk describes the radiation induced coefficients calculated as $dk = [ln(T_b/T_a]/d$, where T_b, T_a are the optical transmittances before and after irradiation, correspondently, and d is the thickness of the PWO sample. LY is the abbreviation for light yield, which is measured in units of photoelectrons per MeV, phe/MeV, deposited energy.

Parameter	Limit [a]
optical transmittance at 360 nm	≥ 35 %
optical transmittance at 420 nm	≥ 60 %
optical transmittance at 620 nm	≥ 70 %
dk (420 nm)	< 1.1 m^{-1}
LY at T= +18° C within 100 ns integration time gate	≥ 16 phe/MeV
LY (100 ns)/LY(1000 ns) at T= +18° C	> 90 %

The aim of the R&D program was to obtain full size (up to 20 cm) lead tungstate crystals produced again by Czochralski method according to the PANDA EMC specifications as listed in table 1. This paper reports on results of the ongoing investigations.

2. Survey based on small samples

In the first stage, 11 series of lead tungstate ingots were grown under different conditions with respect to the atmosphere and the appropriate selection and concentration of dopants. In each case two small samples were cut from the top and bottom part of the ingot to study the homogeneity of the crystal quality. The samples have different shapes but the thicknesses are close to 1 cm with the two opposite sides optically polished to perform transmittance and light yield measurements. Besides the optical and luminescence properties the radiation hardness was investigated by determining the change of the transmission after γ-irradiation (^{60}Co) with an integral dose of 30 Gy at a dose rate of ~ 2 Gy per minute measured at room temperature using a Varian Cary 4000 spectrophotometer. The light yield was measured using a calibrated Hamamatsu R2059-01 photomultiplier at a temperature of +18°C and for selected samples at -25°C, respectively. TYVEK paper was used as reflector material.
As expected from previous experiences, the growth in N_2 + O_2 atmosphere and the doping with a sufficient amount of La/Y ions provided the best transmission characteristics near and below the emission wavelength of 420 nm as well as a good radiation hardness.

3. First results of large size crystals

As a crucial test of the new developed and installed furnace, full size ingots have been grown. The very first ingot showed a yellow coloring due to over-doping but the next ones were clear and visually transparent but slightly distorted due to an insufficient guiding of the pulling rod. The next obtained ingots have been used to produce the first large size crystals with dimensions of 20x20x200 mm^3 or 16x16x100mm^3, respectively. The tests were primarily focusing on the reliability of the furnace. The crystals were initially grown in argon and finally in a N_2 atmosphere. As raw material not anymore used PWO-II crystals produced at BTCP have been chosen, since the optimization of the crystal production at CRYTUR will be based on the identical raw material selected and already pre-mixed in a specific stoichiometric composition optimized for BTCP. All samples were optically polished. The rectangular shape was chosen to allow a reliable investigation of the homogeneity of the crystal by measuring the optical transmittance and other parameters even transversal at different positions along the length of the crystal.

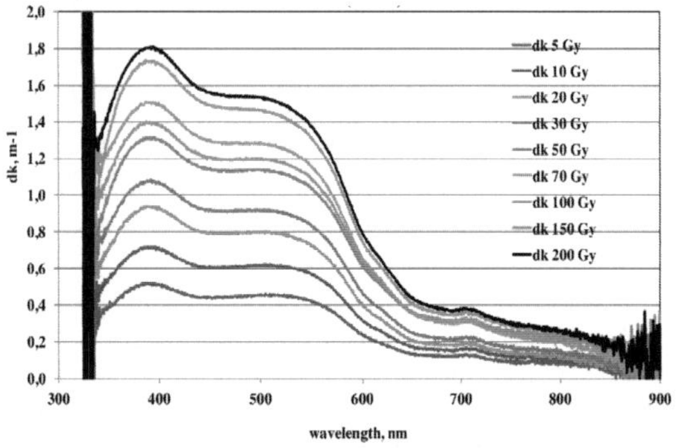

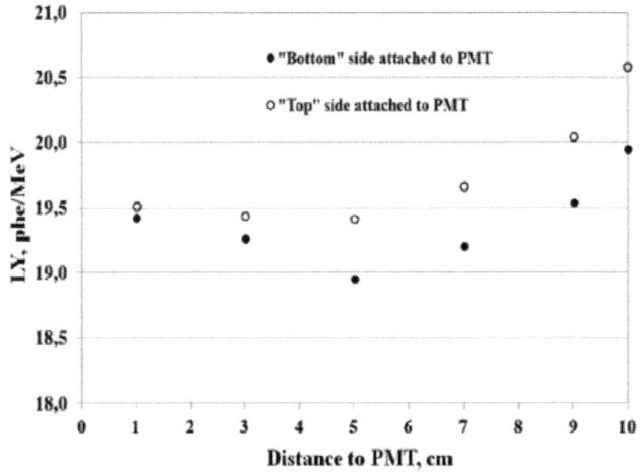

Figure 1. (left) Distribution of the longitudinal induced absorption coefficient of a 100 mm PWO crystal exposed to an integrated dose of up to 200 Gy. (right) The light yield measured as a function of position relative to the photo sensor for both crystal orientations. The positions "Top" and "Bottom" were randomly assigned to opposing ends of the crystal.

The shown sample of 100 mm length was free of visible imperfections or scattering centers. The optical performance showed only small variations along the crystal. However, the absolute value of the longitudinal transmission is significantly below the typical distribution of a former BTCP crystal. But the position of the absorption edge is nearly identical. The distribution of the induced absorption coefficient shows the typical behavior and the absolute value of dk at 420nm just meets the requirement to stay below 1.1m^{-1}. In order to test the homogeneity and the overall yield of luminescence, the light yield has been measured at room temperature for different locations of the ^{137}Cs source. On average, the observed light yield appears to reach the required limit even when taking into account the shorter crystal length. Figure 1 illustrates the already achieved radiation hardness and light yield.

4. The present performance of full size rectangular crystals

The most recent ingots show already a significant overall improvement with respect to colouring and overall size, which was achieved by modifications and optimizations of the growing technology and the fine tuning of the doping. Several rectangular crystals of 200 mm length and a cross section of 20x20 mm^2 have been produced in nearly identical quality with respect to all inspected performance parameters.

Figure 2 shows a photograph of one of these ingots confirming the clear transparency and the very symmetric shape. The luminescence yield measured at room temperature reaches a value of 17.7 phe/MeV for an integration gate of 100 ns. This value correspond to more than 92% compared to the collected yield within 1μs. Even the response at a temperature of -25°C does not indicate any slow contributions.

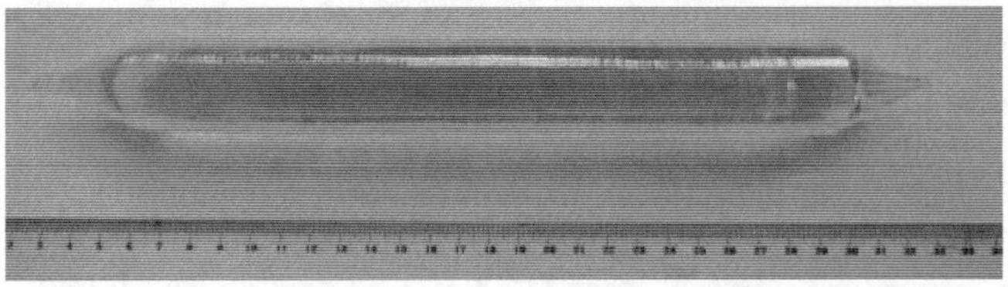

Figure 2. Photograph of one of the most recently grown PWO ingots (serial number 81).

Figure 3 documents the achieved radiation hardness for a sample optimized for the light yield. The distributions show the induced absorption coefficient in longitudinal and transverse direction after imposing an integral dose of 30 Gy (^{60}Co). The transverse measurement confirms as well the homogeneity of the distribution of dopants.

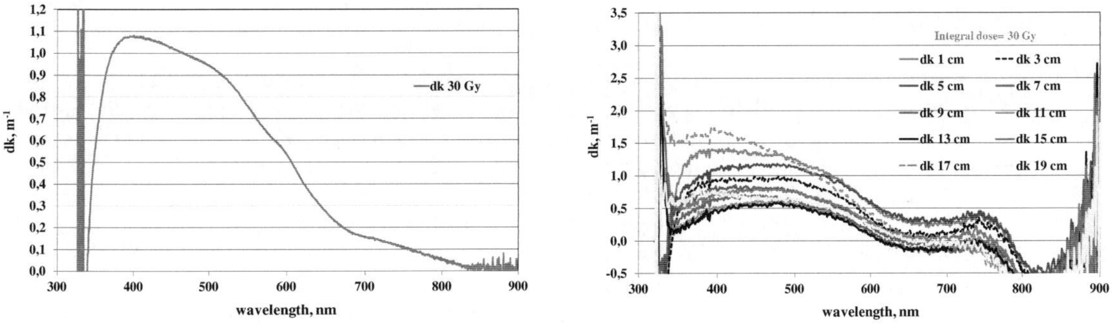

Figure 3. The longitudinal and transverse induced absorption coefficient as a function of wavelength measured at room temperature after imposing an integral dose of 30 Gy (^{60}Co). The results are determined for a 20x20x200 mm^3 full size crystal (# 79).

5. Conclusions and outlook

The development and manufacturing of PWO crystals was successfully restarted by the company CRYTUR. The obtained results confirm the first milestone to produce crystals very close to the PWO-II quality as reached by the former producer BTCP in spite of significant modifications of the technology. As a next step, the production of tapered crystals has been started as required for the barrel section of the PANDA-EMC. A pre-production of ~100 crystals will provide the input for any further optimization and to determine the overall conditions for mass production of the remaining 8500 crystals for PANDA.

References

[1] E. Auffray et al., "Status of PWO crystal production for the electromagnetic calorimeter of CMS and of its construction," Nucl. Instr. and Meth. in Phys. Res. A, vol. 537, Issues 1-2, pp. 373-378, Jan. 21, 2005.

[2] A. Borisevich et al., "Lead tungstate crystal with increased light yield for the PANDA electromagnetic calorimeter," Nucl. Instr. and Meth. in Phys. Res. A, vol. 537, Issues 1-2, pp. 101-104, Jan. 21, 2005.

[3] Deeply Virtual Compton Scattering with CLAS at 6 GeV, Experimental Proposal, 2001 (http://www.jlab.org/exp prog/proposals/01/PR01-113.pdf).

[4] M. Battaglieri, et. al. (CLAS Collaboration), The CLAS12 Forward Tagger, Technical Design Report, 2012 (http://www.ge.infn.it/ batta/jlab/fttdr.2.0.pdf).

IOP Conf. Series: Journal of Physics: Conf. Series **928** (2017) 012032 doi:10.1088/1742-6596/928/1/012032

Performances of Crystal Scintillators in a Severe Radiation Environment Caused by Gamma Rays

F Yang, L Y Zhang and R-Y Zhu
256-48, HEP, California Institute of Technology, Pasadena, CA 91125, USA
E-mail: zhu@hep.caltech.edu

Abstract. Because of their superb energy resolution and detection efficiency crystal scintillators are widely used in high energy and nuclear physics experiments. A crucial issue is radiation damage. We report an investigation on γ-ray induced radiation damage in crystal scintillators of large size, including BaF_2, BGO, CeF_3, pure CsI, LSO/LYSO/LFS and PWO with an integrated dose up to 340 Mrad at a dose rate up to 1 Mrad/h. Optical and scintillation properties of these crystal samples were measured before and after irradiations. The result shows that pure CsI crystals have good radiation hardness below 100 krad. BaF_2, BGO and LYSO crystals show good radiation hardness beyond 1 Mrad. In terms of light output degradation LSO/LYSO/LFS is clearly the best among all scintillation crystals.

1. Introduction

Because of their superb energy resolution and detection efficiency, crystal scintillators are widely used in HEP experiments. The CMS lead tungstate ($PbWO_4$ or PWO) crystal calorimeter, for example, has played an important role in the discovery of the Higgs boson [1]. One crucial issue, however, is their radiation damage in a severe radiation environment, which requires precision monitoring to correct variations in crystal's transparency [2]. We report an investigation on γ-ray induced radiation damage in various crystal scintillators of more than 20 cm length, such as BaF_2 (20×20×250 mm^3), BGO (25×25×200 mm^3), CeF_3 (33×32×191 mm^3), pure CsI (50×50×300 mm^3), LYSO (25×25×200 mm^3) and PWO (28.5^2×30^2×220 mm^3). The 30 cm long pure CsI sample was irradiated to 1 Mrad, and was cut to 20 cm for irradiations beyond 1 Mrad. Gamma-ray irradiations up to 1 Mrad were carried out at two facilities at Caltech: an open ^{60}Co source and a sealed ^{137}Cs source. The former provides dose rates between 2 and 100 rad/h by placing samples at appropriate distances. The later provides a dose rate of about 7 krad/h in 2015 with 5% uniformity along the sample's longitudinal axis when the samples are placed at the center of the irradiation chamber [3]. Gamma-ray irradiations beyond 1 Mrad were carried out at the Total Ionization Dose (TID) facility of Jet Propulsion Laboratory (JPL), where a group of high intensity ^{60}Co sources provides an adjustable dose rate up to 1 Mrad/h in an opening throat of 10×10×13.5 inch. The irradiation at the TID facility was carried out in steps: 9, 90 and several steps of 100 Mrad each to reach 340 Mrad. The dose rate was 180 krad/h for the 9 Mrad irradiation and 1 Mrad/h for the rest.

2. LSO/LYSO/LFS Crystals

LYSO, LSO and LFS crystals by six different vendors are investigated. Figure 1 shows normalized emission weighted longitudinal transmittsnce (EWLT, top) and light output (LO, bottom) as a function of the integrated dose for six LYSO/LSO/LFS samples. EWLT is defined as [4]:

$$EWLT = \frac{\int LT(\lambda)Em(\lambda)d\lambda}{\int Em(\lambda)d\lambda} \qquad (1)$$

IOP Conf. Series: Journal of Physics: Conf. Series **928** (2017) 012032 doi:10.1088/1742-6596/928/1/012032

The losses of EWLT and LO are about 35% and 42% respectively in OET LFS after 340 Mrad. The average LO loss after 10 Mrad is about 18% with a divergence of less than 7.5%, indicating good consistency of radiation hardness for crystals of this type from six vendors.

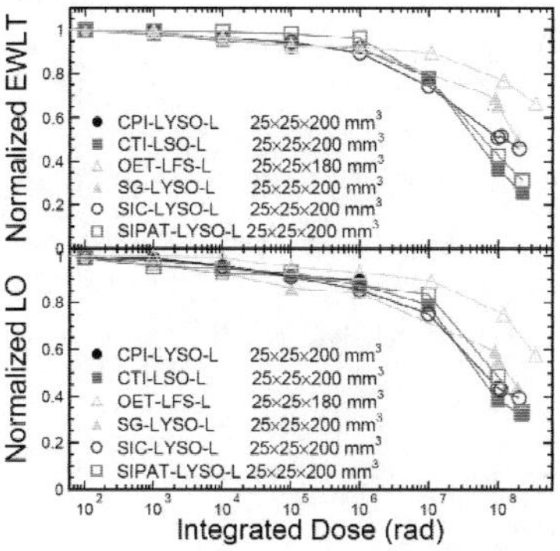

Figure 1. Normalized EWLT and LO are shown as a function of integrated dose up to 340 Mrad for six LYSO/LSO/LFS crystals from different vendors.

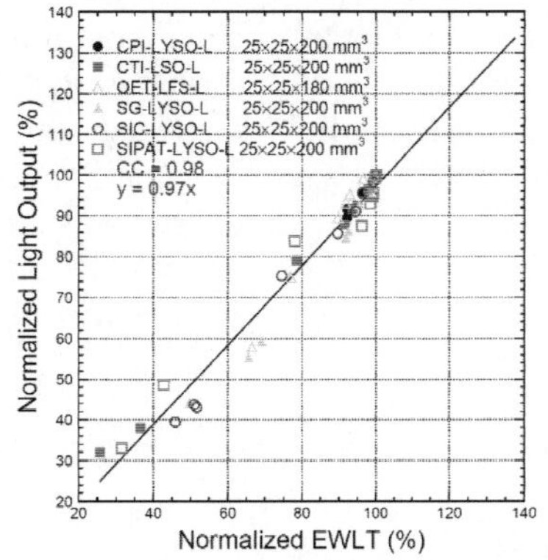

Figure 2. Correlation between normalized EWLT and LO is shown for six LYSO/LSO/LFS crystals from different vendors.

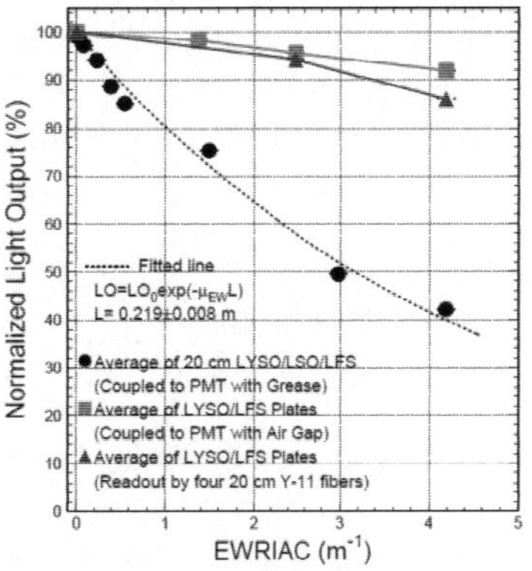

Figure 3. Normalized LO as a function of EWRIAC is shown for 20 cm long LYSO/LSO/LFS crystals and LYSO/LFS plates of 14×14×1.5 mm^3.

Figure 2 shows correlation between normalized EWLT and LO for six LYSO/LSO/LFS crystals. Also shown in the plot is a linear fit and the linear correlation coefficients (CC), defined as [5]:

$$\mathrm{CC} = \frac{\sum(x-\bar{x})(y-\bar{y})}{\sqrt{\sum(x-\bar{x})^2 \sum(y-\bar{y})^2}} \qquad (2)$$

An excellent positive correlation of 97% is observed between normalized EWLT and LO for six LYSO/LSO/LFS crystals from different vendors, indicating that LO loss in crystals of this type is caused by transmittance loss, so may be corrected by using a light pulse based monitoring system. Figure 3 shows normalized LO as a function of emission weighted radiation induced absorption coefficient (EWRIAC) for 20 cm long LYSO/LSO/LFS crystal samples as well as 14×14×1.5 mm^3 LYSO/LFS plates with five holes designed for the proposed LYSO/W Shashlik calorimeter [6]. While the LO of 20 cm long crystals (black dots) was measured by a PMT coupled to the samples directly, the LO of plates was measured by two methods. One is with the plates coupled to a PMT directly (red squares). The other is with the plate coupled a PMT through four Y-11 wavelength shift fibers (blue triangles). After 100 Mrad irradiation, the LYSO/LSO/LFS samples have a EWRIAC value of 3 m^{-1}. The corresponding measured LO loss is about 50% for 20 cm long crystals, and 4% and 6% respectively for the 14×14×1.5 mm^3 plates directly coupled to the PMT and through Y-11 WLS fibers. Compared to 20 cm long crystals, about a ten times better radiation hardness is observed by reducing the light path length in 14×14×1.5 mm^3 plates, which greatly enhances the radiation hardness of the LYSO/W Shashlik calorimeter as compared to a LYSO calorimeter of total absorption. An exponential fit reveals that the average scintillation light path length in 20 cm long crystals is about 22 cm. The data shown in Figure 3 may be used to estimate performance of LYSO/LSO/LFS crystals under γ-ray irradiation.

3. BaF_2, CeF_3, CsI, BGO and PWO crystals

It is well known that γ-ray induced radiation damage in BaF_2 and pure CsI crystals does not recover at room temperature [7, 8], so is dose rate independent. Figure 4 shows normalized EWLT (top) and LO (bottom) of the fast scintillation component as a function of integrated dose for three BaF_2 crystals from different vendors, where the fast component refers to scintillation light peaked at 220 nm with scintillation decay time of less than 1 ns. While the sample SIC2012 was grown at SICCAS in 2012, the samples BGRI 2015 and Incrom 2015 were grown at BGRI and Incrom respectively in 2015. Figure 5 shows the normalized EWLT (top) and LO (bottom) as a function of the integrated dose for two pure CsI crystal samples from different vendors. The 30 cm long samples SIC2013 was grown at SICCAS in 2013. The 23 cm long sample Kharkov 1 was grown at Kharkov in 2015. The sample SIC2013 was 30 cm long when it was irradiated up to 1 Mrad, and was cut to 20 cm for irradiations beyond 1 Mrad at the TID facility of JPL. The samples from SICCAS and Kharkov show consistent radiation hardness. The LO is about 80% after 100 krad, and was reduced to about 30% after 1 Mrad, indicating that pure CsI is radiation hard at low doses up to 100 krad but not beyond. It is also noticed that the radiation hardness of sample SIC 2013 is consistent with the sample Kharkov 2015 and two Kharkov samples grown twenty years ago [8].

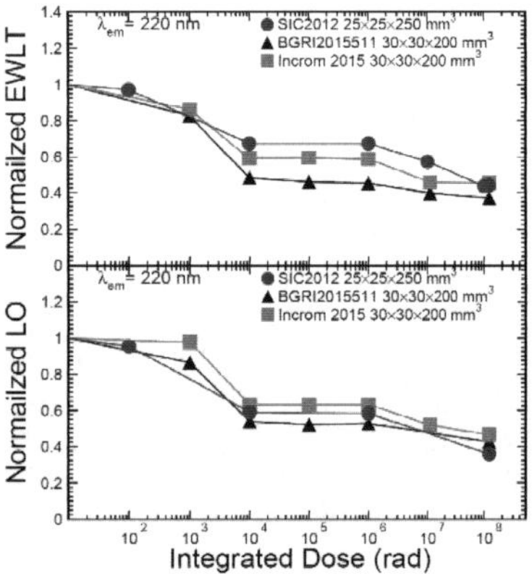

Figure 4. Normalized EWLT and LO of the fast scintillation component as a function of integrated dose for three BaF_2 crystals from different vendors.

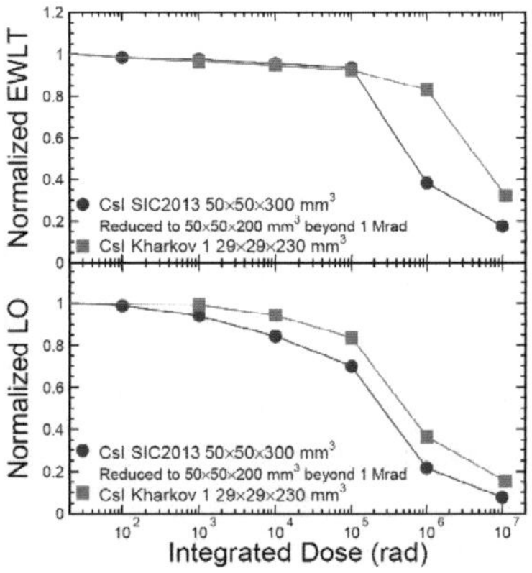

Figure 5. Normalized EWLT and LO are shown as a function of integrated dose for two pure CsI crystals from different vendors.

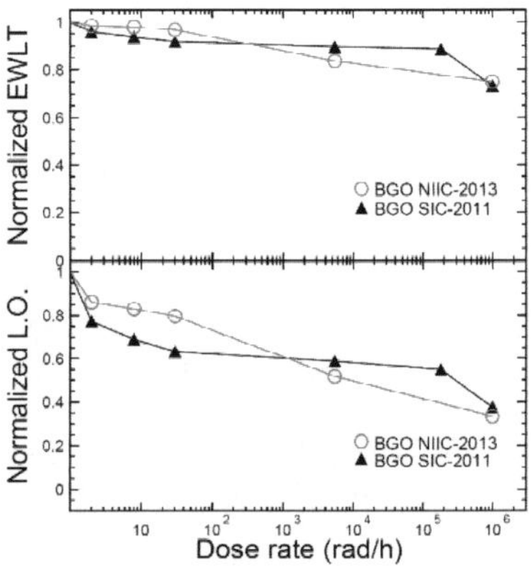

Figure 6. Normalized EWLT (top) and LO (bottom) are shown as a function of dose rate for two BGO crystals from different vendors.

Radiation damage in BGO, CeF_3 and PWO crystals recovers under room temperature, leading to a dose rate dependent damage [9, 10]. Figure 6 shows the normalized EWLT (top) and LO (bottom) in equilibrium as a function of the dose rates for two BGO crystal samples from different vendors. The radiation hardness of these two samples from different vendors is more or less consistent. Early BGO crystals produced for the L3 experiment, however, are not as radiation hard as these samples [10], indicating that crystal quality is improved during mass production for the medical industry.

Figure 7 shows the normalized EWLT (top) and LO (bottom) in equilibrium as a function of the dose rates for two CeF_3 crystal samples. Consistent radiation hardness was observed in these samples. The LO of CeF_3 crystal SIC2014 was too low to be measured under the 1 Mrad/h irradiation. Both samples were grown twenty year ago. Development is required to improve radiation hardness of CeF_3 for this material to be used in a severe radiation environment.

Figure 8 shows the normalized EWLT (top) and LO (bottom) in equilibrium as a function of the dose rates for two PWO-II crystals recently grown by BTCP for Panda and one PWO crystal recently grown by SICCAS for JLAB. Also shown in the plot is a PWO crystal grown about fifteen years ago by SIC for CMS. Previous investigations show that the radiation hardness of PWO crystals against ionization dose is diverse, and has no correlation with their initial optical quality [5]. Although selected samples grown recently are better than that grown for CMS about 15 years ago, quality control is still required to make sure PWO crystals used to construct future calorimeter meet a well-defined specification for crystal's radiation hardness.

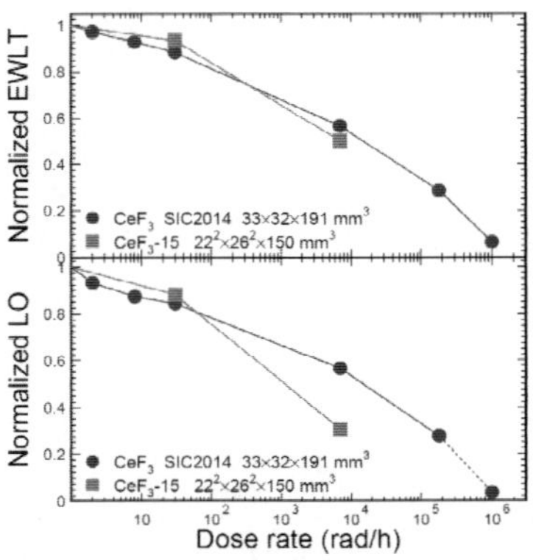

Figure 7. Normalized EWLT (top) and LO (bottom) are shown as a function of dose rate for two CeF_3 crystals

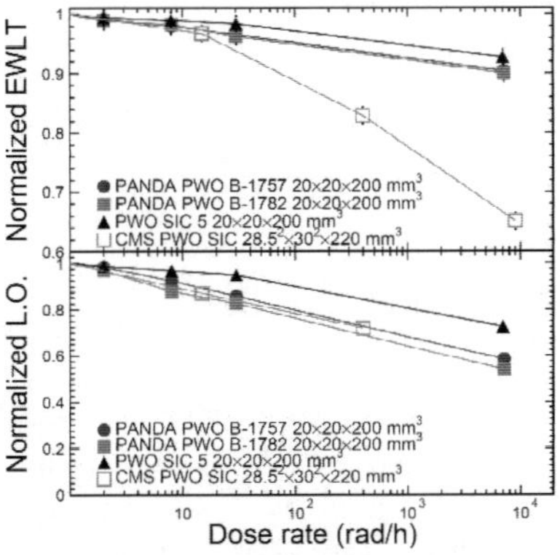

Figure 8. Normalized EWLT (top) and LO (bottom) are shown as a function of dose rate for four PWO crystal samples

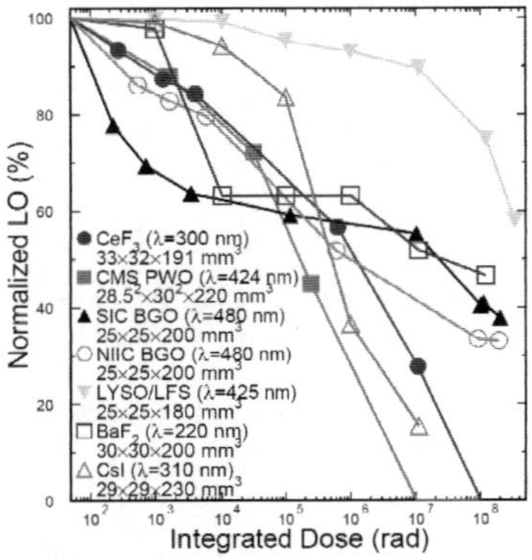

Figure 9. Normalized LO as a function of integrated dose for various crystals

4. Summary

In this investigation no recovery was observed after 200 Mrad irradiation in two LYSO samples, indicating radiation damage in LYSO is not dose rate dependent. Consistent degradation in transmittance and light output was observed for 20 cm long LYSO/LSO/LFS crystals from six vendors, indicating that radiation damage in commercially produced crystals of this type is under control. Because of its high cost due to expensive raw materials and high melting point, however, future crystal calorimeters may consider alternative cost-effective crystals.

Ionization induced radiation damage in LSO/LYSO/LFS, BaF_2 and pure CsI does not recover under room temperature. That in BGO, CeF3 and PWO does, leading to a dose rate dependent damage. A dose rate dependent damage level requires frequent monitoring and thus detector cost.

Ignoring the dose rate dependence, Figure 9 shows the normalized LO as a function of integrated dose for various long crystal samples. In terms of light output loss, LYSO clearly shows the best radiation hardness among all crystal scintillators. The best sample of this type maintains 75% and 60% LO after 100 and 340 Mrad. On the other hand, BGO and BaF_2 crystals also maintain 35% and 45% LO respectively after 200 and 120 Mrad, so may be considered as cost-effective alternatives for future HEP experiments with severe radiation environment. Pure CsI shows good radiation hardness below 100 krad. Because of its low cost, it is a good alternative for future HEP experiments in a modest radiation environment.

Acknowledgement

This material is based upon work supported by the U.S. Department of Energy, Office of Science, Office of High Energy Physics program under Award Number DE-SC0011925.

References

[1] S. Chatrchyan, V. Khachatryan, A. M. Sirunyan, A. Tumasyan, W. Adam, E. Aguilo, *et al.* 2012 *Physics Letters B* **716** 30
[2] R. Y. Zhu 2004 *IEEE T Nucl Sci* **51** 1560
[3] J. M. Chen, R. H. Mao, L. Y. Zhang, and R. Y. Zhu 2007 IEEE T Nucl Sci **54** 1319
[4] J. M. Chen, R. H. Mao, L. Y. Zhang, and R. Y. Zhu 2007 *IEEE T Nucl Sci* **54** 718
[5] J. M. Chen, R. H. Mao, L. Y. Zhang, and R. Y. Zhu 2007 *IEEE T Nucl Sci* **54** 375
[6] L. Y. Zhang, R. H. Mao, F. Yang, and R. Y. Zhu 2014 *IEEE T Nucl Sci* **61** 483
[7] R. Y. Zhu 1994 *Nucl Instrum Meth A* **340** 442
[8] Z. Y. Wei and R. Y. Zhu 1993 *Nucl Instrum Meth A* **326** 508
[9] R. Y. Zhu 1998 *Nucl Instrum Meth A* **413** 297
[10] F. Yang, R. H. Mao, L. Y. Zhang, and R. Y. Zhu 2012 *J. Phys.: Conf. Ser* **404** 012025

Hadron calorimetry test bench

Dmitri Kotchetkov[1], Justin Frantz[1], Nowo Riveli[1], Victor Mayatsky[2], Vladimir Brovin[2], Valentina Logunova[2], Marina Alexeeva[2], Nadezhda Dmitrienko[2], Boris Kosyanchuk[3], Yuri Pankov[3], Alexander Kozlov[3] and Terry Awes[4]

[1] Department of Physics and Astronomy, Ohio University, Athens, OH 45701, U.S.A.
[2] OOO Uniplast, Bolshaya Nizhegorodskaya St. 77, Vladimir, 600016, Russia
[3] ZAO Polizell, Bolshaya Nizhegorodskaya St. 77, Vladimir, 600016, Russia
[4] Physics Division, Oak Ridge National Laboratory, 1 Bethel Valley Rd., Oak Ridge, TN 37831, U.S.A.

E-mail: dmitrikotchetkov@gmail.com

Abstract. A reconfigurable sampling hadron calorimetry test bench was built. Different scintillator-absorber stack configurations can be easily constructed within the test bench for extensive detector performance studies. Three configurations, with absorber layers of uniform thicknesses of 48 mm, 24 mm, and 16 mm, and with scintillator layers of uniform thicknesses of 4 mm, were studied with electron and hadron beams.

1. Test benches as research and development tools

Sampling calorimetry test benches are among tools that allow diversified, yet inexpensive, research and development projects. Using a test bench, one can carry extensive studies that lead to better optimization of a final detector design, meanwhile reducing the cost of the detector and the cost of R&D. Configurations of a scintillator-absorber stack can be tuned to optimal constant or varying sampling fractions, in order to achieve the best detector performance for the energies of interest. Compensation, metrological, and mechanical engineering options can be optimized with the test bench, as well. Different absorber materials, readout devices and readout electronics can be tried out. Reconfigurable and expandable test benches can offer wealth of studies before one actually progresses to building a prototype detector.

2. Reconfigurable test bench

We designed and fabricated a reconfigurable test bench for hadron calorimetry studies. The components and corresponding quantities of the test bench set are listed in table 1.

The box-case (figure 1), in which the scintillator-absorber stack is placed, is assembled from steel components. Its support is made of a 30 mm thick steel sheet welded to one I-beam and twelve C-beams. Two frames made of sheets with milled windows are each attached to the front and rear ends of the support. Two side walls are attached to the C-beams of the support and to the front and rear frames. Each side wall has milled slots that run with a step of 120 mm along the length of the wall. The scintillator-absorber stack is assembled by placing absorber plates and scintillator tiles on their edges, orthogonally to the support and to the side walls of the box-case. The slots in the opposing walls of the box-case allow fixing the stack for a certain chosen total thickness, using a bracket inside the box-case. If needed, extra steel plates are placed at the end of the scintillator-absorber stack to eliminate a gap between the stack and the bracket.

Table 1. Test bench set.

Component	Quantity
steel box-case	1
scintillator tile (350 x 350 x 4 mm^3)	210
scintillator tile (350 x 350 x 5 mm^3)	14
non-scintillator tile (350 x 350 x 2 mm^3)	232
antimonial lead alloy plates (350 x 350 x 2 mm^3)	380
steel plates (350 x 350 mm^2, various thicknesses)	5
36-fiber bundle (clear waveguides, 2 m)	6
12-fiber bundle (clear waveguides, 2.25 m)	2
4-fiber bundle (clear waveguides, 2 m)	2
2-fiber bundle (clear waveguides, 1 m)	2

The thin antimonial lead alloy plates (antimony content is 10%) allow making absorber layers with different thicknesses, which are multiples of 2 mm. The scintillator-absorber configurations can be chosen with uniform or variable thicknesses of the absorber layers through the depth of the test bench.

The scintillator tiles (figure 2) are made of 98.3% polystyrene, 1.7% p-Terphenyl and 0.01% POPOP. The tiles are enveloped with aluminum foil which, in turn, is enveloped with thin plastic film. Wall plastic paper is put on both surfaces of each tile to protect the foil and thin film envelopes from scratches.

Figure 1. Steel box-case.

Figure 2. Scintillator tiles.

The non-scintillator polystyrene tiles were made to research mechanical design options for real hadron calorimeters. Such non-scintillator tiles can be placed between the scintillator tiles and the absorbers to see how the energy resolution and compensation might be affected by such placement.

Kuraray Y-11 wavelength shifting fiber with a diameter of 1.2 mm is glued, with optical cement, in grooves made in the scintillator tile. Two fiber ends exit the tile through steel tubes glued at both ends of the groove. Equal quantities of the "left" (figure 3) and the "right" (figure

4) scintillator tiles were made (dimensions in the figures are in milimeters). The fiber layout maps of the "left" and the "right" tiles are mirror images of each other; the "left" and the "right" tiles alternate in the scintillator-absorber stack.

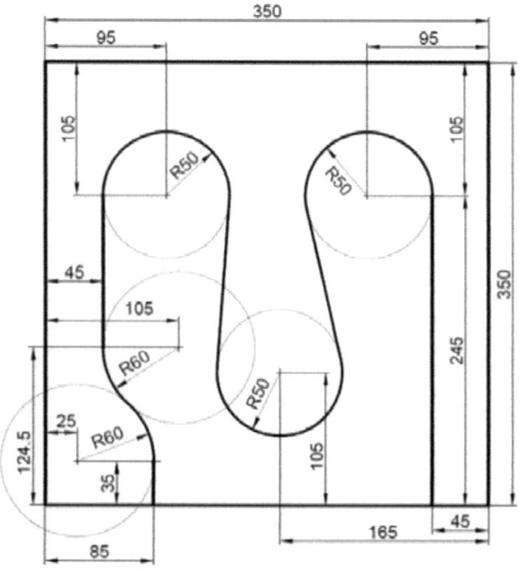

Figure 3. "Left" scintillator tile.

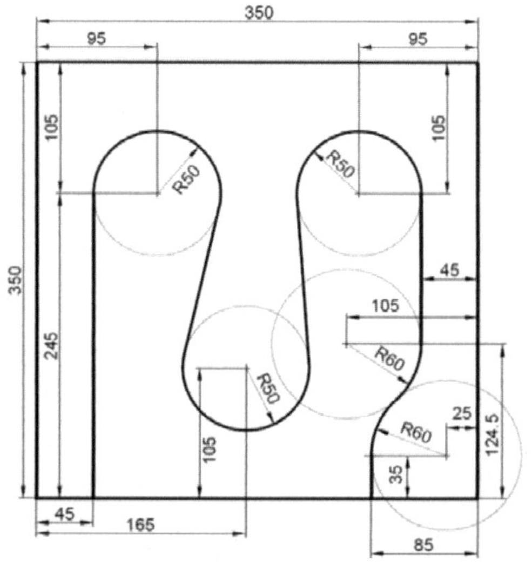

Figure 4. "Right" scintillator tile.

Upon exiting the tile, the steel tube, with the inserted wavelength shifting fiber, is coupled with a clear waveguide fiber that transmits the light to a photomultiplier tube (PMT). Several clear waveguide fibers are bundled together, using a mechanical collet (figure 5), so the cross-sectional surfaces of the clear waveguide fibers can be directly coupled with the PMT window (figure 6). This way, every PMT can serve several scintillator tiles. Depending on chosen thicknesses of the absorber layers, one or several photomultipliers detect the light from the scintillator-absorber stack.

Figure 5. Fiber bundle collet.

Figure 6. Bundles coupled to PMTs.

The 5 mm thick scintillator tiles were made to study lateral shower leaks. For such studies, these tiles can be attached to the side walls of the box-case.

3. Test beam studies

Performance of the test bench was preliminarily studied at T9 beam line of CERN Proton Synchrotron with beam momenta ranging from 1 GeV/c to 10 GeV/c. Energy signals from hadron (negative pion) and electron beams were measured with three test bench configurations. In the first configuration, 15 scintillator-absorber layers were used. The thickness of every absorber layer was 48 mm. A total of 16 scintillator tiles (one tile was placed at the end of the stack) were read out by 32 fibers of 36-fiber bundle, using one PMT. In the second configuration, the number of scintillator-absorber layers increased to 31. The thickness of each absorber layer was reduced to 24 mm. 32 scintillator tiles (one at the end of the stack) were read out by 64 fibers of two 36-fiber bundles, using two PMTs. In the third configuration, the data were taken with 47 scintillator-absorber layers. The thickness of every absorber layer was 16 mm. 48 scintillator tiles (again, one tile was placed at the end of the stack) were read out by 96 fibers of three 36-fiber bundles with three PMTs. The gains of the PMTs (Philips XP2264/H04) were kept the same for all measurements except for the test of the third configuration with the hardon beam, when the gains were reduced by about a factor of 4. The electromagnetic and hadronic energy resolution functions for all three configurations of the test bench are plotted in figure 7.

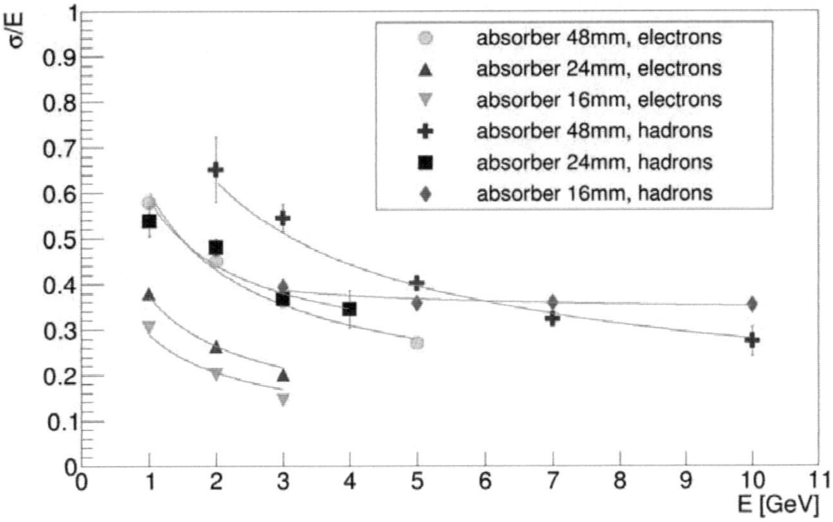

Figure 7. Energy resolutions of the signals from electron and hadron beams.

The energy resolutions as functions of the beam energies were fitted with either $A/\sqrt{E} \bigoplus B$ or $A/\sqrt{E}$. In the configuration with the 48 mm thick absorbers, the electromagnetic energy resolution was found to be $0.601/\sqrt{E} \bigoplus 0.077$, while the hardonic energy resolution was $0.886/\sqrt{E}$. With the 24 mm thick absrobers, the electromagnetic energy resolution was $0.374/\sqrt{E}$ and the hadronic energy resolution was $0.548/\sqrt{E} \bigoplus 0.211$. With the 16 mm thick absorbers in the stack, the electromagnetic and the hadronic energy resolutions were $0.290/\sqrt{E}$ and $0.337/\sqrt{E} \bigoplus 0.335$, correspondingly.

Acknowledgments
This work was supported through U.S. National Science Foundation grants No. 1308299 and No. 0969986, and a special grant from Ohio Univeristy Insitute of Nuclear and Particle Physics. Brookhaven National Laboratory (Upton, NY, U.S.A.) funded the production of the test bench, as well as hardware shipping costs, through the contract No. 202065 with Ohio University. ATLAS group of Brookhaven National Laboratory contributed in funding the test beam experiment.

IOP Conf. Series: Journal of Physics: Conf. Series **928** (2017) 012034 doi:10.1088/1742-6596/928/1/012034

Study of the New Glass and Glass Ceramic Stoichiometric and Gd^{3+}-loaded $BaO*2SiO_2$ (DSB:Ce) Scintillation Material for Future Calorimetry

R. W. Novotny[1], K.-T. Brinkmann[1], A. Borisevich[2], V. Dormenev[1], M. Korjik[2], H.-G. Zaunick[1], and S. Zimmermann[1]

[1]Justus-Liebig-Universität Giessen, II. Physikalisches Institut , Heinrich-Buff-Ring 16 D 35392 Giessen, Germany
[2]Institute for Nuclear Problems, Bobruiskaya11, 220030, Minsk, Belarus

Email: r.novotny@exp2.physik.uni-giessen.de

Abstract. In the last forty years, application of crystalline materials in ionizing radiation detectors has played a crucial role in the discovery of matter properties and promoted a continuous progress in the detecting technique. Further concepts of the detectors at HEP experiments will require an unique combination of the material features, particularly in case of collider experiments. Crucially important becomes a minimal level of radiation damage effects under the electromagnetic part of ionizing radiation and energetic hadrons as well: low deterioration of the optical transmission, low level of afterglow and low level of radioluminescence due to radio-nuclides being generated due to secondary nuclear reactions in the detector material itself. A systematic study of the radiation hardness of inorganic optical and scintillation materials have been performed. We concluded that both oxide and fluoride crystals which consist of atoms with atomic number less than 60 will be reasonably survivable in the irradiation environment of future experiments at colliders. In this study we focused on the study of cheap, capable for a mass production glass ($BaO*2SiO_2$) and DSB: Ce glass ceramics obtained from this glass. We also made this glass more heavy by admixing gadolinium oxide into the matrix. Glass with Gd^{3+} admixture possesses two times larger light yield than pure (BaO*2SiO2) glass and glass ceramics. Both types of the materials were produced as fibre and blocks of larger volume.

1. Introduction

In the last forty years, the application of crystalline materials in ionizing radiation detectors has played a crucial role in the discovery of matter properties and promoted a continuous progress in the detecting technique. Further concepts of the detectors at HEP experiments will require an unique combination of the material features, particularly in case of collider experiments. Crucially important becomes a minimal level of radiation damage effects in particular due to high fluences of energetic hadrons: low deterioration of the optical transmission, low level of afterglow and low level of radio-luminescence due to radio-nuclides being generated in nuclear reactions within the detector material. A systematic study of the radiation hardness of inorganic optical and scintillation materials have been performed during last three years [1,2]. We concluded that both oxide and fluoride crystals which consist of elements with atomic numbers below 60 will reasonably survive in the irradiation environment of future experiments at colliders. In this study we focus on material which is cheap and capable for mass production. We have explored glass ($BaO*2SiO_2$) and DSB: Ce glass ceramics, which was obtained in a further refining process [3,4] as well as heavier modifications by adding heavy rare earth ions. Transparent glass ceramics contains nano-sized particles of Ba_2SiO_5 which improve the scintillation properties. The phase diagram of the $BaO*SiO_2$ indicates several compounds. The melting point increases with the Ba content. However, to form glass the content of SiO2 should not be below 30 mol%. Moreover, special attention should be paid to keep up the stoichiometry of $BaO*2SiO_2$ which

IOP Conf. Series: Journal of Physics: Conf. Series **928** (2017) 012034 doi:10.1088/1742-6596/928/1/012034

can be obtained in the pure form in a narrow range of the SiO_2 content. The glass produced from stoichiometric compositions is radiation hard to hadronic irradiation [1]. The Ba-Si system is rather flexible to incorporate trivalent ions such as the rare earth ions Lu, Gd and Yb. However, trivalent ytterbium is known to be a strong quencher of the scintillation process in Ce^{3+} doped materials. Gadolinium Gd is not expensive and its presence in the crystalline material promote development of bright scintillation and a higher efficiency for absorption of γ-rays.

2. The investigated samples

DSB glass and glass ceramics are obtained by glass production technology with successive thermal annealing. They can be produced in bulk and fibre shapes. The "mother" glass is prepared from the constituents Ba and Si and produced in a mold from the molten glass. Specification for purity of the initial ingredients is similar to a specification for crystalline scintillators. Afterwards it is annealed according to an optimized heating procedure to improve its properties. Gd salt is added in the glass at the stage of the raw material preparation.

Table 1. Some selected properties of the developed scintillation material. (*depends on the Gd content in the final glass; ** for maximum Gd content)

material	ρ g/cm^3	Z_{eff}	X_0 cm	λ_{max} nm	cutoff undoped material nm
$(BaO*2SiO_2)$:Ce glass	3.7	51	3.6	440, 460	310
DSB:Ce	3.8	51	3.5	440, 460	310
$(BaO*2SiO_2)$:Ce glass heavy loaded with Gd	4.7 - 5.4*	58	2.2**	440, 460	318

Creation of the nano-crystallites occurs due to homogeneous seeding during thermal annealing. The composition does not contain special dopants acting as nucleation agents. Nano-structuring of the glass can be obtained at temperature in the range of 800-900 °C in a short treatment and a sudden crystallization of the glass occurs. Further thermal treatment leads to an increase of the dimensions of crystallites up to microns and the ceramics becomes non-transparent. Table 1 summarizes some of the relevant parameters of the investigated different materials.

3. Experimental results

The spectroscopic and scintillation properties were measured with a Cary1E Varian spectrophotometer, a custom made luminescence spectrometer SDL-2 and a start-stop scintillation kinetics spectrometer. The spectra of the signal amplitudes using various commercial photomultipliers, such as XP20202 or R2059-1, were recorded with commercial electronics and DAQ systems.

3.1. The luminescence properties

Some relevant physical properties of the investigated materials are listed in table 1. One can state that the luminescence spectra of the three materials are similar and consist of two bands peaking near 440 and 460 nm. The intensities of these bands vary from material to material. However, the band at the longer wavelength dominates in glass materials whereas the component at shorter wavelength shows an increase of the relative intensity when the material is partly crystallized. The similarity of the luminescence spectra confirms that the luminescent centers have the same origin in all samples. It indicates that even in a glass heavy loaded with Gd^{3+} and surrounding the Ce^{3+} the luminescence centers remain the same as in $BaO*2SiO_2$ glass. Table 2 summarizes the measured kinetics of the luminescence measured at room temperature for the three different materials. The Gd loaded samples show slower kinetics in case of a smaller content of Gd. However, the kinetics is accelerated with the increase of concentration. At large Gd content it becomes an efficient transporter of the excitations avoiding capture of carriers by unlinked boundaries which occurs in the glass. To our expectation further increase of the Gd^{3+} content will lead to a shortening of the scintillation kinetics.

Table 2. Scintillation kinetics of the measured samples

Material	Decay constants and their fractions in the kinetics		
	Fast ns (%)	Middfast ns (%)	Slow ns (%)
$(BaO*2SiO_2)$:Ce glass	22 (12)	72(50)	450(38)
$(BaO*2SiO_2)$:Ce glass loaded with 7 weight% of Gd oxide		86(40)	330(60)
$(BaO*2SiO_2)$:Ce glass loaded with 20 weigth% of Gd oxide	50(19)	120(39)	400(40)

The light output of first samples of DSB:Ce measured with low energy α-and γ-sources [3, 4] was on the level of 100 phe/MeV using a standard photomultiplier with bialkali photocathode and an integration time of 4μs. Similar tests with samples loaded with Gadolinium Oxide show a significantly increased light yield and efficiency for obtaining even a photo peak for γ-rays. As an example, figure 1 shows the response to a ^{137}Cs source for two samples. The sample #1 contains 10 weight percent Gd_2O_3 and sample #2 the double loading. In both cases 0.5 weight percent of Ce are added. Both samples have a cross section of 10x10 mm^2 and a thickness of 5 mm

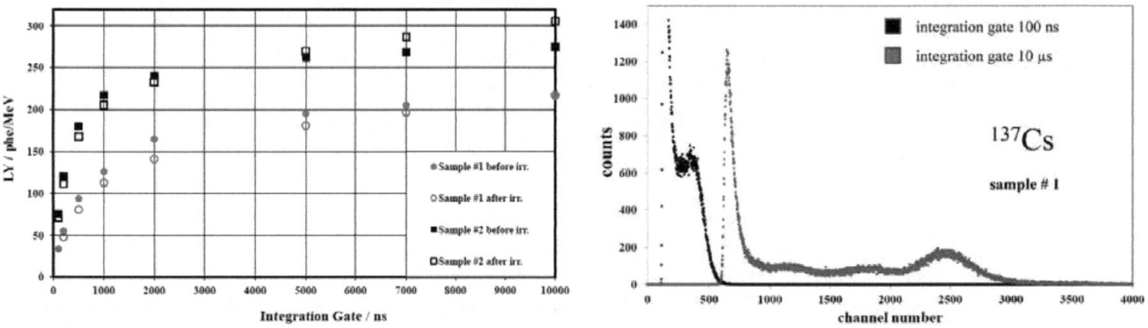

Figure 1. (left) Measured light output for two Gd loaded samples as a function of the integration gate of the output of the photomultiplier with bialkali photocathode. The response to a ^{137}Cs source was measured at room temperature. (right) Response of sample #1 to γ-rays from a ^{137}Cs source. The signals from the photomultiplier are integrated over a gate width of 100 ns or 10 μs, respectively.

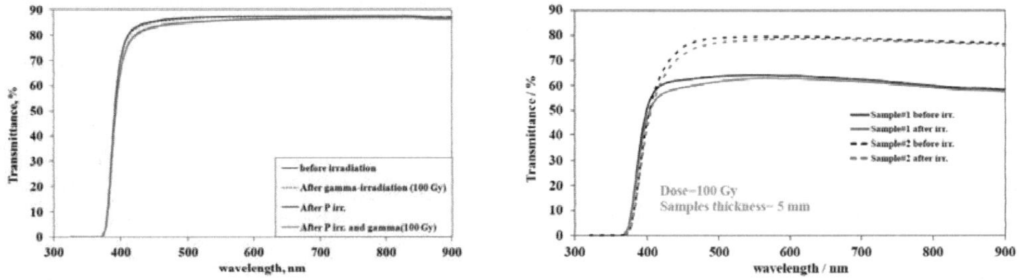

Figure 2.
(left) Optical transmission of a DSB:Ce sample before and after irradiation with γ-rays and 150 MeV protons. (right) Change of the optical transmission of two Gd loaded samples after γ-irradiation.

3.2. The radiation hardness

There have been done extensive studies of the different samples with respect to radiation hardness using either γ-rays (^{60}Co) or even a high fluence of 150 MeV protons provided by the KVI at the University of Groningen (The Netherlands). In particular, DSB:Ce after thermal treatment shows an excellent radiation hardness. The samples have been irradiated with an integral dose of 100 Gy of γ-rays and an integral fluence of $5 \cdot 10^{13}$ protons/cm^2. The impact on the optical transmission is documented in figure 2 (left). The change of optical transmission after γ-irradiation of the two samples differently loaded with Gd is shown in figure 2 (right).

3.3. First production of DSB:Ce fibers and large size blocks

There have been first experiences in producing up to 200 mm long fibers as well as blocks up to 120 mm length. Both geometries show the expected luminescence. However, inhomogeneities and cracks, in particular in case of the fibers, limit light collection and cause short attenuation lengths. First samples are shown as photographic pictures in figure 3.

Figure 3. (left) Photograph of a bundle of 200 mm long fibers made of DSB:Ce with cross sections of ~ 1mm. (right) First large volume sample made of DSB:Ce. The rectangular block with a volume of 23x23x120 mm^3 has been polished on all surfaces is is illuminated with a UV-lamp.

4. Conclusions and outlook

The status report on the development on new radiation hard scintillator material presents even at a very early stage promising features for future applications in calorimetry. The first results of samples loaded with Gadolinium indicate new options for optimizing the efficiency, the light yield and the decay kinetics. However, there are still significant improvements of the manufacturing process necessary to avoid inhomogeneities, defects and inclusions of bubbles.

References

[1] K.-T. Brinkman, et al., *Radiation Damage and Recovery of Medium Heavy and Light Inorganic Crystalline, Glass and Glass Ceramic Materials after Irradiation With 150MeV Protons and 1.2MeV Gamma-Rays*, Conference Record IEEE Nuclear Science Symposium MIC, Washington State Convention Center, Seattle WA, USA, 8-15 November, 2014, N0 6-3.

[2] V. Dormenev, V. et al., *Comparison of Radiation Damage Effects in PWO Crystals Under 150 MeV and 24 GeV High Fluence Proton Irradiation*, IEEE Trans. on Nucl. Sci. 61 (2014) 501-506.

[3] E. Auffray, et al., *DSB:Ce^{3+} scintillation glass for future applications*, Journal of Physics: Conference Series 587 (2015) 012062.

[4] A. Borisevich et al., *Optical transmission radiation damage and recovery stimulation of DSB:Ce^{3+} inorganic scintillation material*, Journal of Physics: Conference Series 587 (2015) 012063.

IOP Conf. Series: Journal of Physics: Conf. Series **928** (2017) 012035 doi:10.1088/1742-6596/928/1/012035

Study of properties of the plastic scintillator EJ-260 under irradiation with 150 MeV protons and 1.2MeV gamma-rays

V. Dormenev[1], K.-T. Brinkmann[1], M. Korjik[2], and R. W. Novotny[1]

[1]Justus-Liebig-Universität Giessen, II. Physikalisches Institut , Heinrich-Buff-Ring 16 D 35392 Giessen, Germany
[2]Institute for Nuclear Problems, Bobruiskaya11, 220030, Minsk, Belarus

Email: v.dormenev@exp2.physik.uni-giessen.de

Abstract. One of the most critical aspects for the application of a scintillation material in high energy physics is the degradation of properties of the material in an environment of highly ionizing particles in particular due to hadrons. There are presently several detector concepts in consideration being based on organic scintillator material for fast timing of charged particles or sampling calorimeters. We have tested different samples of the organic plastic scintillator EJ-260 produced by the company Eljen Technology (Sweetwater, TX, USA). The ongoing activity has characterized the relevant parameters such as light output, kinetics and temperature dependence. The study has focused on the change of performance after irradiation with 150 MeV protons up to an integral fluence of $5\cdot10^{13}$ protons/cm^2 as well as with a strong ^{60}Co γ-source accumulating an integral dose of 100 Gy. The paper will report on the obtained results.

1. Introduction

Plastic scintillators play a significant role in the construction of large area detectors in high energy physics (HEP) experiments. Future concepts for detectors, particularly at collider facilities, will require an unique combination of the material features and affordable price. Crucially important will be a minimal level of radiation damage effects under the electromagnetic part of the ionizing radiation as well as due to high energy hadrons. It requires a tolerable deterioration of the optical transmittance, a moderate level of the damage of the scintillation mechanism itself and a minor contribution of the radio-luminescence due to radioactive nuclides which might be generated by nuclear reactions in the material of the detector itself.

A systematic study of the radiation hardness of inorganic optical and scintillation materials has been performed by us since several years [1-7]. We compared damage effects in a variety of inorganic scintillators after irradiation with γ-quanta of a ^{60}Co source, 150 MeV protons (KVI, Groningen, The Netherlands) and 24 GeV/c protons (PS, CERN, Switzerland). It resulted in a detailed understanding of the damaging processes in different inorganic samples, in particular in self-activated, Ce^{3+} doped and cross-luminescent crystalline materials.

2. The investigated samples

Here we report on first results of the study of commercially produced plastic scintillation material EJ-260 (ELJEN Technology, Sweetwater, USA). This plastic represents a bright, fast, green light emitting scintillator with properties similar to other products available on the market such as BC428 and NE103, respectively. The relevant parameters are given in table 1. Three different samples with a

IOP Conf. Series: Journal of Physics: Conf. Series **928** (2017) 012035 doi:10.1088/1742-6596/928/1/012035

thickness of 1.25 cm were investigated before and after irradiation with γ-quanta (^{60}Co source, $E_\gamma \approx$ 1.2 MeV), 150 MeV protons, respectively.

Table 1. Some selected properties of the investigated scintillation material EJ-260 produced by ELJEN Technology, Sweetwater, USA.

Light Yield	Luminosity	Decay Time	Light attenuation	H/C ratio	Softening Point
% Antracene	nm	ns	cm		°C
60	490	9.2	350	1/1.1	75

3. Experimental results

The samples were investigated with respect to their optical transmission and luminescence properties before and after irradiation. The initial light output was measured with a calibrated photomultiplier tube (Hamamatsu R2059-01) with bialkali photocathode using different integration gates. Figure 1 illustrates the obtained results.

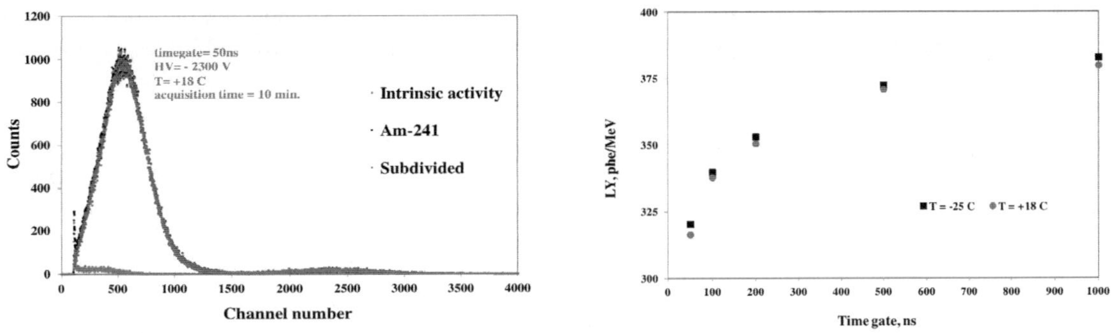

Figure 1. (left) Response to low energy γ-rays of a ^{241}Am source (59.5keV) including the contribution due to background. (right) Light output as a function of integration gate and temperature measured with a ^{241}Am γ-source. The used photomultiplier had a quantum efficiency of 18% at the relevant emission wavelength.

3.1. The radiation damage due to γ-rays

No damage of the material properties was observed after the irradiation with γ-quanta up to an absorbed dose of 100 Gy which is rather consistent with similar studies. The samples did not show any changes in the optical transmittance (see figure 2), scintillation kinetics or light output.

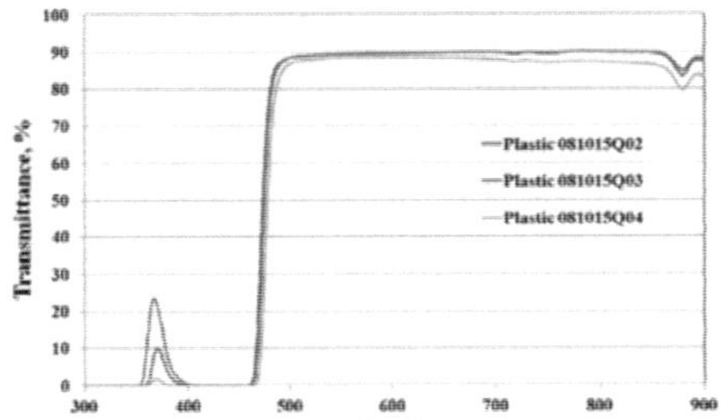

Figure 2. Optical transmission of the investigated samples before and after irradiation with γ-rays of an integral dose of 100 Gy.

3.2. The irradiation with 150 MeV protons

Figure 3(left) shows the results of the transmittance measurement of two samples obtained before and two months after the exposure to 150 MeV protons with a total fluence of $5{\cdot}10^{13}$p/cm^2. The samples had to be stored cooled due to a too high radioactivity. There is a significant reduction of the optical transmittance in the spectral range of the luminescence spectrum.

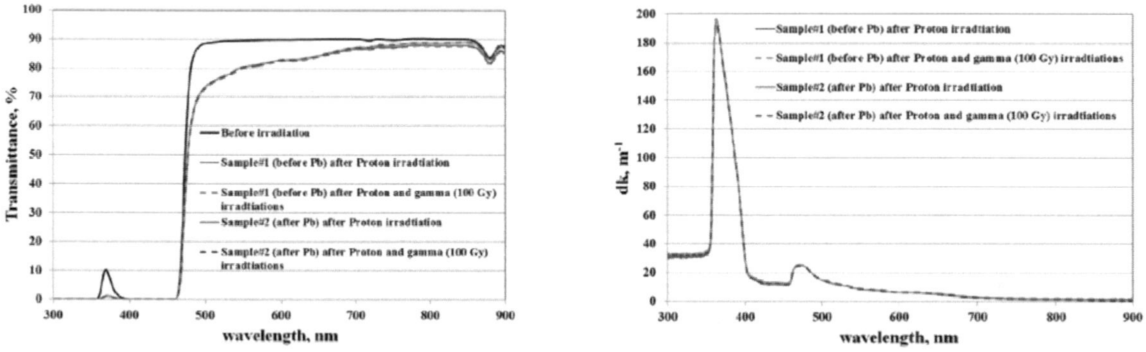

Figure 3. (left) Optical transmittance spectra of two pieces of the EJ260 plastic sample before and after irradiations with 150 MeV protons and low energy γ-rays. (right) Spectra of the radiation induced coefficient of the two pieces of the EJ-260 plastic sample after irradiations with 150 MeV protons and low energy γ-rays.

Figure 3(right) shows for a more quantitative picture the spectrum of the induced absorption coefficient. The spectral range of 380-480 nm is blocked for the observation due to the absorption band of the green emitting organic dye. However, a even larger induced absorption is observed in the gap between the absorption bands of the dyes (350-380 nm). It indicates that most of the color centers which are created in the plastic polymer matrix are created in the UV range. We assume that the damaging mechanisms of the matrix under high energy protons are similar to those we recognized in cases of inorganic materials [1-2, 4]. Even in a light matrix like a polymer high energy protons initiate a variety of nuclear reactions like (p,^{12}C) which lead to the production of light fragments: deuterium, tritium, ^{3}He and ^{7}Be. The presence of ^{7}Be radio-nuclides was confirmed by us in the proton-irradiated samples by characteristic γ-rays identified with a Ge-detector four months after the irradiation took place.

It is worth to note, that the recovery of the optical transmittance damage after the irradiation with protons is rather slow. The additional irradiation with γ-quanta of the proton irradiated samples did not show any additional damage (see Fig. 3). Most probably, similar to inorganic compounds, a creation of the clusters of the defects occurs in polymers as well. Contrary to color centers, which are usually based on a single defect, the color centers in clusters are recombining slowly [1, 2].

Figure 4 shows the dependence of the light yield of the sample irradiated by 150 MeV protons versus the gate width of the integration of the signal. Results were obtained at two temperatures of 18° C and -25° C, respectively.

We did not observe a significant difference in the scintillation kinetics measured at two temperatures. Moreover, we did not observe any slow components in the scintillation kinetics. The integrated light yield reaches a plateau after 500 ns. However, we observed a slowing down of the kinetics in the range below 500 ns. This could be caused by a damage of the organic dye and a change of the transfer efficiency from the UV to the green light emitting dye. Also a trapping of excitations created by additional colour centres cannot be excluded.

As a next step, a comparison of the damage effects between protons of 150 MeV and 24 GeV/c is already foreseen in collaboration with the group of E. Auffray et al. at CERN, Geneva.

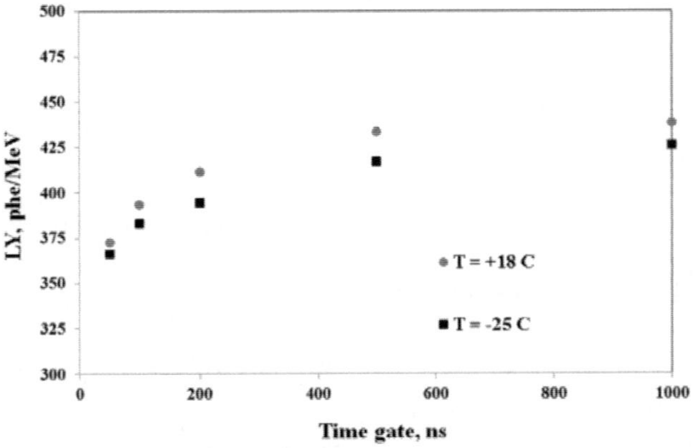

Figure 4. Dependence of the scintillation light yield on the integration gate of the PMT signal measured at two different temperatures for a sample after being irradiated with 150 MeV protons

References

[1] V. Dormenev, M. Korjik, T. Kuske, V. Mechinski, and R. W. Novotny, *Comparison of Radiation Damage Effects in PWO Crystals Under 150 MeV and 24 GeV High Fluence Proton Irradiation*, IEEE Trans. on Nucl. Sci., 61 (2014) 501-506.

[2] E. Auffray et al., *Radiation Damage of Oxy-Orthosilicate Scintillation Crystals Under Gamma and High Energy Proton Irradiation*, IEEE Trans. on Nucl. Sci. 61 (2014) 495-500.

[3] E. Auffray, A. Fedorov, M. Korjik, D. Kozlov, M. Lucchini, V. Mechinsky, and K. Pauwels, *Comparative Study of Damage Effects Induced by Gamma and High Energy Protons Irradiation in YAP:Ce, YAG:Ce and DSB:Ce Scintillating Materials*, Conference Record IEEE Nuclear Science Symposium MIC, Washington State Convention Center, Seattle WA, USA, 8-15 November, 2014, N55-8.

[4] R. W. Novotny, K.-T. Brinkmann, A. Borisevich, V. Dormenev, V. Kalinov, M. Korjik, D. Kozlov, A. Voitovich, H.-G. Zaunick, *Radiation Damage and Recovery of Medium Heavy and Light Inorganic Crystalline, Glass and Glass Ceramic Materials after Irradiation with 150 MeV Protons and 1.2 MeV Gamma-Rays,* Conference Record IEEE Nuclear Science Symposium MIC, Washington State Convention Center, Seattle WA, USA, 8-15 November, 2014, N06-3.

[5] M. Korjik, E. Auffray, *Limits of Inorganic Scintillating Materials to Operate under a High Dose Rate Irradiation Environment in Collider Experiments*, O6-1, (Invited talk) Inorganic Scintillators and Their Applications, SCINT2015, 7-12 June 2015, Berkeley, USA.

[6] E. Auffray, A. Borisevitch, A. Gektin, I. Gerasymov, M. Korjik, D. Kozlov, D. Kurtsev, V. Mechinsky, O. Sidletskiy, R. Zoueyski, *Radiation damage effects in Y_2SiO_5: Ce scintillation crystals under γ-quanta and 24 GeV protons,* Nucl. Instr. and Meth. in Phys. Res. A 783 (2015) 117-120.

[7] A. Borisevich, V. Dormenev, M. Korjik, D. Kozlov, V. Mechinsky, and R. W. Novotny, *Optical transmission radiation damage and recovery stimulation of DSB: Ce^{3+} inorganic scintillation material,* Journal of Physics: Conference Series, 587 (2015) 012063.

IOP Conf. Series: Journal of Physics: Conf. Series **928** (2017) 012036 doi:10.1088/1742-6596/928/1/012036

Hadrons in the CALICE silicon-tungsten electromagnetic calorimeter

Sviatoslav Bilokin[1], Naomi van der Kolk[2], Roman Pöschl[1] on behalf of the CALICE Collaboration

[1]Laboratoire de l'Accélérateur Linéaire, Centre Scientfique d'Orsay, Bât. 200, 91898 Orsay, France

[2] Max-Planck-Institut für Physik, Föhringer Ring 6, 80805 München, Germany

E-mail: poeschl@lal.in2p3.fr (Corresponding author)

Abstract. A detailed study of hadronic interactions is presented using data recorded with the highly granular CALICE silicon-tungsten electromagnetic calorimeter. The predictions of several Monte Carlo GEANT4 models are compared with experimental data, taken at FNAL in 2008. The contribution recaps results published in 2015 and a set of new results available since the beginning of 2016. The published results present a detailed analysis of hadronic showers in terms of radial and longitudinal hit and energy distributions. For the most recent analysis a simple track-finding algorithm was developed to study tracks left by secondary particles, emerging from hadronic interactions in the highly granular electromagnetic calorimeter prototype. Present Monte Carlo simulations provide a good description of the experimental data in terms of new observables, available through the detailed analysis of secondary particles; the Monte Carlo predictions are within 20% of the data, and for many observables much closer.

1. Introduction

The physics at future high-energy lepton colliders requires jet energy reconstruction with unprecedented precision. Detector concepts for the International Linear Collider (ILC) and the Compact Linear Collider (CLIC) rely on Particle Flow Algorithms [1, 2] to achieve the necessary precision. This event reconstruction technique requires highly granular calorimeters to deliver optimal performance. Such calorimeters are developed and studied by the CALICE collaboration.

To develop realistic Particle Flow Algorithms, the interactions of hadrons must be modelled reliably in Monte Carlo simulations and the detector response to hadrons must be well-understood. Highly granular calorimeter prototypes provide a unique means to test and to further develop models of hadronic cascades.

The response of the CALICE silicon-tungsten electromagnetic calorimeter prototype (Si-W ECAL) [3] is used to test hadronic shower models at low energies. The depth of the Si-W ECAL corresponds to approximately one interaction length (λ_{I}), which means that, although the complete shower is not recorded, the first hadronic interaction can be studied in great detail because of the fine longitudinal and transversal sampling. The Si-W ECAL was operated in a test beam at Fermi National Accelerator Laboratory (FNAL) with negatively charged pions (π^-) in the energy of range $2-10\,\mathrm{GeV}$. The majority of charged pions and other hadrons within high energy jets have energies in this range and therefore it is of considerable interest to validate

IOP Conf. Series: Journal of Physics: Conf. Series **928** (2017) 012036 doi:10.1088/1742-6596/928/1/012036

the performance of Monte Carlo simulations. Results of a study of global observables such as radial and longitudinal hit and energy distributions of hadronic showers, have been published in [4]. As an extension of this analysis this article presents new results that become available due to the high granularity of the Si-W ECAL [5].

2. The CALICE silicon-tungsten electromagnetic calorimeter

The Si-W ECAL prototype consists of a sandwich structure of 30 layers of silicon as active material, alternating with tungsten as the absorber material. The active layers are made of silicon wafers segmented into 1×1 cm^2 pixels (or pads). Each wafer consists of a square of 6×6 pixels and each layer contains a 3×3 matrix of these wafers, resulting in an active zone of 18×18 cm^2.

The Si-W ECAL is divided into three modules of ten layers each. The tungsten thickness per layer is different in each module, increasing from 1.4 mm in the first module (layers 1–10), to 2.8 mm in the second (layers 11–20) and 4.2 mm in the third (layers 21–30). The total thickness corresponds to 24 radiation lengths (X_0) and approximately one interaction length. More than half of the hadrons traversing the Si-W ECAL prototype undergo a primary interaction within its volume.

3. Global observables of hadronic showers

Test beams were conducted in May and July of 2008 at the Fermilab Test Beam Facility at FNAL. The presented analysis uses data from runs with π^- mesons at energies of 2, 4, 6, 8 and 10 GeV. Monte Carlo simulations corresponding to the recorded test beam data have been produced using the simulation tool kit GEANT4 (version 9.6 patch 1). The full geometry of the CALICE test beam set-up is taken into account in the simulation via the MOKKA framework which provides the geometry interface to GEANT4. Four different physics lists, FTFP_BERT, FTFP_BERT_HP, QGSP_BERT and QBBC, are compared to the data in order to study the different hadronic models which are implemented in the physics lists for different energy ranges.

As a first step in the analysis a pure data sample is selected. Residual contamination from electron events and events with multiple incoming particles is corrected for. Subsequently interacting events are selected based on the deposited energy in individual layers of the Si-W ECAL. The starting layer of the hadronic shower can be reconstructed with an accuracy of $\pm$ 2 layers at an efficiency of at least 50% at 2 GeV and 87% at 10 GeV. The same event selection criteria as used for the data are also applied to the simulated events. The predictions from the simulations are then compared to the data. The fraction of interacting events and the total deposited energy are studied, as well as radial and longitudinal shower profiles, the size and energy density of the interaction region, and the distribution of secondary particles from the first hard interaction. As an example for the results Fig. 1 shows the longitudinal shower profile and the mean hit energy per layer for π^- mesons with an energy of 10 GeV compared with different implementations of the physics list FTFP_BERT [6]. While the older version 9.3 of GEANT4 describes the data well, a successor version 9.6 fails to describe the data. The situation improves slightly with a more recent version 10.1 of GEANT4.

4. Studying the structure of the first hadronic interaction

After the study of general observables of the hadronic shower as outlined in the previous sections the analysis goes one step further demonstrating the outstanding potential of the CALICE Si-W ECAL physics prototype to obtain a detailed picture of the interactions of hadrons with matter. Figure 2 shows the event display of a primary pion of 10 GeV that interacts in the Si-W ECAL. Visible are the incoming primary hadron, the interaction zone and secondary particles emerging from it.

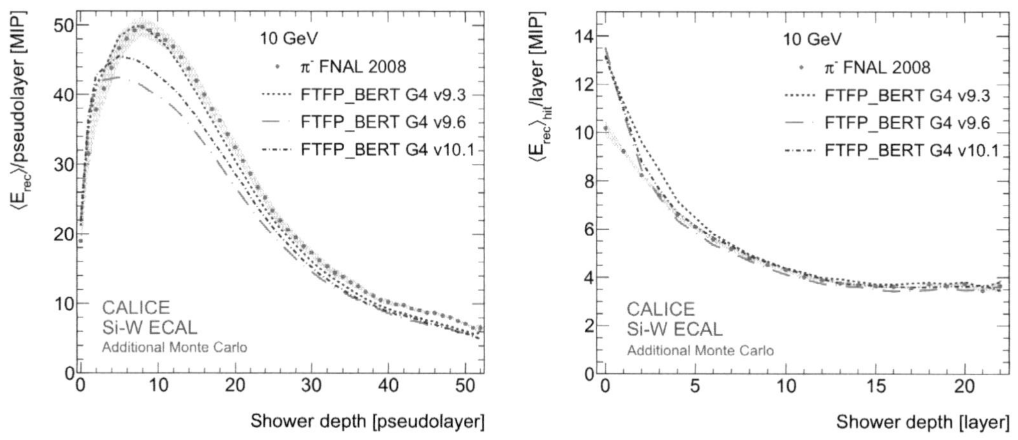

Figure 1. *Longitudinal shower profile (left) and Mean hit energy per layer (right) for 10 GeV pions compared with predictions from the FTFP_BERT physics list in different versions of* GEANT4.

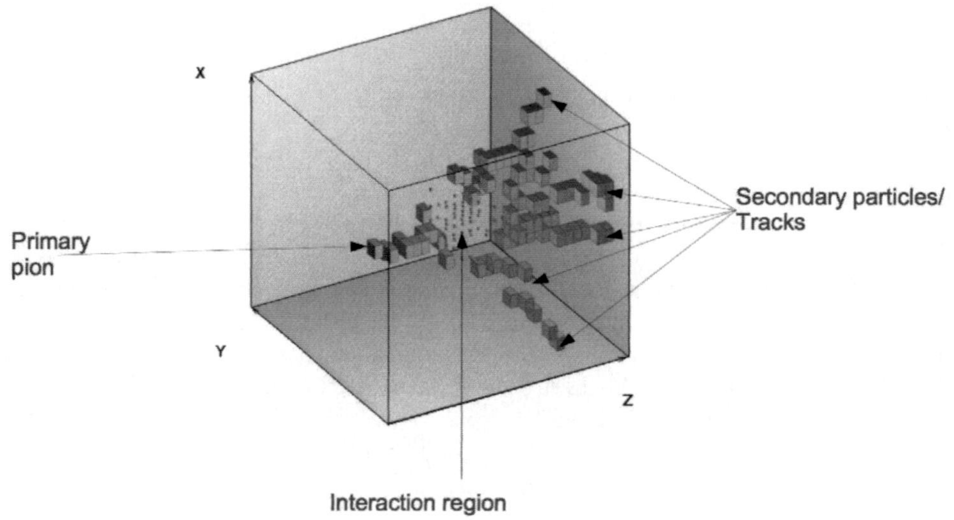

Figure 2. *Event display of the interaction of a primary pion with an energy of 10 GeV in the Si-W ECAL. The interaction leads to a dense interaction region and secondary particles that can be reconstructed as outgoing tracks.*

The following results are obtained with a new simple track-finding algorithm for the Si-W ECAL. This algorithm allows for the reconstruction of tracks produced by secondary particles created in the interaction of hadrons with the absorber material, and hence to study the interaction region of hadronic showers in the Si-W ECAL. The track-finding algorithm produces a new set of differential observables, based on reconstructed tracks of secondary particles and the interaction region of the hadronic cascades. The results are stable w.r.t. small variations of the main parameter of the track-finding algorithm.

The accuracy with which the simulation describes the data varies with the beam energy and the chosen physics observable. In most of the cases data and Monte Carlo agree within 10%

without revealing a clear preference for one of the chosen physics lists. The largest source of discrepancy between data and Monte Carlo is the energy and radius of the interaction region. As shown in the left part of Fig. 3, the measured energy deposition in the interaction region is up to 20% higher than predicted by the Monte Carlo simulation.

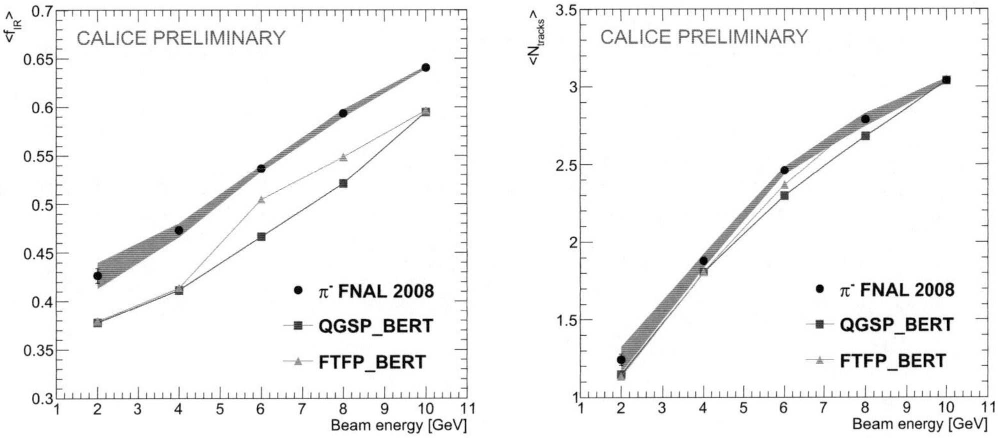

Figure 3. *Left: Relative energy deposition in the interaction region of hadronic interactions in the Si-W ECAL. Right: Number of secondary tracks emerging from hadronic interactions in the Si-W ECAL.*

The distributions of the number of secondary tracks, see right part of Fig. 3, and the number of hits per track for data are well described by the tested physics lists. The polar angles of reconstructed tracks in the Monte Carlo simulation agree with data within 8% on average and the distribution of azimuthal angles is well reproduced by the Monte Carlo simulations even in view of the non-trivial detector geometry.

With respect to a more general outlook future work aims at transferring the insights about the interaction region and the secondaries emerging from it to the optimisation of Particle Flow Algorithms. The comparatively simple algorithm as presented in this contribution will allow to judge on the performance of more involved algorithms such as PandoraPFA [2] or ARBOR [7].

A tighter track selection leads to clean MIP-like tracks. The detector response is stable to about 1-2% over the tested energy range with an expected good agreement with Monte Carlo simulations. This observation can be exploited in the future as a starting point for a study on the possibility of an in-situ calibration or at least a regular monitoring of the detector by means of the selected tracks.

References

[1] Brient J C and Videau H 2001 *eConf* **C010630** E3047 (*Preprint* hep-ex/0202004)
[2] Thomson M 2009 *Nucl.Instrum.Meth.* **A611** 25–40 (*Preprint* 0907.3577)
[3] Repond J *et al.* (CALICE Collaboration) 2008 *JINST* **3** P08001 (*Preprint* 0805.4833)
[4] Bilki B *et al.* (CALICE) 2015 *Nucl. Instrum. Meth.* **A794** 240–254 (*Preprint* 1411.7215)
[5] *CALICE Analysis Note* URL https://twiki.cern.ch/twiki/pub/CALICE/CaliceAnalysisNotes/can055.pdf
[6] *CALICE Analysis Note* URL https://twiki.cern.ch/twiki/pub/CALICE/CaliceAnalysisNotes/CAN050a.pdf
[7] Ruan M and Videau H 2014 (*Preprint* 1403.4784)

IOP Conf. Series: Journal of Physics: Conf. Series **928** (2017) 012037 doi:10.1088/1742-6596/928/1/012037

DHCAL with minimal absorber: measurements with positrons

B. Freund[1,4], B. Bilki[1,3], F. Corriveau[4], C. Neubüser[1,2], Y. Onel[3], and J. Repond[1]

On behalf of the CALICE Collaboration

[1] Argonne National Laboratory,
9700 S. Cass Avenue, Argonne, IL 60439, U.S.A
[2] DESY,
Notkestrasse 85, D-22607 Hamburg, Germany
[3] University of Iowa,
203 Van Allen Hall, Iowa City, IA 52242-1479, U.S.A
[4] McGill University,
3600 University Street, Montreal, QC H3A2T8, Canada

E-mail: benjamin.freund@cern.ch

Abstract. In special tests, the active layers of the CALICE Digital Hadron Calorimeter prototype, the DHCAL, were exposed to low energy particle beams, without being interleaved by absorber plates. The thickness of each layer corresponded approximately to 0.29 radiation lengths or 0.034 nuclear interaction lengths, defined mostly by the copper and steel skins of the detector cassettes. This paper reports on measurements performed with this device in the Fermilab test beam with positrons in the energy range of 1 to 10 GeV. The measurements are compared to simulations based on GEANT4 and a standalone program to emulate the detailed response of the active elements.

1. Introduction

This paper reports on special tests performed in the Fermilab test beam using the detector cassettes of the Digital Hadron Calorimeter prototype, the DHCAL [1], without absorber material interleaved between the active layers. The active layers of the DHCAL contained thin Resistive Plate Chambers (RPCs) with a readout featuring 1 x 1 cm^2 pads.

In its configuration without absorber plates, the so-called Min-DHCAL [2] provided the opportunity to study electromagnetic and hadronic showers with extremely fine segmentation, especially longitudinally, spreading the showers over the entire depth of the stack. This paper presents measurements with positrons in the energy range of $1-10$ GeV. The experimental results are compared to detailed Monte Carlo simulations based on GEANT4. These comparisons provide an ideal tool to gain deeper insights into the response of an RPC-based calorimeter and into the physics of low energy showers.

IOP Conf. Series: Journal of Physics: Conf. Series **928** (2017) 012037 doi:10.1088/1742-6596/928/1/012037

2. Description of the DHCAL with minimal absorber

The DHCAL used Resistive Plate Chambers (RPCs) [3, 4] as active elements. The area of each plane was approximately 1 x 1 m^2 and was equipped with three 32 x 96 cm^2 chambers. The readout boards contained pads of 1 x 1 cm^2 and were placed on the anode side of the chambers. The chambers and the readout boards in turn were inserted into a cassette structure with a 2 mm thick steel front cover and a 2 mm thick rear copper cover. The thickness of each cassette was about 12.5 mm and corresponded to 0.29 radiation lengths X_0 or 0.034 nuclear interaction lengths λ_I. The DHCAL in this configuration with minimal absorber, consisted of 50 cassettes, spaced 2.54 cm apart. The total number of readout channels was 460,800. The electronic readout provided a single bit for each channel, corresponding to a single threshold. To first order, the energy of incoming particles was reconstructed as being proportional to the number of hits. This method works, as the noise rate in the stack was negligibly small [2]. For further details on the design, operation, and commissioning of the chambers and the electronic readout system, see ref. [1, 5].

3. Data collected at the Fermilab Test Beam Facility

The Min-DHCAL was exposed to the test beam at the Fermilab Test Beam Facility, FTBF [6]. The facility provides a primary 120 GeV proton beam and momentum selected secondary beams in the range of 1-66 GeV/c . The latter are a mixture of electrons, muons and pions, where the fraction of electrons is dominant at momenta below 5 GeV/c and tapers off for momenta above 32 GeV/c. The beamline included two Cerenkov counters for particle identification and two scintillator paddles (19 x 19 cm^2), located approximately two meters upstream of the Min-DHCAL. The data acquisition was triggered by the coincidence of these two paddles.

The data on which this paper is based were collected in November 2011. Runs were taken with a selected momentum in the range of 1-10 GeV/c.

4. Simulation of the test beam set-up

The simulation of the beam tests is based on the GEANT4 program [7] version 10.02. The simulated set-up includes the active elements with their cassette covers, resistive plates, the gas, and electronic readout boards. Any energy deposition generated by the simulation in the gas gap of the RPCs is used as a seed for the simulation of an avalanche. The latter is simulated by a standalone program, called RPC_sim [2], which is governed by six parameters. Five of these are tuned by comparing the distribution of the number of hits per layer in both measured and simulated muon track events. The remaining parameter, a distance cut, d_{cut}, introduced to suppress close-by avalanches, was tuned using the 3 and 10 GeV positron data [2].

Initially, the GEANT4 program utilized the FTFP_BERT physics list. However, this led to a unsatisfactory description of, among other measurements, the energy resolution, indicating a deficit in the generation of initial ionizations in the gas gap of the RPCs. A migration to the Option 3 or _EMY [8] based electromagnetic physics list, which is particularly appropriate for low energies, resulted in a significant improvement of the description of the experimental data, in particular, the energy resolution.

5. Hit and Event Selection

To ensure the high quality of the data, several loose cuts were applied to the selection of both hits and events. By requiring a signal in the upstream Cerenkov counter, the fraction of muon and pion induced events was effectively reduced to zero, due to the negligible rate of accidental hits in the counter. The acceptance for positrons was approximately 97.5% and independent of energy.

6. Equalization of the RPC responses

Through-going muons are used to measure and equalize the response of the 150 different RPCs in the stack. The equalization procedure is performed for each run individually using the admixture of muons in the beam.

The efficiency ϵ, the average pad multiplicity μ, and their product $\epsilon\mu$ are determined for each chamber individually and also averaged over the entire stack and all runs. For a given chamber, i, the calibration factor is calculated as the ratio of the product averaged over the entire stack, $\epsilon_0\mu_0$, and the product, $\epsilon_i\mu_i$, as measured for chamber i.

7. Systematic errors

The following systematic errors associated with uncertainties in the measurements have been considered: the residual non-uniformity in the calibrated response (dominant), the rate limitation of RPCs [9], contamination of the positron sample by muons and pions, and contributions from accidental noise hits in the chambers. All systematic errors pertaining to the experimental data are assumed to be independent and uncorrelated between energies, and are therefore added in quadrature.

The differences observed between the FTFP_BERT and FTFP_BERT_EMY physics lists point to some uncertainty in the simulation of electromagnetic showers in GEANT4. Additionally, uncertainties in the emulation of the RPC response lead to further systematic errors in the simulated results. However, since the 3 and 10 GeV measurements were used to tune the distance cut parameter of the RPC_sim program, d_{cut}, the simulation lost most of its predictive power and an assignment of systematic errors to the simulation of positrons has therefore become problematic.

8. Results

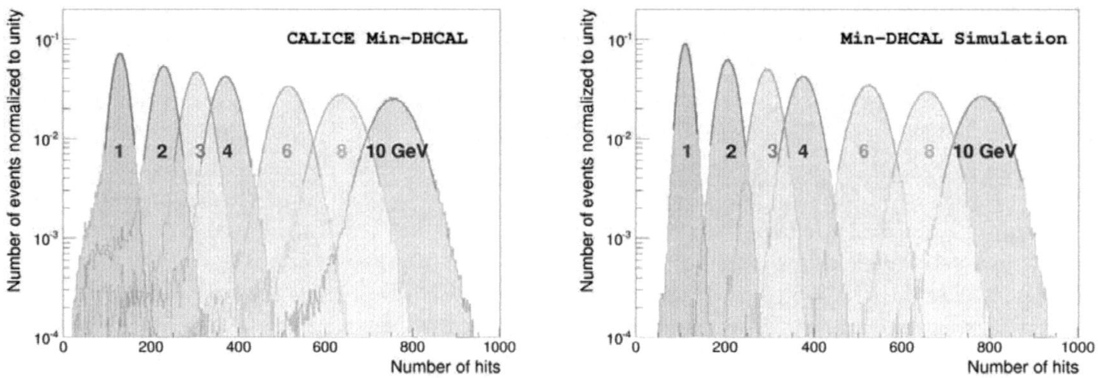

Figure 1. Distribution of the number of hits for all selected positron events for data (left) and simulation based on the FTFP_BERT_EMY physics list (right). The distributions are plotted separately for each beam momentum setting (1, 2, 3, 4, 6, 8, 10 GeV/c) and are each normalized to unity. The distributions are fit with a Gaussian function in the range of ± 2 standard deviations. The results of the fits are shown as solid lines.

Figure 1 shows the distribution of the number of hits for all selected positron events for both data and simulation. The response curves have been normalized to unity for each momentum

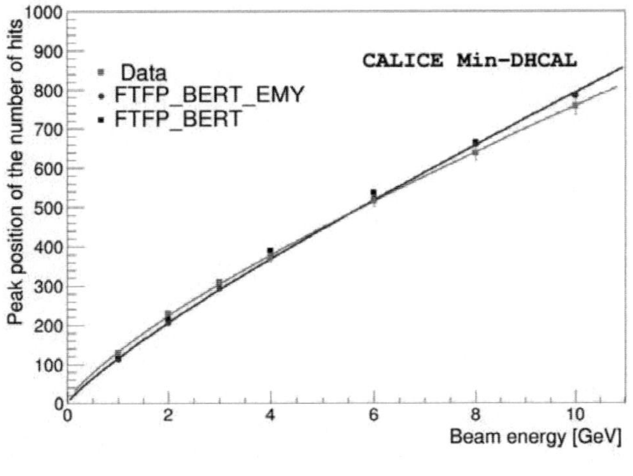

Figure 2. Peak position of the number of hits versus positron beam energy for both data (red) and simulation based on FTFP_BERT (black) and FTFP_BERT_EMY (blue). The experimental data points and the simulation based on FTFP_BERT_EMY have been fitted to a power law shown as solid lines. The error bars of the data include both the statistical and systematic (dominant) errors. The statistical error bars of the simulation are smaller than the marker size.

selection and are well described by fits of a Gaussian function in the range of $\pm 2\sigma$ around the peak value (determined iteratively).

The mean values obtained from the Gaussian fits are shown as function of beam energy in Fig. 2. The data are compared to the results of the Monte Carlo simulation based on both the FTFP_BERT and the FTFP_BERT_EMY physics lists. Both are seen to be in good agreement with the data. The data/simulation are fitted to a power law

$$N_{hit} = a_0 (E_{beam}/GeV)^m \tag{1}$$

where the exponent m is a measure of the non-linearity (saturation) of the response. A value of unity would indicate a perfectly linear response. A value of $m = 0.76 \pm 0.02(0.836 \pm 0.001)$ is obtained for data (simulation based on FTFP_BERT_EMY), indicating a strong saturation of the response. The saturation is mostly due to the large pad size compared to the density of particles in the core of electromagnetic showers. The simulation based on the FTFP_BERT physics list produces similar results as the ones based on FTFP_BERT_EMY, as indicated by the black squares in Fig. 2. The inverse of the power law is utilized to reconstruct the energy of the positrons.

Figure 3 shows the resulting relative resolutions as function of beam energy for both data and simulation (based on both physics lists). The measured resolutions are approximately 15% better than the corresponding resolutions obtained by the simulation based on the FTFP_BERT physics list, indicating a possible deficit in the number of ionizations in the gas gap. On the other hand, the simulation based on the FTFP_BERT_EMY physics list reproduces the measurements quite well, but are in average about 6% better than the data. The energy resolution versus beam energy was fitted to the standard parametrization with a constant and a stochastic term.

Table 1 summarizes the results of the fits, showing a reasonable agreement between the stochastic terms of data and simulation.

The imaging capabilities of the DHCAL provide an unprecedented tool for the detailed study of the shape of showers. As an example, Fig. 4 shows the distribution of the radial distance of

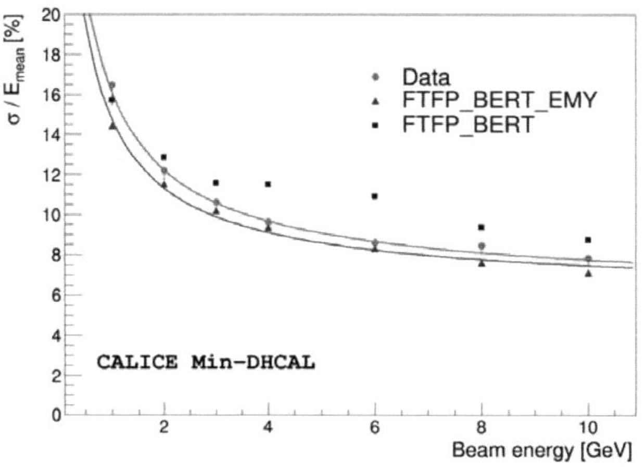

Figure 3. Energy resolution versus positron beam energy for data (red) and simulation based on FTFP (black) and FTFP_BERT_EMY (blue). The experimental points were corrected for the known momentum spread of the beam. The error bars of the data include both statistical and systematic uncertainties. The error bars of the simulation indicate the statistical uncertainty only. The curves are the results of fits to the quadratic sum of a constant and stochastic term.

	c [%]	α [%]
Data	6.3 ± 0.2	14.3 ± 0.4
Simulation (FTFP_BERT_EMY)	6.2 ± 0.1	13.4 ± 0.2

Table 1. Fit parameters for the constant and stochastic terms of the energy resolution for positrons.

each hit to the shower axis, as measured for 6 GeV positrons. The accelerated decrease in entries above a radius of 50 cm is an artefact of the square shape of the detector planes with dimensions of 96 x 96 cm^2. Excellent agreement between data and simulation is observed over the entire range of radii apart from a small depletion at small radii in the data. Note that the number of hits varies by six orders of magnitude over the entire range in radii. Both the statistical and systematic uncertainties of the data are very small and mostly invisible in the plots.

9. Conclusion

The Digital Hadron Calorimeter (DHCAL) detector planes without absorber plates, the Min-DHCAL, was exposed to particles in the Fermilab test beam. The response of the individual Resistive Plate Chambers in the calorimeter stack was equalized using through-going muon tracks. The response of the Min-DHCAL to positrons, its energy resolution and various electromagnetic shower shapes were measured in the energy range of 1 to 10 GeV. The results of a Monte Carlo simulation based on GEANT4 and a standalone program, RPC_sim, to emulate the response of the RPCs, were compared to the data. The RPC_sim program was tuned to reproduce the measured response to muons and to reproduce the measurements obtained with 3 and 10 GeV positrons. Due to the tuning process the simulation lost its predictive power for both muons and positrons.

The GEANT4 simulation utilized either the FTFP_BERT and the FTFP_BERT_EMY physics lists. The latter provides higher accuracy, in particular for the simulation of electromagnetic processes in thin layers. Despite tedious efforts of tuning of the RPC_sim parameters

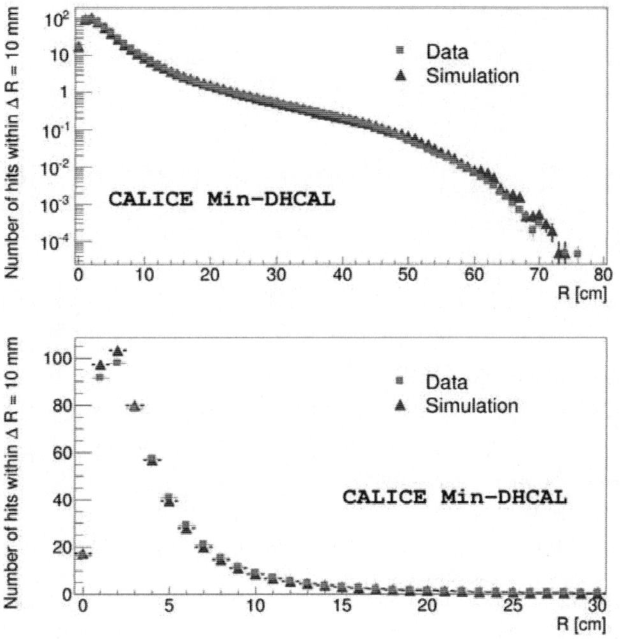

Figure 4. Distribution of the radial distance of hits from the shower axis for 6 GeV positrons. The upper (lower) plot uses a logarithm (linear) y-scale. The areas of both plots are normalized to one event. The error bars include the statistical and systematic uncertainties for the data and statistical uncertainties only for the simulation.

to reproduce the measurements, only a poor description of the data was obtained with FTFP_BERT, suggesting a deficit of ionizations in the gas gap of the RPCs. A significant improvement is seen with the use of FTFP_BERT_EMY, leading to a good to excellent agreement with the data.

References

[1] C. Adams et al., JINST 11 (2016), P07007.
[2] B. Freund et al., JINST 11 (2016), P05008
[3] R.Santonico and R.Cardarelli, Nucl. Instr. and Meth. **187** (1981)377-380.
[4] G.Drake et al., Nucl. Instr. and Meth, **A578**, 88-97, 2007.
[5] C. Adams et al., *Design, Construction, and Testing of the Digital Hadron Calorimeter (DHCAL) Electronics*, arXiv:1603.01654 [phys.ins-det](to be published).
[6] `http://ftbf.fnal.gov`
[7] E. Longo and I. Sestili, Nucl. Instrum. Methods **128**, 283 (1975)
[8] V.N. Ivanchenko et al.,*Recent Improvements in Geant4 Electromagnetic Physics Models and Interfaces* , Prog. Nucl. Sci. Technol. **2** (2011) 898.
[9] B. Bilki et al.: *Measurement of the rate capability of Resistive Plate Chambers*, JINST 4 P06003 (2009).

IOP Conf. Series: Journal of Physics: Conf. Series **928** (2017) 012038 doi:10.1088/1742-6596/928/1/012038

Comparison of Energy Reconstruction Schemes and Different Granularities in the CALICE Scintillator-Steel Analogue Hadron Calorimeter

Coralie Neubüser

Deutsches Elektronen Synchrotron (DESY), Notkestraße 85, 22607 Hamburg
now at CERN - European Laboratory for Particle Physics, CH-1211 Geneve, Switzerland

E-mail: `coralie.neubuser@cern.ch`

Abstract. The CALICE collaboration develops different high-granularity hadronic calorimeter technologies for a future linear collider. These technologies differ in active material, granularity and their readout and thus their energy reconstruction schemes. The Analogue Hadron Calorimeter (AHCAL), based on scintillator tiles with Silicon Photomultiplier readout, measures the signal amplitude of the energy deposition in cells of at most $3 \times 3\,\mathrm{cm}^2$ size. The Digital, Resistive Plate Chamber (RPC) based, HCAL (DHCAL) detects hits above a certain threshold by firing pad sensors of $1 \times 1\,\mathrm{cm}^2$. A 2 bit readout is provided by the, also RPC based, Semi-Digital HCAL (SDHCAL), which counts hits above three different thresholds per $1 \times 1\,\mathrm{cm}^2$ cell. All three calorimeter concepts have been realised in a $1\,\mathrm{m}^3$ prototype with interleaved Steel absorber and tested at various test beams. This study investigates the impact of the readout, granularity and active medium on the energy resolution individually by applying the reconstruction procedures on AHCAL data, that can also be processed in a way which emulates a (semi-) digital readout system. The difference in granularity is studied via simulations of an AHCAL with $1 \times 1\,\mathrm{cm}^2$ cell sizes.

Additionally, a so-called Software Compensation algorithm is developed to weight hits dependent on their energy content and correct for the difference in the response to the electromagnetic and hadronic sub-showers ($\frac{e}{h} \neq 1$) and thus reduce the influence of fluctuations in the π^0 generation. The impact on the energy resolution will be discussed and compared to the other energy reconstruction schemes.

1. Introduction

For a future linear electron-positron collider such as ILC or CLIC, the desired jet energy resolution of $3-4\,\%$ for a wide range of jet energies can be achieved by using Particle Flow Algorithms for the jet reconstruction. Within the CALICE collaboration, several concepts for a hadron calorimeter (HCAL) optimised for Particle Flow are studied and have been tested with large, $\sim 1\,\mathrm{m}^3$ prototypes: the so-called analogue, digital and semi-digital HCAL concepts. The concepts differ in active material for the shower detection, granularity, readout technology and reconstruction method. This makes it difficult to disentangle the influence of each of these components to the energy resolution of jets as well as of individual particles. Since the analogue HCAL prototype has a larger cell size than the other two concepts, and the digital and semi-digital HCAL prototypes do not provide analogue hit size information, it is impossible to study all different aspects in test beam data. For the data taken with the analogue HCAL prototype, a direct comparison of the reconstruction methods is possible, and within the simulation the cell size can be altered to the S- and DHCAL cell size of $1 \times 1\,\text{cm}^2$. The effect of the active media can be studied only in simulation. For reliable results from the $1 \times 1\,\text{cm}^2$ AHCAL simulation it is important to validate the simulation of hadronic showers by comparing them to the measured test beam data, especially for the quantities that are relevant for the energy reconstruction.

2. Analogue Hadronic Calorimeter in the CERN 2007 test beam

For this analysis the AHCAL test beam data from 2007 with steel absorber is chosen. The 2007 CERN test beam setup at the SPS consisted of 30 layers of CALICE silicon-tungsten ECAL, 38 layers of the scintillator-steel analogue HCAL and 16 layers of the scintillator-steel tail catcher and muon tracker (TCMT). The absorber plate thickness for the HCAL was $\sim$ 2 cm. A detailed description of the test beam setup can be found in [1].
The run list and event selection follows the published analysis [2] and consists of 29 runs at 11 energies for pions between 10 and 80 GeV. Negative pion events are selected by requiring the showers to start in the second to sixth HCAL layer and by rejecting muons that have a smaller energy deposit than 150 MIP in the HCAL. One MIP is defined as the response of a minimum-ionising particle like a 0.4 GeV muon in a single AHCAL cell. The value of a MIP is measured within calibration runs in ADC counts. For a detailed description of the event selection and the whole analysis see [4]. Additionally, the test beam runs are simulated using the software packages GEANT4 version 9.6 patch 1.The physics lists FTFP_BERT and QGSP_BERT from GEANT4 show best performance for hadrons and were therefore chosen for comparisons in this analysis.

3. Energy Reconstruction

The goal of this analysis is first a direct comparison of the four reconstruction methods: analogue, digital, semi-digital and analogue software compensation, applied to the same AHCAL data and second a study of the impact of the granularity on the reconstruction performance. This includes using the same methods to extract the mean energy and the resolution. Since the distributions of the reconstructed energies from the number of hits are expected to show a non-gaussian tail, the following procedure is used to fit the distributions and to extract the mean and the width. In a first step, the Novosibirsk function $f(x) = A \cdot \exp\left(-\frac{1}{2}\left(\frac{\ln^2[1+\Lambda\tau(x-\mu)]}{\tau^2} + \tau^2\right)\right)$ with $\Lambda = \frac{\sin\left(\tau\cdot\sqrt{\ln 4}\right)}{\sigma\cdot\tau\cdot\sqrt{\ln 4}}$ [3] is used to fit the distributions. The fit range is limited to $\mu \pm 3\,\sigma$ from a pre-fit with a Gaussian. The fit consistently provided a χ^2/ndf value better than 3. In order to extract the mean and the width of this fit function, a histogram is filled with random values generated according to the Novosibirsk function with the extracted fit parameters. The mean and RMS of the histogram are used as response and resolution for the studied energy.
For each method the energy is reconstructed on an event-by-event basis.
Systematic uncertainties due to the uncertainty on the beam energy, the MIP to GeV conversion and the detector stability is taken into account. For the simulations the systematic uncertainty is estimated from the uncertainty on the light crosstalk between neighbouring cells.

3.1. Analogue

From the linear hit energy sum E_{sum} above 0.5 MIP threshold in the AHCAL the analogue energy is reconstructed as

$$E_{\text{rec,analogue}} = 0.3232\,\text{GeV} + \frac{e}{\pi} \cdot \omega \cdot E_{\text{sum}} \cdot c \qquad (1)$$

with the correction for the non-compensation with $e/\pi = 1.19$, the electromagnetic calibration factor ω for the conversion from MIP to GeV scale and the scaling c, which is determined by a fit to $c = 1.04$ and for the $1 \times 1\,\mathrm{cm}^2$ AHCAL to $c_{1\times1} = 1.01$.The approximately constant contribution of the track in the ECAL is taken into account using a constant value of 0.3232 GeV.

3.2. Digital

Within the digital energy reconstruction the non-linear digital response is linearised as follows. The mean number of hits is fitted with a power law $\langle N_{hits}\rangle = a \cdot (E_{beam} - m)^b$. By constraining the digital reconstructed energy to show a linear behavior $E_{rec,digital} = E_{beam}$, the energy is reconstructed with the fit parameters of the power law to

$$E_{rec,digital} = m + \sqrt[b]{\frac{N_{hits}}{a}}. \tag{2}$$

3.3. Semi-Digital

The semi-digital energy reconstruction follows

$$E_{rec,semi-digital} = \alpha N_1 + \beta N_2 + \gamma N_3, \tag{3}$$

with N_1 the number of hits with energies above the first threshold t_1 and below the second t_2, N_2 the number of hits above t_2 and below the third threshold t_3 and N_3 the number of hits above t_3. These thresholds are optimised for the energy reconstruction performance [4]. α,β and γ weight the hits depending on their energy content. Hadronic showers change their structure and compositeness with energy, which is taken into account by parameterising α,β and γ as quadratic polynomials of the total number of hits $N_{hits} = N_1 + N_2 + N_3$. To find the best parameters a χ^2 function of the form $\chi^2 = \sum_{i=1}^{N} \frac{\left(E^i_{beam} - E^i_{rec}\right)^2}{E^i_{beam}}$ is minimised, where i runs over all events.

3.4. Analogue Software Compensation

The analogue software compensation algorithm follows the principle of the semi-digital energy reconstruction by using the same χ^2 minimisation technique, to determine different weights ω for different classes of hits, dependent on the energy e_j and the total measured energy E_{sum} in the event. Thus the energy is reconstructed by

$$E_{rec,SC} = \sum_{j=0}^{N_{hits}} \omega\left(e_j, E_{sum}\right) \cdot e_j. \tag{4}$$

Hereby the hit energy spectrum is divided into 8 energy ranges.
An example of the energy distributions is given in Figure 1, which shows the analogue reconstructed energy distributions in comparison to the reconstructed energies using the analogue software compensation algorithm for data and simulation.

4. Results

The resolutions obtained with the different reconstruction methods applied to the $3 \times 3\,\mathrm{cm}^2$ AHCAL data and the $1 \times 1\,\mathrm{cm}^2$ AHCAL simulations are compared in Figure 2a and 2b. In addition, the best resolution reached with AHCAL data, by applying software compensation techniques [2], is indicated. For the comparison one should keep in mind that in the earlier

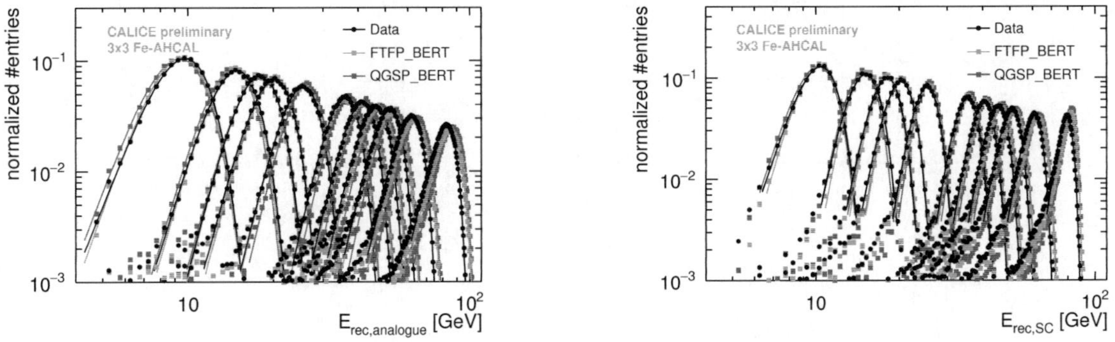

(a) Analogue (b) Analogue Software Compensation

Figure 1: a) Analogue reconstructed energy distributions and b) the reconstructed energies after the use of the analogue software compensation algorithm for the beam energies from 10 to 80 GeV; The black dots show the testbeam data, the orange squares show the FTFP_BERT and the blue squares the QGSP_BERT simulated $E_{rec,analogue}$ and $E_{rec,SC}$ distributions. The corresponding Novosibirsk fits are represented by solid lines.

analysis, the TCMT is fully included and the track in the ECAL considered in the energy reconstruction, while here a simplified treatment of the ECAL is used, the TCMT contribution is neglected and the widths are determined using a Novosibirsk function. The non-linearities of the four methods studied in this analysis are shown in the upper parts of Figure 2.
For the AHCAL with a granularity of $3 \times 3\,\mathrm{cm}^2$, the analogue and digital reconstruction procedures show rather similar resolutions at the lowest energies. For larger energies, the resolution of the analogue reconstruction method continues to decrease, while the digital resolution degrades dramatically. The semi-digital reconstruction and the software compensation both apply weights to the energy depositions in a shower depending on the hit energy. The semi-digital reconstruction achieves a resolution similar to the software compensation for the lowest energy, 10 GeV. For higher beam energies the resolution follows a similar shape as for the software compensation but with absolute values 1-2 % worse. The best resolution of all four methods for the whole energy range is found using the analogue software compensation algorithm.
The simulated AHCAL with $1 \times 1\,\mathrm{cm}^2$ cell size shows an improved resolution for the semi-digital and digital readout schemes, see Figure 2b. Compared to the classical analogue energy reconstruction the digital reconstruction shows better results for beam energies below 35 GeV. This improvement despite the reduction of information can be explained by the shape of the analogue cell signal, which follows a Landau distribution that is characterised by a long tail to high values. By counting cells above a certain signal amplitude, the signal fluctuations to high values are removed and thus the energy reconstruction can be improved. A degradation due to saturation effects of the digital resolution is observed above 25 GeV.
The increase of the number of thresholds from 1 to 3, digital to semi-digital, results in a large improvement of the energy resolution of the $1 \times 1\,\mathrm{cm}^2$ AHCAL simulation and in an even larger improvement for the AHCAL with $3 \times 3\,\mathrm{cm}^2$ cells.

In conclusion, the best resolution is achieved by applying either a weighting by the software compensation algorithm or by the semi-digital energy reconstruction. This is understood because both methods apply energy dependent weights, which are determined by a χ^2 minimisation that optimises the resolution. For both methods a decreasing resolution with increasing beam energy is observed. The $1 \times 1\,\mathrm{cm}^2$ AHCAL simulation needs the semi-digital energy reconstruction to

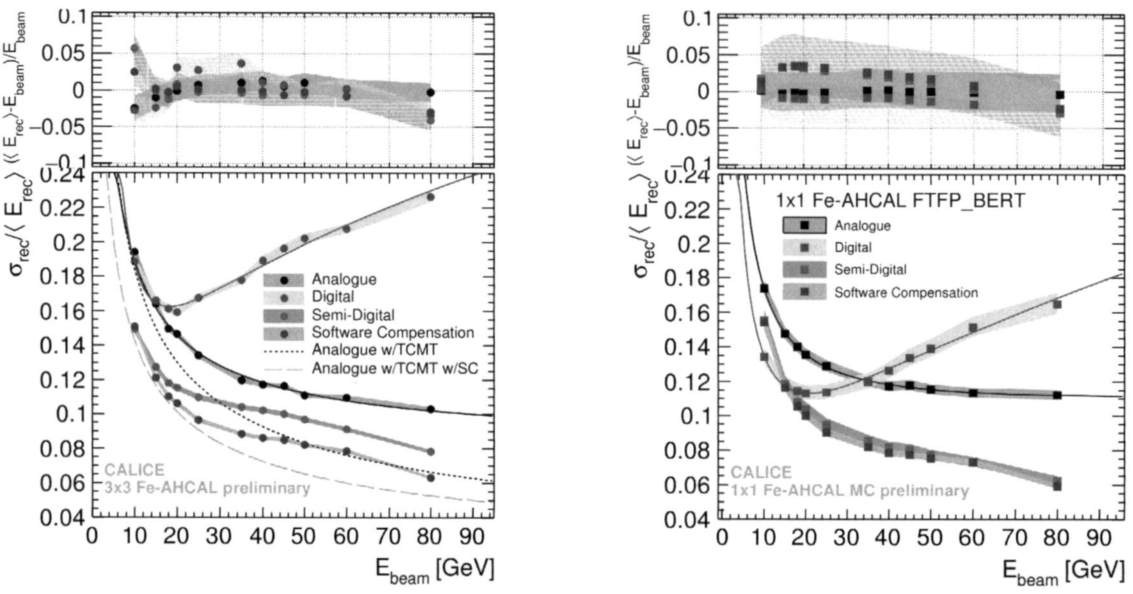

(a) $3 \times 3\,\text{cm}^2$ (b) $1 \times 1\,\text{cm}^2$

Figure 2: Energy dependence of the relative energy resolution of the AHCAL testbeam data in (a) and the simulation with $1 \times 1\,\text{cm}^2$ granularity and the FTFP_BERT physics list in (b), obtained using different approaches for the energy reconstruction of pions: analogue (black), digital (green), semi-digital (red) and applying the analogue software compensation algorithm (blue). The dashed and dotted curves in (a) show the resolution achieved in [2] with and without software compensation techniques, using the energy deposits in the TCMT and in the ECAL in addition to the AHCAL. The plots on the top show the residuals to the beam energy with the bands indicating the systematic and statistical uncertainties. The statistical errors are smaller than the markers.

achieve the best possible energy resolution, while the analogue software compensation algorithm has to be used for the energy reconstruction in the $3 \times 3\,\text{cm}^2$ AHCAL.
A first comparison [4] of the digital and semi-digital energy resolutions of the $1 \times 1\,\text{cm}^2$ AHCAL simulations to the resolution achieved by the DHCAL and SDHCAL prototypes shows an advantage of the scintillator-tile calorimeter especially at low energies. However this effect can have many different explanations: the different sampling fraction of 5 mm scintillator thickness versus 1.15 mm gas gap; a different threshold setting; the difference in the signal distributions (a Landau distribution and a Polya function, that is very sensitive to high voltage variations); or the difference in the ionising energy loss in the scintillator and the gas mixture. It needs further investigations to disentangle the effects of the differences mentioned above.

References

[1] CALICE collaboration, Validation of GEANT4 Monte Carlo models with a highly granular scintillator-steel hadron calorimeter, 2013 JINST 8 P07005, doi:10.1088/1748-0221/8/07/P07005
[2] CALICE collaboration, Hadronic energy resolution of a highly granular scintillator-steel calorimeter using software compensation techniques, 2012 JINST 7 P09017, doi:10.1088/1748-0221/7/09/P09017
[3] H. Ikeda et al., NIM A441 (2000), p. 401 (Belle Collaboration)
[4] C. Neubüser, Comparison of Energy Reconstruction Schemes and Different Granularities in the CALICE AHCAL, CALICE Analysis Note CAN-049a, May 2014 https://twiki.cern.ch/twiki/pub/CALICE/CaliceAnalysisNotes/CAN-049a.pdf

IOP Conf. Series: Journal of Physics: Conf. Series **928** (2017) 012039 doi:10.1088/1742-6596/928/1/012039

Towards a Technological Prototype for a High-granularity Electromagnetic Calorimeter for Future Lepton Colliders

Taikan Suehara[1] on behalf of CALICE SiW-ECAL group[123456]

[1]Department of Physics, Faculty of Science, Kyushu University, 744 Motooka, Nishi-ku, Fukuoka, 819-0395 Japan
[2]Laboratoire Leprince-Ringuet, Ecole Polytechnique, 91128 Palaiseau Cedex, France
[3]Laboratoire de Physique Nucleaire et de Hautes Energies, UPMC, Case courrier 200, 4 place Jussieu, 75005 Paris, France
[4]Laboratoire de l'Accelerateur Lineaire, Centre Scientifique d'Orsay, 91898 ORSAY Cedex, France
[5]OMEGA Microelectronics Center, Ecole Polytechnique, 91128 Palaiseau Cedex, France
[6]Department of Physics, Graduate School of Science, The University of Tokyo, 7-3-1 Hongo, Bunkyo-ku, Tokyo, 113-0033 Japan

E-mail: `suehara@phys.kyushu-u.ac.jp`

Abstract. The CALICE collaboration is preparing large scale prototypes for highly granular calorimeters for detectors to be operated at a future lepton collider. Currently a prototype of a silicon-tungsten electromagnetic calorimeter SiW-ECAL will be assembled which in terms of dimensions and layout meets already most of the requirements given by the lepton collider physics programme and hence the detector design. In particular the front end electronics will be embedded into the layer structure of the calorimeter and have to fit within alveolar layers with less than 1 cm in height. In this paper the design of the prototype and the steps towards the realisation are presented. The presented technology plays also a key role in the upgrades of the LHC Experiments CMS and ATLAS.

1. Introduction

Particle flow is considered to be the key concept of calorimetry in future energy-frontier lepton colliders such as International Linear Collider (ILC). Jet energy in the particle flow is measured with momentum of tracks and energy of neutral clusters. Since the momentum resolution of tracks is much better than the energy resolution of clusters in general, this gives better energy resolution of jets.

The particle flow requires a highly granular calorimeter system to separate particles, especially in electromagnetic calorimeter (ECAL). The reference design of ECAL for International Large Detector (ILD), one of two detector concepts for ILC, has 30 layers of tungsten absorbers and sensors with the total depth of 24 radiation lengths, as described in the Detailed Baseline Design report [1]. Silicon sensor technology (SiW-ECAL) is one of the promising options for ILD ECAL, because of the easy implementation of fine segmentation and the good maturity of technology used for various detector applications. One of the big challenges of the SiW-ECAL system is to realize a readout system of a huge number ($\sim 10^8$) of channels with embedded ASICs in between the sensor and absorber layers.

The concept of SiW-ECAL system was first tested with a "Physics Prototype" [2], which proved the concept of particle flow calorimetry in good agreement with the Monte Carlo simulations with several test beam campaigns in 2005-2011.

The goal of the second "Technological Prototype" is to establish a system which is ready to be used as a real detector as ILD ECAL, such as:

- Embedded electronics with dedicated ASICs with 1-3 mm electronics layers
- Smaller cell size of around 5×5 mm^2
- Power pulsing capability (bias currents of front end electronics enabled and shut down in synchronization with the ILC beam structure)
- Buffering up to 15 event frames with self-triggering capability
- Multiple ASICs connected on the bus structure

The first technological prototype was fabricated with one 5×5 mm^2-segmented sensor and four ASICs and tested with an electron beam in DESY in 2012-13 [3]. Based on the experience of physics and technological prototypes, we have developed a new technological prototype which is aimed to be even closer to the to ILD ECAL.

The SiW-ECAL development has been conducted mainly within the framework of CALICE collaboration, which aims to develop highly granular calorimeters to be operated at future colliders in a consistent manner. Strong synergies exist with the recently raised effort of developing high-granular calorimeter to be installed in the CMS for the HL-LHC upgrade [4].

2. Sensor studies

Figure 1 shows a silicon sensor. The biggest technical issue of the sensors is square-shaped events caused by hits on the sensor edge. This is considered to be due to a floating guard-ring that surrounds the sensor. We are investigating sensors without guard rings as well as optimizing the spacing on the edge. Our study with laser photons hitting on the edge showed that the sensors without guard rings do not suffer from the square events.

Figure 2 shows the comparison of the dark current of sensors without guard rings with different edge spacings and thicknesses. The sample with smaller spacing and 500 μm thickness gives significantly lower breakdown voltage than others. Other samples have acceptable leakage current and breakdown voltage.

We also studied the sensors with neutron irradiation. Data were taken at the neutron beamline of up to 7.8 MeV at a tandem accelerator in Kobe university. The increase of the dark current observed after the irradiation shows that the effect scaled to the estimated irradiation of ILC calorimeter is in an acceptable range, as shown in Fig. 3.

3. Electronics and Assembly

SKIROC2 (Silicon Kalorimeter ReadOut Chip 2) is a 64 channel readout ASIC designed for SiW-ECAL. It has self-triggering capability, 15 analog memories, 12-bit ADC and 12 bit counter for bunch tagging of each event in a ILC train. It features further two slow shapers of different gain and power pulsing capability. Performance of the SKIROC2 chip is shown in [5].

Figure 4 shows a schematic of our "short slab". It is equipped with four 256-pad silicon sensors and 16 ASICs packaged in BGA400. (Alternative chip-on-board implementation of the ASICs is also being studied.) It also has a kapton film to apply bias voltage (100-120 V) to sensors, cooling and electric shielding, detector interface (DIF) to send data to the DAQ system, and a big capacitance for power pulsing. The sensors are attached by conductive glue with a auto-dispensing robot to the printed circuit board (PCB). Four slabs have been prepared for the first test beam.

Figure 1. A sample of the silicon sensor for SiW-ECAL.

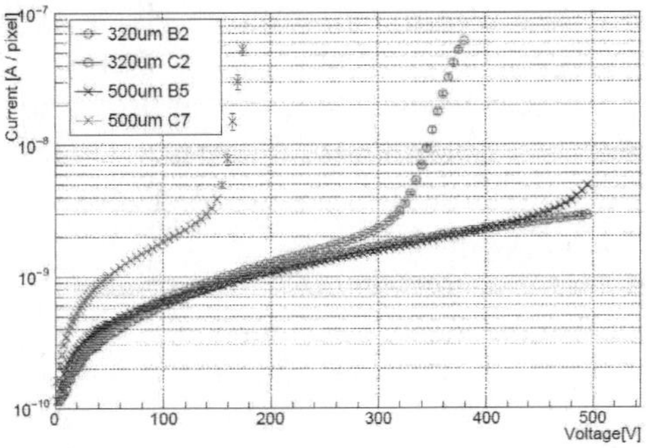

Figure 2. Leakage current of sensors without guard rings.

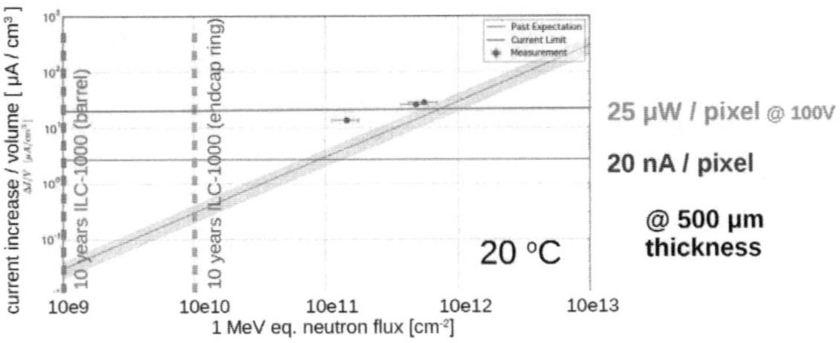

Figure 3. Results of irradiation (red points) with expectation to ILC. The result is nearly consistent with past results of various silicon sensors (inclined line with a band). The dashed lines show the estimated flux of ILC 1 TeV operation. The horizontal lines show our limitation caused by power consumption (upper) and by current electronics (lower).

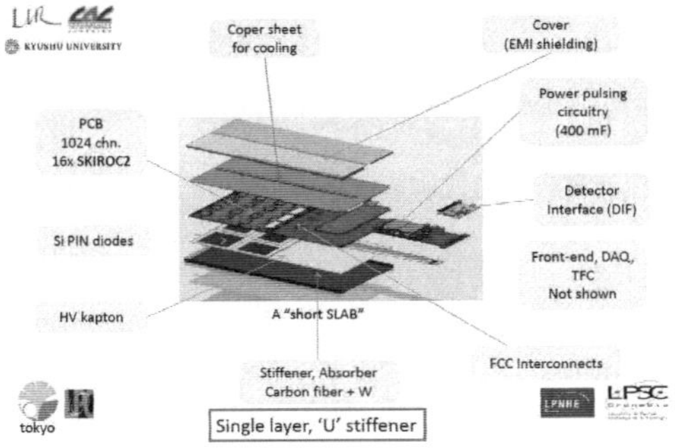

Figure 4. Schematics of the slab designed as the second prototype.

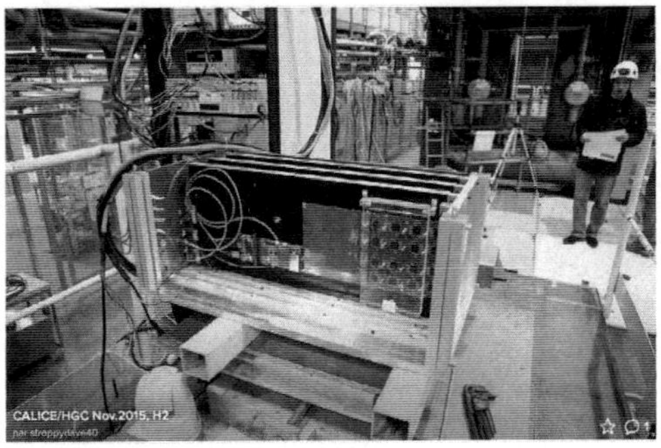

Figure 5. Setup of the test beam at CERN SPS with three active slabs.

4. Test beam at CERN

Beam test was conducted at CERN SPS H2 beam line at November 2015. Three slabs with 12 sensors and 48 ASICs were operated during the test beam. The setup (without shielding outside) is shown in Figure 5. Events with pions, electrons and muons with 10-150 GeV energy were taken.

Figure 6 shows a sample of a hitmap measured by a slab. The beam spot was clearly visible, with some number of channels with hits missing, because they are masked due to larger trigger rate due to higher noise. This is caused by misconfiguration of tuning trigger threshold on each channel, which is planned to be corrected in the next ASIC production (SKIROC2a).

Figure 7 shows distribution of signal-to-noise (S/N) ratio obtained on each channel (except inactive channel) by calculating 60 ADC count divided by root mean square of the pedestal distribution of the channel on several types of muon runs. 60 ADC counts correspond approximately to the maximum probability of minimum ionizing particle (MIP). It shows that S/N ratio of between 15 and 20 is obtained for most of channels, which is big enough to have essentially no accidental noise with 0.5 MIP trigger threshold.

The test beam was operated with a 5 times higher gain of preamplifier than the gain assumed for the real ILC operation to assure dynamic range up to more than 1000 MIPs. (The feedback capacitance of the preamplifier was set to 1.2 pF instead of 6.0 pF.) Reducing gain currently lowers the S/N ratio, but the high S/N ratio at the high gain shows noise level low enough at the preamplifier input, which is the most essential part for the noise.

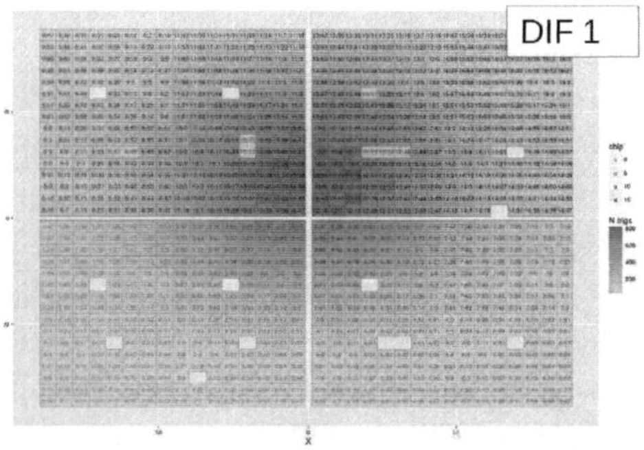

Figure 6. A hitmap of one slab with four sensors.

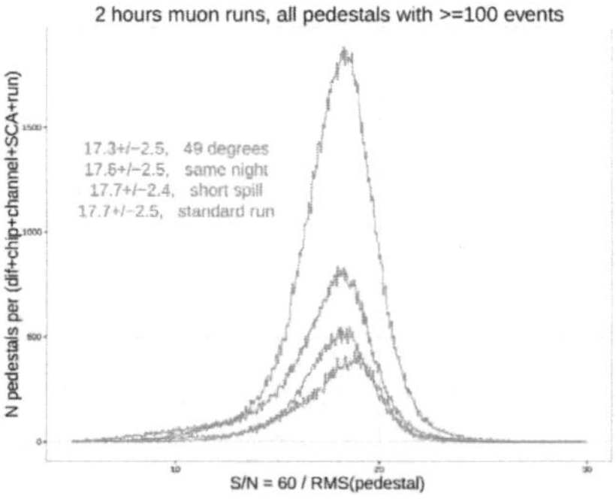

Figure 7. Distribution of signal-to-noise ratio in each channel (preliminary).

5. Summary and Plan

This contribution reports on progress of the CALICE SiW-ECAL prototype with ILC compatible integrated electronics and silicon sensors with $\sim 5 \times 5$ mm^2 pixelization. Sensor studies have shown that the sensor without guard ring is a good candidate for SiW-ECAL. The improved slabs of technological prototype with BGA-packaged SKIROC2 have been fabricated and first test beam has been done in CERN SPS beam line. The preliminary result shows a good signal-to-noise ratio.

We plan to fabricate more layers, including those with sensors without guard rings, and conduct a series of test beam in 2016. The new slabs will be also equipped with the new SKIROC2a for comparison with the previous version. We will also fabricate "long slabs" which are chains of ten slabs to fit the structure of ECAL assembly of the real detectors for future colliders.

References

[1] Behnke T *et al.* 2013 The International Linear Collider Technical Design Report - Volume 4: Detectors (*Preprint* 1306.6329)
[2] Adloff C *et al.* 2009 *Nucl. Instrum. Meth.* **A608** 372–383
[3] Amjad M S *et al.* 2015 *Nucl. Instrum. Meth.* **A778** 78–84
[4] Butler J, Contardo D, Klute M, Mans J, Silvestris L and on behalf of the CMS C (CMS Collaboration) 2015 CMS Phase II Upgrade Scope Document URL http://cds.cern.ch/record/2055167
[5] Callier S, Dulucq F, de La Taille C, Martin-Chassard G and Seguin-Moreau N 2011 *JINST* **6** C12040

IOP Conf. Series: Journal of Physics: Conf. Series **928** (2017) 012040 doi:10.1088/1742-6596/928/1/012040

Impact of Non-Uniformity in Light Collection on the Energy Resolution of the PANDA Electromagnetic Calorimeter at Photon Energies Below 1 GeV

Stefan Diehl, Kai-Thomas Brinkmann, Peter Drexler, Valery Dormenev, Rainer W. Novotny, Christoph Rosenbaum and Hans-Georg Zaunick for the PANDA-Collaboration

2nd Physics Institute, Justus-Liebig-University Giessen, Heinrich-Buff-Ring 16, 35392 Giessen, Germany

E-mail: `stefan.diehl@exp2.physik.uni-giessen.de`

Abstract. The electromagnetic calorimeter (EMC) of the PANDA detector at the future FAIR facility comprises more than 15,000 lead tungstate (PWO) crystals. The barrel part will consist of 11 crystal geometries with different degree of tapering, which causes a non-uniformity in light collection as an interplay between the focusing and the internal absorption of the light. For the most tapered crystals the detected light is enhanced by 40%, if the scintillation process is created in the front part of the crystal. Due to the shower development and its fluctuations the non-uniformity leads to a reduction of the energy resolution. To reduce this effect, one lateral crystal side face has been de-polished to a roughness of 0.3 μm. Measurements confirm an increase of the light yield in the rear part of the crystal. In contrast, only a slight decrease can be observed in the front part. The overall non-uniformity is significantly reduced below 5%. This paper will discuss the experimental studies based on GEANT4 and optical simulations to understand the impact of a de-polished side face on the light collection. For consequences on the future performance, a 3x3 sub-array of de-polished crystals was directly studied using a tagged photon beam in the energy range from 50 MeV up to 800 MeV, respectively, performed at the tagged photon facility at MAMI, Mainz. The comparison to an array composed of polished crystals confirms a significant improvement of the constant term of the energy resolution from above 2 % down to 0.5 % and only a small increase of the statistical term. The results can be reproduced in GEANT4 simulations.

1. Introduction

The PANDA detector will be one of the central experiments of the future Facility for Antiproton and Ion Research (FAIR) which is currently under construction near Darmstadt in Germany. PANDA will be a fixed target experiment using a cooled antiproton beam of up to 15 GeV to study new aspects in the fields of hadron spectroscopy, nucleon structure investigations, the modification of hadrons in matter and the search and characterization of Hypernuclei [1]. The electromagnetic calorimeter (EMC) of the PANDA detector, which will be one of the central components to achieve the physical goals, will consist of more than 15,000 lead tungstate (PWO) crystals operated at -25 °C to increase the light yield [1]. To reach the physics goals, a detection of photons down to 10 MeV is mandatory.

2. Origin and reduction of the non-uniformity in light collection

The barrel part of the target EMC will be composed of 11 crystal geometries with a different degree of tapering. The read out will be performed with two rectangular large area avalanche photo diodes per crystal with an active area of 1 cm^2 each. Due to the tapering the crystals show a non-uniformity in light collection as a result of an interplay between the focusing and the internal absorption of the scintillation light within the crystal (see Fig. 1). The non-uniformity has been determined using low energy γ-rays, cosmic muons and a 80 MeV proton beam leading to comparable results (see also [2]). The measured non-uniformity curves for the two most tapered crystal types (1 and 2) and for a crystal with an average degree of tapering (type 6) are shown in Fig. 2.

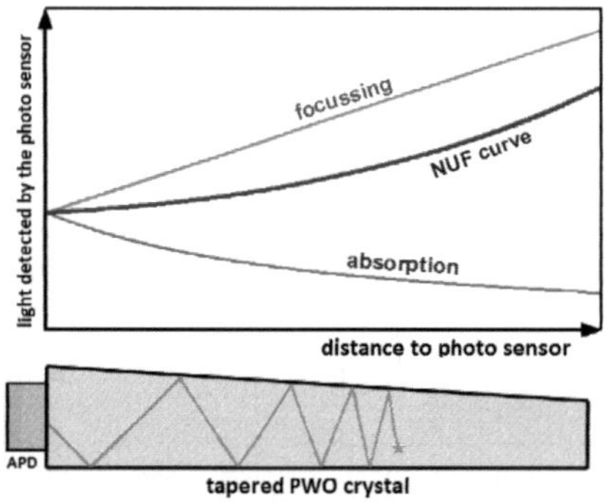

Figure 1. Illustration of the effects which contribute to the non-uniformity of the light collection.

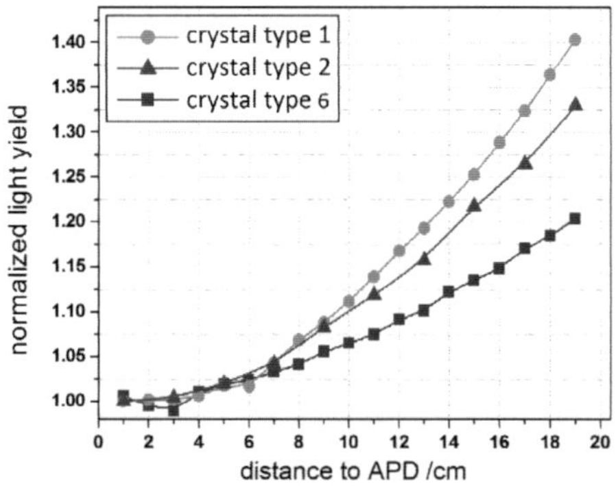

Figure 2. The measured non-uniformity curves of crystals with different degree of tapering.

While crystals with an average degree of tapering (type 6) show a light yield difference between creation of the scintillation light in the front and the rear part of the crystal of 20 %, this value increases for the most tapered crystals to more than 40 % with a slope in the front part of almost 3 %/cm. Due to the spread of the electromagnetic shower within the crystal and due to its fluctuations, this effect causes a smearing of the response leading to a reduction of the energy resolution, in particular the constant term. This problem has already been considered by the CMS-ECAL collaboration and was solved by de-polishing crystals on one lateral side [3]. Since the PANDA barrel EMC is focusing on much lower energies ranging from 10 MeV up to a few GeV, besides the uniformity of the light collection, also the absolute light yield has to be considered, if the crystal surface is modified. To investigate the impact of a de-polished side face on these parameters, the experiences from CMS [3] have been used and one lateral side face of a set of 25 crystals has been de-polished at the CMS setup to a roughness of $R_a = 0.3$ μm.

Figs. 3 and 4 show the absolute and relative position dependent light yield of six typical type 2 crystals wrapped with mirror reflective VM2000 before and after the de-polishing procedure. As an effect of the de-polishing, a significant increase of the absolute light yield in the rear part of the crystal can be observed combined with a slight decrease in the front part. As a direct result, the non-uniformity is reduced down to a typical value of 5 % with a nearly homogeneous response in the front and central part of the crystal. The observed behaviour can be reproduced with the ray tracing model of GEANT4 (see Fig. 5). The increase of the light yield in the rear part of the crystal can be explained by photons which change their direction on the way to the front face of the crystal, which is illustrated by the significant increase of photons with path lengths between 1 cm and 39 cm in a crystal with a de-polished side face in Fig. 6.

IOP Conf. Series: Journal of Physics: Conf. Series **928** (2017) 012040 doi:10.1088/1742-6596/928/1/012040

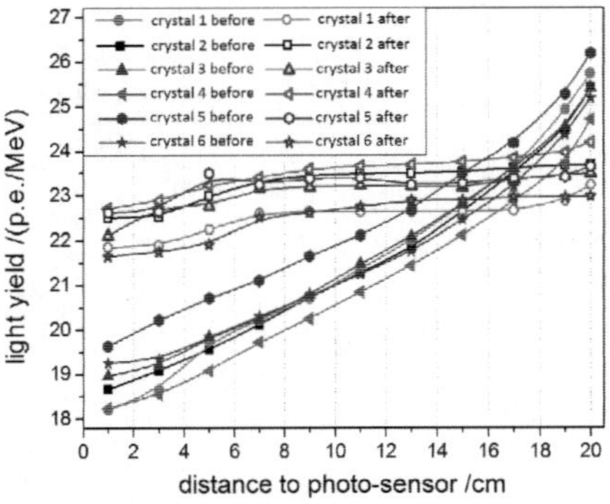

Figure 3. Absolute light yield of six typical type 2 crystals before and after de-polishing of one lateral side face.

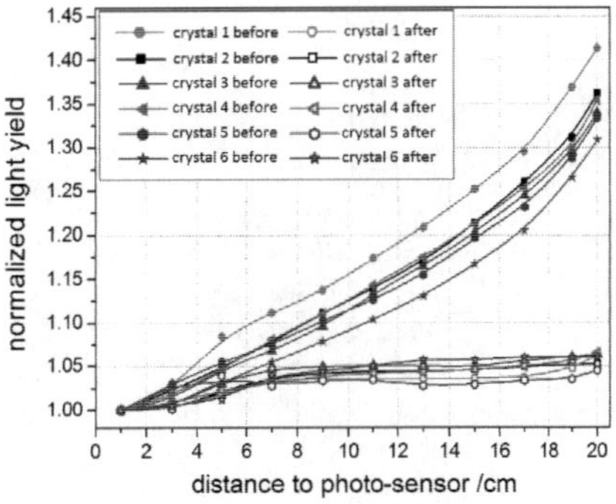

Figure 4. Relative light yield of six typical type 2 crystals before and after de-polishing of one lateral side face.

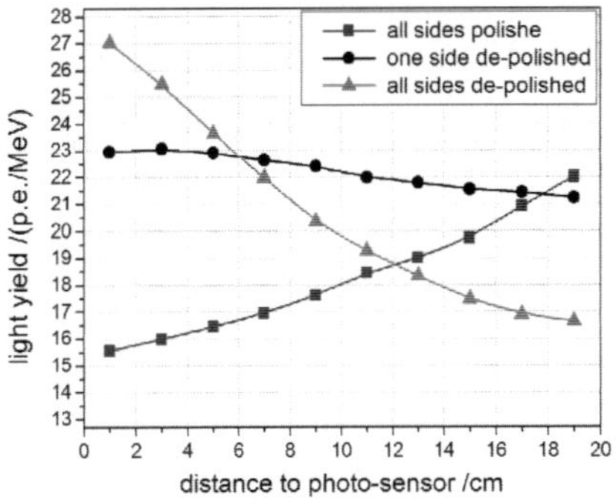

Figure 5. Simulated light yield of a completely polished crystal and a crystal with one / all lateral side faces de-polished. The crystal is wrapped with mirror reflective VM2000 in all cases.

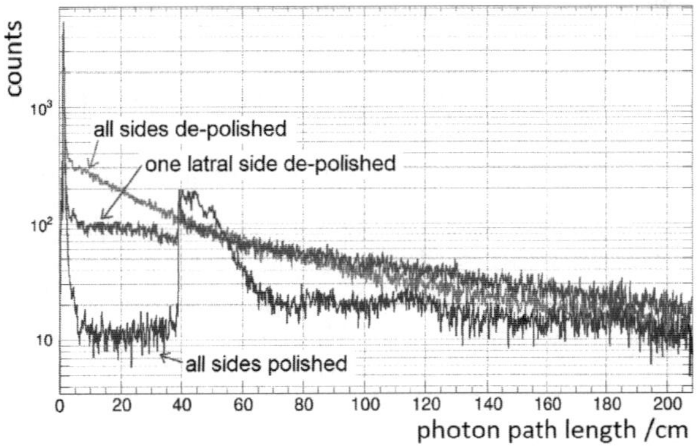

Figure 6. Path length distribution of photons which reach the photo-sensor emitted from a source positioned 1 cm away from the crystal rear face for different surface concepts.

3. Impact of the non-uniformity on the energy resolution of the PANDA barrel EMC

For the first time a 3x3 sub-matrix of de-polished crystals has been implemented in the current close to final barrel EMC prototype (see Fig. 7) and compared with an identical array of completely polished crystals. The response to tagged photons in the energy range between 50 MeV and 800 MeV, respectively, has been measured for a central interaction within the 3x3 array at the MAMI facility in Mainz (Germany). Fig. 8 shows the obtained energy resolution for the 3x3 sub-array of de-polished crystals in comparison to a similar matrix composed of polished crystals. The experimental results show a significant improvement of the energy resolution for energies down to 200 MeV if crystals with one de-polished side face are used, while the energy resolution is comparable for both arrays in the region between 50 MeV and 200 MeV, respectively. As a consequence, the constant term of the relative energy resolution is significantly reduced from above 2 % down to 0.5 % while the statistical term increases only slightly. The

IOP Conf. Series: Journal of Physics: Conf. Series **928** (2017) 012040 doi:10.1088/1742-6596/928/1/012040

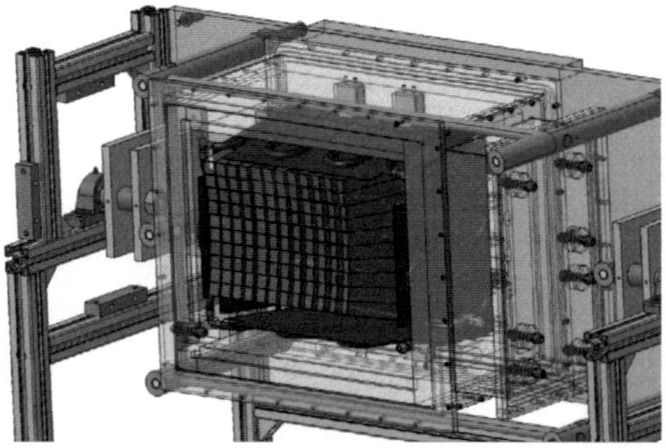

Figure 7. Schematic drawing of the current PANDA barrel EMC prototype PROTO120.

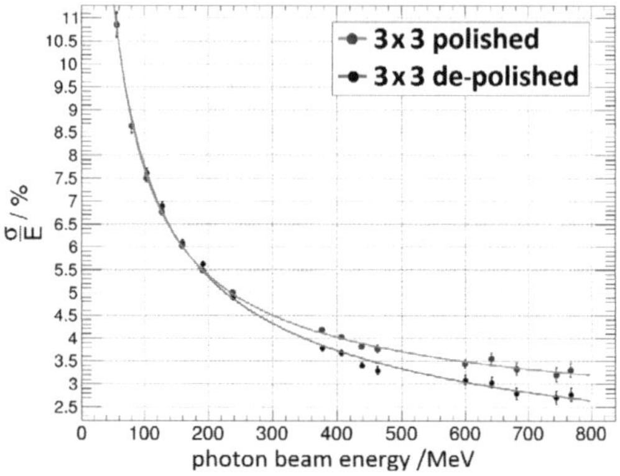

Figure 8. Relative energy resolutions of 3x3 arrays of de-polished and completely polished crystals operated at a temperature of -25°C implemented into PROTO120 [5].

shown results can be well reproduced with a specially developed GEANT4 model, including the non-uniformity and other empirical properties of the crystals and their readout. More details on the experimental and simulation procedure can be found in [4] and [5].

4. Conclusion and Outlook

The de-polishing of one lateral side face provides a nearly uniform response combined with an only moderate light loss in the front part of the crystal. On the other hand the light yield from the rear part of the crystal even increases due to a direction change of a certain amount of photons, initiated by the de-polished side face. Especially at high energies a significant improvement of the energy resolution can be achieved if the light collection is uniform. Due to the only moderate light loss in the front part of the crystal, no change of the energy resolution can be observed in the energy region between 50 MeV and 200 MeV, in the experimental data as well as in GEANT4 simulations.

5. Acknowledgements

The project is supported by BMBF, GSI and HIC for FAIR. The authors would like to thank Etiennette Auffray from CERN for providing the de-polishing of the crystals.

References

[1] PANDA collaboration, Technical Design report for: The PANDA Electromagnetic Calorimeter, Darmstadt 2009, arXiv:0820.1216.

[2] D. Bremer, Measurement and Simulations on Position Dependencies in the Response of Single PWO Crystals and a Prototype of the PAND EMC, PhD thesis (Giessen, 2014).

[3] E. Auffray et al., Crystal Conditioning for High Energy Physics Detectors, Nucl. Instrum. Meth. in Phys. Res. A 486, 22-34 (2002).

[4] S. Diehl, Optimization of the Influence of Longitudinal and Lateral Non-Uniformity on the Performance of an Electromagnetic Calorimeter, PhD thesis (Giessen, 2015), *http://geb.uni-giessen.de/geb/volltexte/2016/11998/*.

[5] S. Diehl, D. Bremer, P. Drexler, V. Dormenev, T. Eissner, T. Kuske, S. Nazarenko, R. W. Novotny, C. Rosenbaum, H.-G. Zaunick, Influence of the Light Collection Non-Uniformity in Strongly Tapered PANDA PWO Crystals on the Energy Resolution of the PANDA Electromagnetic Calorimeter at Energies below 1 GeV, IEEE Transact. on Nucl. Science vol. 63, iss. 2, 564-568, doi: 10.1109/TNS.2015.2493341 (2015)

Radiation hardness test of un-doped CsI crystals and Silicon Photomultipliers for the Mu2e calorimeter

S Baccaro[1], A Cemmi[1], M Cordelli[2], E Diociaiuti*[2,3], R Donghia*[2,3], S Giovannella[2], S Loreti[4], S Miscetti[2], M Pillon[4] and I Sarra[2]

[1] ENEA FSN, Casaccia R.C., Rome, Italy
[2] Laboratori Nazionali di Frascati - INFN, Frascati, Italy
[3] Universitá degli Studi Roma Tre, Rome, Italy
[4] Department of Fusion and Technology for Nuclear Safety and Security - ENEA C.R., Frascati, Italy

E-mail: eleonora.diociaiuti@lnf.infn.it, raffaella.donghia@lnf.infn.it

Abstract. The Mu2e calorimeter is composed by 1400 un-doped CsI crystals coupled to large area UV extended Silicon Photomultipliers arranged in two annular disks. This calorimeter has to provide precise information on energy, timing and position. It should also be fast enough to handle the high rate background and it must operate and survive in a high radiation environment. Simulation studies estimated that, in the hottest regions, each crystal will absorb a dose of 300 Gy and will be exposed to a neutron fluency of 6×10^{11} n/cm^2 in 3 years of running.

Test of un-doped CsI crystals irradiated up to 900 Gy and to a neutron fluency up to 9×10^{11} n/cm^2 have been performed at CALLIOPE and FNG ENEA facilities in Italy. We present our study on the variation of light yield (LY) and longitudinal response uniformity (LRU) of these crystals after irradiation. The ionization dose does not modify LRU while a 20% reduction in LY is observed at 900 Gy. Similarly, the neutron flux causes an acceptable LY deterioration ($\leq 15\%$). A neutron irradiation test on different types of SIPMs (two different array models from Hamamatsu and one from FBK) have also been carried out by measuring the variation of the leakage current and the charge response to an ultraviolet led. We concluded that, in the experiment, we will need to cool down the SIPMs to 0 °C reduce the leakage current to an acceptable level.

1. Introduction

The Mu2e [1] calorimeter is composed by 1400 pure CsI crystals coupled to large area UV extended Silicon Photomultipliers (SiPM), arranged in two annular disks [2][3]. It has to have a good energy, timing and position resolution and to operate in a high radiation environment [2][3].

Simulation studies shows that in the hottest regions each crystal will absorb 30 krad (300 Gy) of ionization dose and will be exposed to a neutron flux of 6×10^{11} n/cm^2 in three years of running [4]. For this reason, we have tested the variation of the light yield (LY) and the longitudinal response uniformity (LRU) of un-doped CsI crystals irradiated with a dose and a neutron flux exceeding the one expected in the experiment lifetime.

Concerning SiPMs, they have to withstand a fluency of 3×10^{11} n/cm^2, with neutrons at 1 MeV equivalent energy, and absorb a dose up to 20 krad in three years of running [4].

Irradiation tests with a ionization dose have been performed at the ENEA CALLIOPE facility [5], where a ^{60}Co source is used to produce γ's with an energy of 1.25 MeV. The activity of the source during our tests was 0.35×10^{15} Bq, allowing to reach 5 Gy/h at about 5 m distance.

The neutron irradiation tests have been performed at the ENEA FNG facility [6], where a nearly isotropic 14 MeV neutron flux is produced. The maximum neutron intensity is 0.5×10^{11} n/s, close to the target, with a uniform production and a dependence on the distance, R, as $1/R^2$. The desired neutron intensity is reached by positioning the crystal at the needed distance.

2. Pure CsI radiation hardness measurements

In both irradiation tests, the crystals parameters, LY and LRU, have been measured at different steps of the irradiation program. This characterization is done using a low intensity collimated ^{22}Na source. The source is placed between the crystals and a small monitor system, constituted by a $(3\times3\times10)$ mm^3 LYSO crystal, readout by a (3×3) mm^2 MPPC. One of the two back-to-back 511 keV photons produced by the source is tagged by this monitor while the second photon is used to test the crystal, that is readout by means of a 2" UV-extended 9813QB photomultiplier (PMT) from ET Enterprises. The LY has then been evaluated as:

$$\frac{Np.e.}{MeV} = \frac{\mu_{Q[pC]}}{G_{PMT}\cdot E_\gamma[MeV]\cdot q_{e^-}[pC]}, \tag{1}$$

where Q is the charge obtained integrating and converting in pC the signal amplitude and μ_Q is the mean value provided by the gaussian fit applied, $G_{PMT} = 3.8\times10^6$ is the PMT gain, $q_{e^-} = 1.6\times10^{-19}$ pC is the charge of the electron and $E_\gamma = 511$ keV is the energy of the annihilation photon.

A $(3\times3\times18)$ cm^3 CsI crystal from SICCAS has been irradiated at the CALLIOPE facility, by positioning it parallel to the incoming γ's, so that the same dose was delivered along the crystal axis. The total dose absorbed by the crystal was $\sim$ 90 krad (900 Gy, two days at 2.23 Gy/h and seven days at 4.86 Gy/h), that corresponds to about three times the total dose expected in the hottest region of the calorimeter, i.e. the innermost ring around the beam axis, in three years running [4].

As shown in Figure 1 (left), a negligible LY reduction is observed for doses up to 20 krad, while for the last irradiation step the measured decrease is about 20%. The slope for the LRU remains similar for the irradiated and not irradiated crystals.

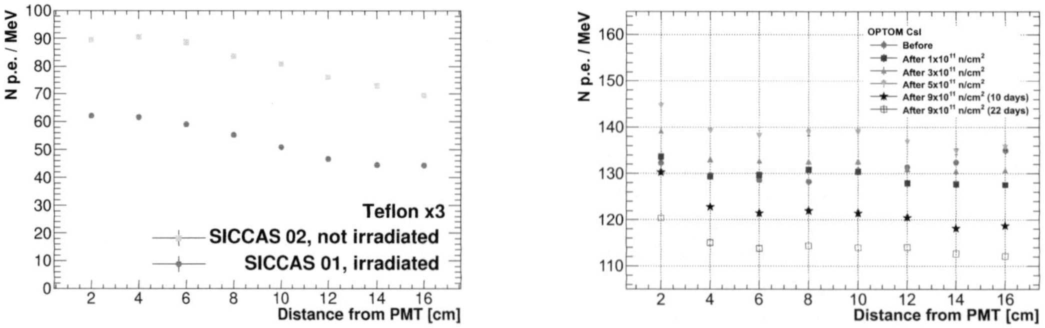

Figure 1. Crystal LY as a function of the source distance from the PMT at different steps of the irradiation programs at CALLIOPE (left) and FNG (right).

Three $(3\times3\times20)$ cm^3 pure CsI crystals from different vendors have been tested with neutron dose at FNG: one from SICCAS (China), one from ISMA (Ukraine) and one from OPTO

MATERIALS (Italy). The total flux delivered, 9×10^{11} n/cm^2, which corresponds to about 1.5/3 times the maximum total flux expected for the first/second calorimeter disk in three years of running [4]. LY and LRU for Tyvek wrapped crystals and PMT optically coupled in air have been measured before each day of irradiation. The measurements have also been repeated 10 and 22 days after the end of the irradiation test.

In Figure 1 (right), an example of the LY of one crystal tested is reported as a function of the source distance from the PMT. The LY increases with the neutron flux because of fluorescence effects and activation. Measurements performed several days after the irradiation test show a decrease of the LY as related to the reduction of these effects. Comparing the first to the last measurements, a decrease of 10-20% is visible for the undoped CsI crystal from OPTO MATERIALS, while no deterioration is present for the ISMA one. The SICCAS crystal has a completely different behavior along the crystal before and after irradiation, with a strong deterioration of the uniformity slope. Moreover, all crystals, except for SICCAS one, after irradiation show good LRU performance, with a total uniformity well below 10%, and negligible deterioration. LRU for CsI from SICCAS changes from 5% to 15% after neutron irradiation.

3. Silicon Photomultiplier radiation damage induced

During the irradiation campaign different models of SiPM have been tested: two SiPM from Hamamatsu [8] and a SiPM from FBK [9]. The (12×12) mm^2 Hamamatsu SiPMs are made by the same array of 16 (3×3) mm^2 cells, but they have different protection material: one SiPM is covered with a silicon protection layer (SPL) while the other one with a Micro-Film (MF). The FBK SiPM is instead a monolithic (6×6) mm^2 SiPM.

To control and monitor the Hamamatsu devices, a pulsed UV-led and two optical fibers have been used to illuminate the SiPM and a PMT positioned 2 m far away from the radiation source. We acquired the signal response to the led pulse of just one cell and the leakage current of another cell. The PMT response has been used as reference in order to obtain a precise measurement of the light input. Since the FBK SiPM is monolithic only its leakage current has been recorded.

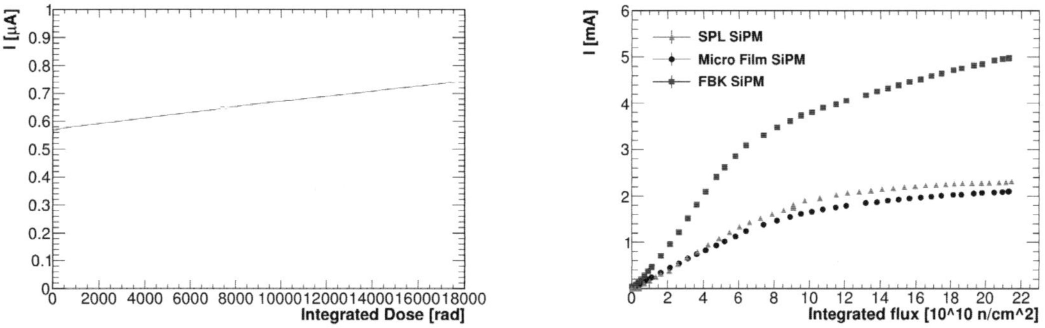

Figure 2. Leakage current as a function of the integrated dose absorbed by the SPL SiPM at CALLIOPE (left) and as a function of the neutron integrated flux at FNG (right).

The three SiPMs were first irradiated with neutrons at the FNG facility. They were positioned 7 cm far away from the source and irradiated, for less than 4 hours, with 14 MeV neutrons up to 2.2×10^{11} n/cm^2, which corresponds to a neutrons flux equal to 2.2 times that expected in the experimental lifetime. We observed that the signal peak of SPL SiPM decreased from ~ 250 mV to ~ 30 mV, while the MF one decreased from $\sim$ 400 mV to $\sim$ 50 mV. In Figure 2 (right) the leakage current of all the SiPMs tested as a function of the integrated flux is reported. To compare results from SiPMs with different cell dimensions, the FBK current has been corrected by a factor 4 because of the greater active area. A current increase is clearly visible in both SiPMs: the leakage current of MF SiPM increased from $\sim$ 16 μA to $\sim$ 2 mA, the one of SPL SiPM from $\sim$ 100 μA to $\sim$ 2.2 mA and of FBK one from $\sim$ 21 μA to $\sim$ 5 mA. Even if the hall temperature was quite stable during irradiation the drop on the gain was mostly dominated by the temperature increment of the SiPM.

Later, another SPL SiPM has been irradiated at CALLIOPE, with a total dose $\sim$ 20 krad in 3 days. The dose effect on SiPM performance is negligible both in term of leakage current and signal amplitude. As shown in Figure. 2 (left) the leakage current increased from $\sim$ 0.6 μA to $\sim$ 0.75 μA while the signal amplitude remained unchanged after the irradiation.

Other irradiation tests with SiPM kept at stable temperature are foreseen.

4. Conclusions

The determination of the LY and LRU changes for un-doped CsI crystals after irradiation with a large ionization dose and with neutron fluency provides an important benchmark for the Mu2e calorimeter, where a high radiation environment is foreseen. Our tests show that doses up to 200 Gy do not modify LY and LRU for a $(3 \times 3 \times 18)$ cm^3 unwrapped CsI crystals from SICCAS coupled in air to an UV-extended PMT. After a total dose of 900 Gy, a 20% reduction in LY is observed instead. Crystals from many vendors have also been irradiated with a neutron flux of 9×10^{11} n/cm^2, corresponding to about 2 times the total flux expected for the hottest calorimeter regions in three years of running. At the end of the irradiation test, an acceptable deterioration for the LY has been observed and the LRU is maintained well below 10%.

We have also tested the radiation damage of the Hamamatsu SPL and Micro Film SiPMs, as well as FBK SiPMs, with neutrons and photons by measuring their change in response and leakage current. The total neutron flux (2.2×10^{11}) causes a decreas of the signal peak (and gain) and a large increase of the leakage current. A dose up to 200 Gy causes a negligible effect. Changes are still acceptable for the running conditions in the experiment when cooling down the SiPM to a running temperature of $\sim$ 0 °C.

Acknowledgments

This work was supported by the EU Horizon 2020 Research and Innovation Programme under the Marie Sklodowska-Curie Grant Agreement No. 690835.

References

[1] L. Bartoszek et al., (Mu2e experiment), "Mu2e Technical Design Report" arXiv:1501.05241
[2] S. Miscetti for the Mu2e collaboration, "Design and status of the Mu2e experiment", FCCP 2015, EPJ Web Conferences, DOI:10.1051/epjconf/2016800001
[3] N. Atanov et al., "Design, status and test of the Mu2e crystal calorimeter", CALOR 2016, J.Phys.Conf.Ser.
[4] G. Pezzullo and B. Echenard, "Study of the radiation dose and neutron flux on the calorimeter", Mu2e-doc-2853
[5] S. Baccaro, A. Cemmi, G. Ferrara and S. Fiore, "Calliope gamma irradiation facility at ENEA-Casaccia R.C. (Rome)", ENEA Report RT/2015/13/ENEA (2015)
[6] M. Martone. M. Angelone and M. Pillon, "The 14 MeV Frascati neutron Generator", Journal of Nuclear Materials 212-215 (1994) 1662-1664
[7] IOP Publishing is to grateful Mark A Caprio, Center for Theoretical Physics, Yale University, for permission to include the `iopart-num` BIBTEXpackage (version 2.0, December 21, 2006) with this documentation. Updates and new releases of `iopart-num` can be found on `www.ctan.org` (CTAN)
[8] `http://www.hamamatsu.com/us/en/index.html`
[9] `https://srs.fbk.eu`

Low Light Image Detection Using Multi-channel Silicon Photomultiplier Camera

Jin-A Jeon[1], H. Y. Lee[1], Il H. Park[1,2], Jik Lee[1,2]

[1]Institute of Science and Technology in Space, Sungkyunkwan University, Suwon, Korea
[2]Department of Physics, Sungkyunkwan University, Suwon, Korea

E-mail: `jajeon@skku.edu`

Abstract. We developed a silicon photomultiplier (SiPM) which is a semiconductor photo-multiplication sensor operating in the Geiger mode. The SiPM can measure a single photon due to its high amplification characteristics as photomultiplier tubes (PMT). SiPMs have a number of advantages, such as low power consumption, small and robust structures, low operating voltages, and insensitivity to magnetic fields. Up until now, arrays of single-channel SiPMs have been used in medical applications and scientific experiments. We fabricated not only single-channel SiPMs, but also multichannel SiPMs using the multichannel-SiPMs. We bulit a 1024-channel SiPM sensor. Finally we assembled a pinhole camera aquipped with the readout electronics. In this paper, we present the 1024-channel SiPM sensor, a pinhole camera, and the preliminary image taken by the camera.

1. Introduction

An SiPM could be called a silicon PMT[1-2]. It is a device with the potential to replace the PMT in a variety of scientific and industrial applications, such as medical devices. Each channel of an SiPM is made of a plurality of micro-pixels, and the micro-pixels are Geiger-mode diodes with a high electric field of a P/N-junction[2-3]. Additionally, micro-pixels are enabled by a reverse bias over the breakdown voltage of the junction, which operates sensitively on/off by a single photon. One channel can count as many photons as the number of micro-pixels coupled in parallel, and the micro-pixel has as high of a gain as a Geiger-mode diode (around 10^5)[3]. Such high optical performance characteristics are obtained in place of the PMT due to the applicability of optical detection in the low-light environment. Despite the problem of damage when a conventional PMT is exposed to high-intensity lights, the SiPM is strong, and it is possible to mass-produce based on a silicon semiconductor process[4-5]. Consequently, it is more suitable to apply as an optical sensor for detecting low-light images. In recent years, development of the SiPM has been divided, largely due to the development of a large matrix of channels for image detector applications and an increase in low-wavelength light detection efficiency of the UV band for medical applications. Multi-channel optical cameras, which consist of 11,328 channel SiPMs, were developed for Cherenkov Telescope Array (CTA) experiments [5]. This X-ray camera is fabricated by arranging a 3×3 mm single channel SiPM to an 8×8 array. It is difficult to assemble and has a large gap between channels. It is important to integrate the multi-channel sensor into a camera, and we have developed a multi-channel SiPM device that has 64 channels arranged as an 8×8 with wafer-level fabrication; these channels have a high degree of integration. To apply the developed 64-channel device on the low-light detecting camera, we arranged sixteen 64-channel elements to be placed in a 4 × 4 grid. This produced a device with 1024 channels. The first step to assembling the camera was the concept based on the pinhole camera for extreme low-light images. The device was tested. The device successfully obtained an image from a weak LED light source.

2. Development of the multi-channel SiPM camera

We designed and manufactured a 64-channel SiPM device that consists of an 8 × 8 array with 1 mm^2 channels, operating at around 25 V. Each channel can be measured by about 10^3 photons by a 32 × 32 array of micro-pixels, which has a size of 35 × 35 μm. The development and characteristics of 64-channel SiPMs will be discussed in more detail in another paper. Our 64-channel SiPM has been designed and manufactured with an interval of several tens of micrometer levels in each channel where the wafer-level processing. This multi-channel processing can minimize the dead area, increasing the number of channels by arranging single channels. The margin of each edge around the hundreds of micrometer is necessary when dicing and assembling each channel. We arranged the device as a 4 × 4 for an image detection camera with the 1024-channel SiPM. It was assembled on the flexible printed circuit boards (FPCB), as seen in Figure 1, for the electrical signal processed by the SiPMs. Four sensors of a 64-channel SiPM are a unit. Each device was assembled with 3 mm intervals in consideration of the securing. Handling the margin of the wiring area for the electrical connections and elements cut margin, we are planning to minimize them.

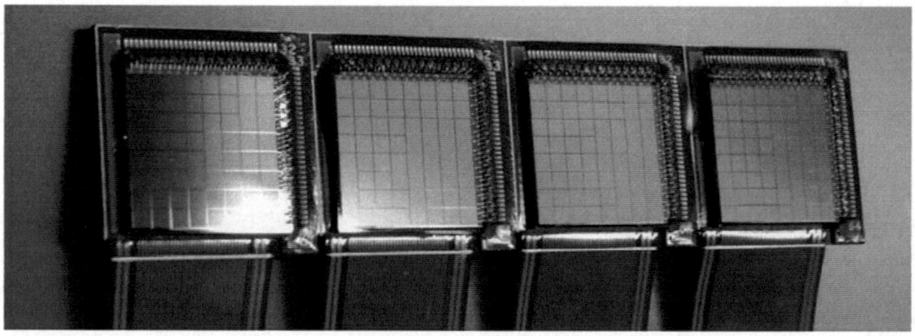

Figure 1. The image of assembled 4 SiPMs on FPCB; 64-channel SiPMs are assembled

A 1024-channel SiPM camera for low light image detection is made by arranging four lines of the main unit, as shown in Figure 2 (Left). The electronics for processing analog signals detected by SiPMs and for applying the bias voltage are connected directly behind the device. The 1024-channel SiPM camera is assembled to open the sensor detection area only for the light-receiving arrangement, shown as a dash circle in Figure 2 (Left), and it was manufactured for stabilization electrically and mechanically. In addition, the total 1024-channel SiPM photo-detector size is about 4 × 4 cm.

3. Performance test of the multi-channel SiPM camera

A pinhole camera that has a 10 cm focal length and a 1 mm diameter was applied to the low-light level optical detection test. The pinhole was attached to a multi-channel SiPM camera in front of the optical receiver in order to reduce the amount of light that would reach the detector, which would be ten thousand times less than image light intensity. We used a pulse of the LED light source that was shorter than 100 ns to make the test environment periodically release the small amount of light, and we arranged LEDs into an A pattern for the light shape detection experiment. The wavelength of the LEDs was about 500 nm, and they were located 30 cm away from the pinhole; the A shape was 10 × 10 cm. All tests were performed in the dark box. In the results, shown in Figure 2 (Right), we were able to find the expected A characters from the images of analog to digital converter (ADC) output counts. A channel consisting of the 32 × 32 micro-pixels showed one of the channel area and exhibited the ADC output value corresponding to the number of photons and painting the color of the area in Figure 2 (Right). The light incident position from the LEDs painted the area as 400 ∼ 800 ADC level, but the pedestal region without the LED was separated as less than zero-level.

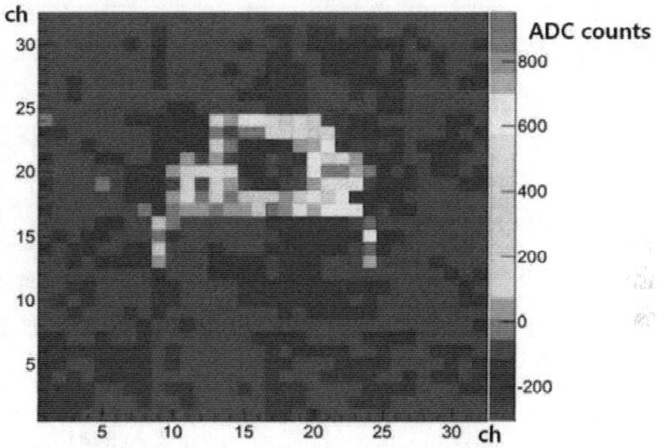

Figure 2. Assembled 1024channel SiPM camera(Left), The image of pulsed LED pattern(Right)

4. Conclusion

The SiPM has been developed as a promising alternative to PMTs in a variety of fields, such as X-ray detectors or cameras. We have designed and produced the 1024-channel SiPM sensor using the developed multi-channel SiPM devices, each device having 64 channels; this design has a higher density than a conventional camera which uses single-channel SiPMs. 64-channel devices were arranger in a 4 × 4 grid to form the 1024-channel SiPM sensor. The sensor was placed in the pinhole camera of the focal length of 10 cm. The camera was able to take the image of weak LED pulses as expected. The developed system is expected to be applied to the next X-ray detection camera. In addition, a camera system with more channels can be integrated because of the expandable design of the system.

References

[1] Saveliev V and Golovin V 2000 *Nucl. Instr. Meth.* A **442** 223
[2] Sadygov Z, Olshevski A, Chirikov A, Zheleznykh I and Novikovd A 2006 *Nuc. Inst. and Meth.* A **567** 70
[3] Buzhan P, Dolgoshein B, Ilyin A, Kantserov V, Kaplin V, Karakash A, Pleshko A, Popova E, Smirnov S, Volkov Yu 2001 *ICFA Inst. Bull.* **21** 28
[4] Hye Young Lee, Jin-A Jeon, Jik Lee, Il H. Park, Sug-Whan Kim and Seul Ki Yang 2012 *New Physics: Sae Mulli* N0. 12 **62** 1364
[5] Otte A. N, Meagher K, Nguyen, Carroll M, Hooper S, McKinney K, Peet S 2015 *Nucl. Instr. Meth.* A **787** 85

Acknoledgments

This work was supported by the Defense Acquisition Program Administration (DAPA) and the Agency for Defense Development (ADD) in Korea (ADDR-117) and the National Research Foundation of Korea (NRF) (2013 R1A1A2064412).

IOP Conf. Series: Journal of Physics: Conf. Series **928** (2017) 012043 doi:10.1088/1742-6596/928/1/012043

The Fabrication and Test of 64-Channel Silicon Photomultipliers

H. Y. Lee[1], Jin-A Jeon[1],Il H. Park[1,2], Jik Lee[1,2]

[1]Institute of Science and Technology in Space, Sungkyunkwan University, Suwon, Korea
[2]Department of Physics, Sungkyunkwan University, Suwon, Korea

E-mail: hylnate@nate.com

Abstract. A silicon photomultiplier (SiPM) is a highly sensitive photo sensor, and it consists of many micropixels that operated in Geiger mode. The SiPM has various preferable characteristics, such as low voltage operation compactness and lightweight. Additionally, the SiPM is robust and insensitive to magnetic fields, and it is capable of the counting of single photons due to the high gain of $10^5 \sim 10^6$. The developments of UV enhanced SiPMs and large area matrices of SiPMs have been actively pursued in the world. The technology can be utilized for applications such as radiation detection, medical science, cameras, and telescopes. In this study, we present the design, fabrication, and test result of the 64-channel SiPM sensor as the first step toward the development of a large matrix of SiPMs.

1. Introduction

The silicon photomultiplier (SiPM) consists of many micropixels and is a next-generation silicon photo sensor with high-speed, low power, low weight characteristics as well as a high sensitivity to light [1]. Each micropixel of the SiPM, called a Geiger mode Avalanche Photo Diode (APD), has the function of single photon counting, which is a characteristic of ultra-sensitivity [2]. As the micropixels of the SiPM are operated in Geiger mode, they have a higher gain (10^6) than a typical APD (10^2). The gain of the SiPM is approximately equal to a photomultiplier tube (PMT), which is frequently used as a photo detector of single photon counting [3]. The SiPM has a short rise time of about 1 ns, a short recovery time of about 20 ns, and a time resolution of about 30 ps/10 photon in a specific way. The sensor also has a low operating voltage of 20 $\sim$ 70 V. The photo detection efficiency (PDE) is at the maximum value of 50 % at peak wavelengths in 400 $\sim$ 900 nm [2]. With these advantages, the SiPM is spotlighted as a photo sensor to replace conventional PMTs. SiPM development throughout the world has been actively proceeding in UV enhancement for medical applications and radiation detection as well as large matrix for application in cameras and telescopes. The SiPM can be applied to all equipment using PMTs in fields of electronics, medicine, military, measurements, and space; they have a very wide range of uses [4-6]. In this paper, we present the design, fabrication, and results of the 64-channel SiPM sensor as the first step toward multichannel sensors.

2. The design and fabrication of the 64-channel SiPM sensor

The SiPM was simulated and designed with the PN junction of a high electric field into an epitaxial layer on p-type substrate, which works at a low operating voltage of about 25 V. The vertical structure of the SiPM has an n+/p/p- diode configuration. Our design of the SiPM sensor is 8 × 8 arrays. Each SiPM consisted of 32 × 32 micropixels with the size of 35 × 35 m^2 in the space of 1 mm^2. We have designed and constructed the 64-channel SiPM sensor on a silicon substrate. Figure 1 (left) shows the designs of the 64-channel SiPM sensor. The multichannel sensor has the 64-readout pads arranged in an L-shape on two sides. Shown in Figure 1 (right) is a picture of the development prototype fabricated as designed.

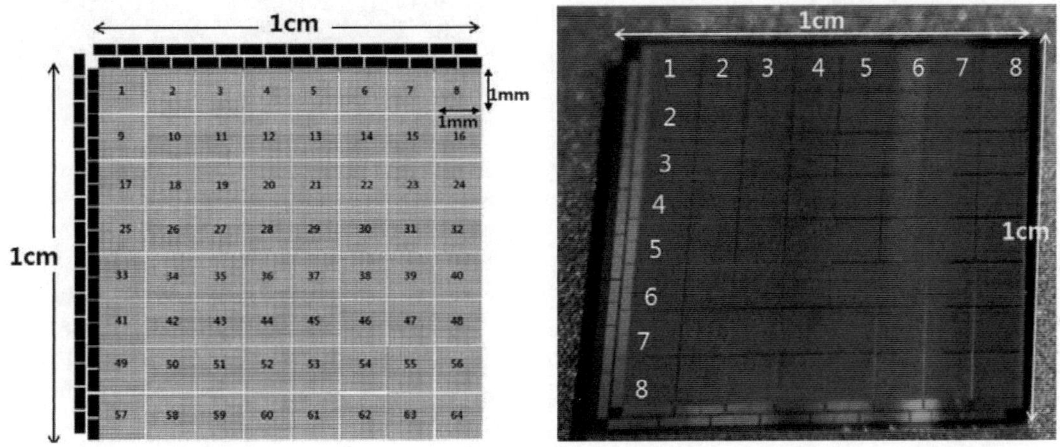

Figure 1. The design (left) and the development prototype (right) of the 64-channel SiPM sensor. The multichannel sensor has the 64 readout pads arranged with an L-shape in two sides. Each SiPM has 32 × 32 micropixels in the size of 1 mm^2

The large matrix is useful for covering a large area with the SiPM. An image sensor can be made by an array of the multichannel SiPM sensors on an electric board. The size of the multichannel sensor has to be in agreement with the yield of fabrication. As shown in Figure 1, the 64-channel SiPM sensor has been designed and fabricated with 8 × 8 arrays in the size of 1 cm^2. The sensor was made with the photo-masks of 8 layers and with approximately 150-step processes in complementary metal-oxide semiconductor (CMOS) fabrication technology. We had the final optimal parameters in the design and fabrication of the multichannel SiPM sensor to make the PN junctions with a high enough electric field for Geiger mode operation. In addition, using 4 × 4 arrays of the 64-channel SiPM, we have also made an image sensor of 1024 channels. The image sensor had the test to take a picture at very low density light, which we will present in another paper.

3. The results of the fabricated 64-channel SiPM sensor

The I-V characteristics of the 64-channel SiPM sensor were measured. The reverse biasing voltage (V) over each SiPM channel was increased up to 40 V in increments of 0.2 V. Only one channel was biased at a time and other channels were unbiased. The I-V characteristics of the sensor are shown in Figure 2 (left). The average leakage current of the 64-channels was about 1 $\sim$ 5 pA before the breakdown voltage. The avalanche breakdown voltages (BV) of all the channels in the sensor were about 24 V, which has very uniform BV characteristics. The leakage current increased rapidly after the BV. The sensor can be operate up to Δ 5 V.

Multiplication process in the SiPM does not discriminate avalanche breakdowns by incident light or thermal excitation. The dark noise in the sensor is caused by thermally generated charge carriers in the silicon. The dark rate of the 64-channel SiPM sensor is 300 $\sim$ 800 kHz at Δ 3 V from the BV. Figure 2 (right) shows the ability of single photon counting of the sensor with low-intensity light in an oscilloscope. As shown in the picture, the single photoelectron (PE) peaks

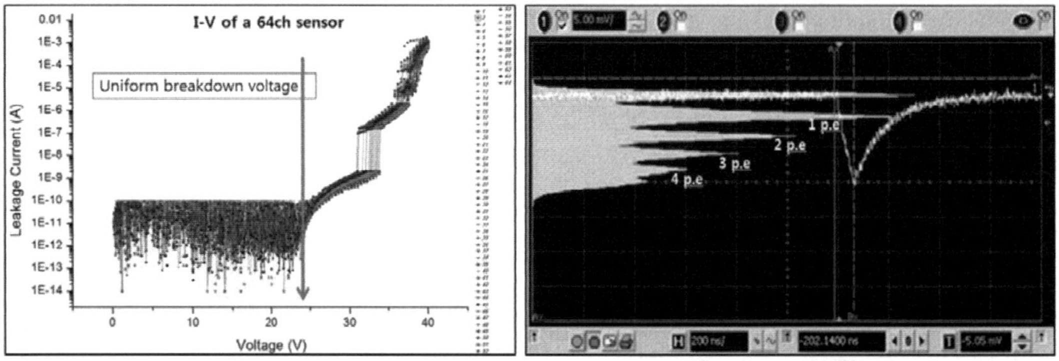

Figure 2. IV curve measured for the fabricated 64-channel SiPM sensor (Left). The peak histogram and signal with a SiPM in an Oscilloscope (Right).

are clearly separated from the pedestal. The single, double, and triple PE peaks are separated from each other in the pulse height histogram. The height of the single PE peak increased with the gain increased by bias voltage. The typical gain in the sensor is $2 \times 10^5 \sim 5 \times 10^5$ at Δ $2 \sim 5$ V. The PDE is about 10 % at a wavelength of 490 nm. Our plan is to study the SiPM improvement of a high PDE at $300 \sim 400$ nm.

4. Conclusion

The 8×8 array sensor of SiPMs was designed with the PN junction of a high electric field on a p-type epitaxial layer for working at low voltage operation. We had the final optimal parameters in the design and fabrication of the SiPM to establish a high electric field enough for Geiger mode operation PN junction. Finally, we fabricated the 64-channel SiPM sensor as a first step toward the fabrication of a large area matrix of SiPMs. We measured dark rate, gain, light response, and PDE on first product of the 64-channel SiPM sensor. The SiPMs turned out to have the capability of single photon counting, the dark rate of $300 \sim 800$ kHz, the gain of $\sim 10^5$, and the PDE of 10 % at 490 nm. The next plan is to improve the PDE of the SiPMs in the wavelength of $300 \sim 400$ nm.

References

[1] A. G. Stewart, V. Saveliev, S. J. Bellis, D. J. Herbert, P. J. Hughes, and J. C. Jackson 2008 *IEEE J. Quantum Electron.* **44** 157

[2] N. Dinu, R. Battiston, M. Boscardin, G. Collazuol, F. Corsi, G. F. Dalla Betta, A. Del Guerra, G. Llos, M. Ionica, G. Levi, S. Marcatili, C. Marzocca, C. Piemonte, G. Pignatel, A. Pozza, L. Quadrani, C. Sbarra, and N. Zorzi 2007 *Nucl. Instrum. Methods Phys. Res.* A **572** 422

[3] M. McClish, P. Dokhale, J. Christian, C. Stapels, and K. S. Shah 2010 *IEEE Trans. Nucl. Sci.* **57** 2280

[4] Hye Young Lee, Jin-A Jeon, Jik Lee, Il H. Park, Sug-Whan Kim and Seul Ki Yang 2012 *New Physics: Sae Mulli* N0. 12 **62** 1364

[5] P. Buzhan, B. Dolgoshein, L. Filatov, A. Ilyin, V. Kantzerov, V. Kaplin, et al. 2003 *Nuc. Inst. and Meth.* A **504** 48

[6] E. Aprile, P. Cushman, K. Ni, and P. Shagin 2006 *Nuc. Inst. and Meth.* A **556** 215

Acknoledgments

The work was supported by the Defense Acquisition Program Administration (DAPA) and the Agency for Defense Development (ADD) in Korea (ADDR-117), the National Research Foundation of Korea (NRF) (2013R1A1A2064412) and the Postdoctoral Research Program of Sungkyunkwan University (2014).